Handbook of Experimental Pharmacology

Volume 288

The *Handbook of Experimental Pharmacology* is one of the most authoritative and influential book series in pharmacology. It provides critical and comprehensive discussions of the most significant areas of pharmacological research, written by leading international authorities. Each volume in the series represents the most informative and contemporary account of its subject available, making it an unrivalled reference source.

HEP is indexed in PubMed and Scopus.

Fabian Pfrengle • Kevin Pagel
Rudolf Tauber
Editors

Complex Carbohydrates in Health and Disease

Springer

Editors
Fabian Pfrengle
Institut für Organische Chemie
BOKU University - Universität für Bodenkultur Wien
Wien, Austria

Kevin Pagel
Institut für Chemie und Biochemie
Freie Universität Berlin
Berlin, Germany

Rudolf Tauber
Institut für Laboratoriumsmedizin
Klinische Chemie und Pathobiochemie
Charité-Universitätsmedizin Berlin
Berlin, Germany

ISSN 0171-2004 ISSN 1865-0325 (electronic)
Handbook of Experimental Pharmacology
ISBN 978-3-032-02582-1 ISBN 978-3-032-02583-8 (eBook)
https://doi.org/10.1007/978-3-032-02583-8

This Springer imprint is published by the registered company Springer Nature Switzerland AG
The registered company address is: Gewerbestrasse 11, 6330 Cham, Switzerland

Preface

This volume of the Handbook of Experimental Pharmacology addresses the highly dynamic field of complex carbohydrates—specifically those attached to glycoproteins and glycolipids—in health and disease. Breakthrough advances in the analysis and synthesis of complex carbohydrates, along with developments in molecular and cell biology as well as clinical studies, have paved the way for translating this field from the natural sciences and basic medicine into clinical application.

Complex carbohydrates—often termed glycans or simply sugars—are usually attached to glycolipids and glycoproteins in cells, the extracellular matrix, and body fluids. They play a central role in many vital biological processes. In conjunction with carbohydrate-specific receptors, glycans mediate cell–cell recognition, binding to the extracellular matrix, and immune responses. They regulate the migration of cells within the organism during embryonic development and tissue repair. Moreover, they contribute to modulating the functional activity of glycosylated proteins, the regulation of intracellular vesicular transport, and the formation of cellular organelles. This wide range of glycosylation functions is based on the structural diversity of glycans, which are generated during oligosaccharide biosynthesis in the endoplasmic reticulum and Golgi apparatus. The generation of glycan structures involves more than 700 genes contributing to glycan biosynthesis, modification, recognition, and catabolism.

Reflecting their important biological roles, glycans are significantly involved in the etiology, pathogenesis, and progression of both inborn and acquired diseases. Congenital disorders of glycosylation (CDG) comprise more than 140 genetic defects in glycoconjugate metabolism, leading to a wide variety of clinical symptoms. In acquired diseases, the role of glycans is particularly prominent in all forms of inflammatory diseases, as well as in bacterial, viral, and parasitic infections, and in cancer.

An understanding of the clinical significance of glycans has led to the use of glycan epitopes in diagnostics and to the development and clinical application of new glycan-based therapeutics and vaccines. Furthermore, because glycosylation critically influences the pharmacodynamics and pharmacokinetics of biologics, tailoring appropriate glycosylation patterns has become a key factor in the design and production of therapeutic antibodies and other glycoprotein drugs.

Glycosylation also impacts the patentability and licensing of glycoprotein therapeutics developed through biotechnology.

This volume, *Complex Carbohydrates in Health and Disease*, was deliberately designed to be interdisciplinary, bringing together chemical, pharmacological, biological, and medical perspectives. The authors hope that this structure will foster cooperation across disciplines and encourage further translation of glycobiology into medical applications.

The chapters in the *first part* focus on the *chemical synthesis of complex carbohydrates and carbohydrate-based drugs*—a key requirement for obtaining homogeneous glycans of defined structure in sufficient quantities. In his chapter, Paul Kosma provides an overview of current methodologies for the chemical synthesis of complex oligosaccharides, outlining the main principles of glycosylation reactions and oligosaccharide assembly, followed by selected recent examples of biomedically relevant glycans. Rakesh Raigawali et al. present strategies for synthesizing glycosaminoglycans, i.e., sulfated carbohydrates ubiquitous in cells and the extracellular matrix. Using heparan sulfate as an example, they describe solution-phase synthesis, one-pot synthesis, universal building blocks, and solid-phase synthesis. The chapter also reviews current knowledge on structure–activity relationships of heparan sulfate-binding proteins such as coagulation factors, growth factors, and chemokines, and discusses prospects for drug development and biosensor applications.

In their eponymous chapter, Yevhenii Radchenko et al. provide insight into the design and synthesis of probes for activity-based profiling of retaining glycosidases—a class of enzymes that catalyze the hydrolysis of glycosidic bonds in complex carbohydrates. The authors describe how activity-based profiling can be applied to monitor glycosidase activity in lysosomal storage disorders and to identify potential glycosidase inhibitors for therapeutic targeting and drug discovery.

The *second part* focuses on *analytical methods*. Niklas Geue et al. summarize state-of-the-art glycan analysis in vacuo, demonstrating how mass spectrometry, tandem mass spectrometry, ion mobility spectrometry, and gas-phase infrared spectroscopy—supported by chromatography and computational modeling—enable structural characterization of isolated glycans and analysis of the complete glycome. Wesley Pietsch et al. provide a comprehensive overview of the emerging use of biological and solid-state nanopores for analyzing the immense structural complexity of glycosaminoglycans. Sergey Samsonov and Mateusz Marcisz describe advances in computational techniques such as molecular docking, molecular dynamics, and free energy calculations, applied to elucidate interactions between glycosaminoglycans and proteins. These approaches offer valuable tools for understanding glycosaminoglycan function and advancing their therapeutic potential.

The *third part*, focusing on *biological functions and medical relevance*, highlights four key areas. Chika Ogura and Shoko Nishihara describe the role of glycans in maintaining stemness and regulating differentiation of pluripotent stem cells—knowledge that is increasingly important for both basic research and innovative clinical applications. Falk Nimmerjahn's chapter explains how the sugar moiety on each immunoglobulin G (IgG) heavy chain modulates IgG's pro- and anti-inflammatory activity. The chapter further details how glycosylation patterns change during inflammation, how these changes can serve as highly specific biomarkers, and how they may inform the optimization or modulation of therapeutic antibodies.

Alexandre Raynor et al. provide a state-of-the-art overview of congenital disorders of glycosylation (CDG), an expanding group of inborn metabolic disorders with more than 170 identified disease types. The authors describe current biomarkers, diagnostic techniques, and therapeutic approaches. Henrique Duarte et al. explore cancer-associated glycosylation alterations found in all types of malignancies studied to date. The chapter discusses their functional roles in invasive growth, metastasis, and tumor immune evasion, as well as their relevance for cancer diagnostics, glycan-based vaccines, and targeted therapies.

We warmly thank all authors for their contributions and their excellent cooperation. We are also deeply grateful to Susanne Dathe, Editor for Neuroscience, Pharmacology & Immunology, and to Alamelu Damodharan, Production Editor (Books) at Springer, for their continuous support, guidance, and patience.

Berlin, Germany — Kevin Pagel
Wien, Austria — Fabian Pfrengle
Berlin, Germany — Rudolf Tauber

Contents

Part I

Synthesis of Complex Carbohydrates and Carbohydrate-based Drugs

Chemical Synthesis of Complex Carbohydrates

Paul Kosma

Contents

Abstract

The chapter provides a brief overview on current methodologies for the assembly of complex oligosaccharides by chemical synthesis. Following an introductory section describing the major factors and variables involved in glycosylation reactions, select examples for advanced approaches towards mammalian and bacterial glycans are discussed illustrating recent progress in the field of glycochemistry.

P. Kosma (✉)
Department of Chemistry, Institute of Organic Chemistry, University of Natural Resources and Life Sciences, Vienna, Austria
e-mail: paul.kosma@boku.ac.at

F. Pfrengle et al. (eds.), *Complex Carbohydrates in Health and Disease*, Handbook of Experimental Pharmacology 288, https://doi.org/10.1007/164_2025_747

Keywords

Glycosylation · Oligosaccharide · One-pot synthesis · Polysaccharide · Sialylation

Abbreviations

ABz	*o*-Alkynylbenzoyl
Ac	Acetyl
All	Allyl
AIBN	2,2′-Azobis(isobutyronitrile)
BAIB	(Diacetoxyiodo)benzene
Bn	Benzyl
Bz	Benzoyl
CA	Chloroacetyl
Cbz	Benzyloxycarbonyl
CPS	Capsular Polysaccharide
DCM	Dichloromethane
DDQ	2,3-Dichloro-5,6-dicyano-1,4-benzoquinone
DIEA	Diisopropylethylamine
DMAP	4-Dimethylaminopyridine
DMDO	Dimethyldioxirane
DMF	*N,N*-Dimethylformamide
DtBSi	Di-*ter*tbutylsilylene
EDCI	1-Ethyl-3-(3-dimethylaminopropyl)-carbodiimide
Fmoc	(9*H*-Fluoren-9-ylmethoxy)carbonyl
Gc	Gas chromatography
HAD	Hydrogen-bonded aglycon delivery
HDTC	Hydrazine dithiocarbonate
IAD	Intramolecular aglycon delivery
IP	*ortho*-Iodophenyl
Kdo	3-Deoxy-D-*manno*-oct-2-ulosonic acid
LOS	Lipooligosaccharide
LPS	Lipopolysaccharide
NeuAc	5-*N*-Acetylneuraminic acid
NIS	*N*-Iodosuccinimide
Lev	Levulinoyl
LG	Leaving group
MCEP	*ortho*-Methoxycarbonylethynylphenyl
MPEP	4-Methoxyphenylethynylphenyl
MP	4-Methoxyphenyl
Nap	2-Naphthylmethyl
NBz	4-Nitrobenzoyl
NPTFA	*N*-Phenyltrifluoroacetimidoyl
PG	Protecting group

Pic	Picolinyl
SPhos	2-Dicyclohexylphosphino-2′,6′-dimethoxybiphenyl
SSEA	Stage-specific embryonic antigen
SSIP	Solvent separated ion pair
Table	2-O-*ortho*-Tosylaminobenzyl
TBAF	Tetrabutylammonium fluoride
TBDPS	*Tert*-butyl-diphenylsilyl
TBPA	Tris(4-bromophenyl)ammoniumyl hexachloroantimonate
TBS	*tert*-Butylsilyl
TCA	Trichloroacetyl
TEMPO	2,2,6,6-Tetramethylpiperidine-1-oxyl
Tf	Trifluoromethanesulfonyl
TFA	Trifluoroacetic acid
THF	Tetrahydrofuran
TMS	Trimethylsilyl
Troc	2,2,2-Trichlorethyl chloroformate
TTBP	2,4,6-Tri-*tert*-butyl pyrimidine

1 Introduction

Besides nucleic acids, proteins and lipids, carbohydrates form an important class of biomolecules mediating a vast spectrum of important biological phenomena in health and disease, such as metabolic utilization, fertilization, cell communication, proliferation and differentiation, interaction with benign and pathogenic microorganisms, their role in hereditary genetic disorders, blood group antigenicity, inflammation or cancer progression, just to name a few examples. The fundamental role of glycans is embedded in the structural variability of saccharides with multiple stereocenters and the formation of linear as well as branched glycans from the disaccharide stage up to macromolecular assemblies. As a comparison, whereas the four nucleotide bases A, C, T, G may be connected to produce 64 trimers and 20 different amino acids could generate 8000 tripeptides, the ten common mammalian sugars could already be interlinked to give 126.000 trisaccharide variants (Turnbull and Field 2007). The chemical space of glycans is further expanded by extended carbon-chain lengths beyond the common hexose units and the presence of additional acidic or basic functions as well as non-carbohydrate appendages such as phosphate, sulfate or carboxylate esters (Werz et al. 2007). Thus, the structural features inherent in the molecular details of sugars harbor an inexhaustible information content to be realized in biological signaling, binding interactions and vaccine development (Varki et al. 2022; Reilly et al. 2019; Meron-Sudai et al. 2023; Ruprecht et al. 2022). As bacterial glycans are structurally the most diverse and challenging to synthesize, these will mostly serve as case studies in this chapter to illustrate progress in glycan assembly. The downside of the structural complexity, however, translates into formidable difficulties for chemical synthesis. Nevertheless, since isolation of complex glycans from natural resources is limited, chemical as well as chemoenzymatic synthesis is still the method of choice to provide structurally defined oligosaccharides, analogues and glycoconjugates in pure form and in usable amounts. A highly dynamic progress in synthetic and analytic methodologies has been witnessed in the past years and provided a robust toolbox for modern glycoscience to study the manifold biological functions of glycans on a molecular level. Whereas the formation of glycosidic linkages in former times has relied mostly on empirical rules and the art of practitioners, recent advances in mechanistic studies and technological platforms have allowed even the construction of extremely highly complex glycans as will be outlined in a few illustrative examples. Still, the design of glycan assembly requires a judicious selection of protecting groups, optimized reactivity and selectivity of both glycosyl donors and glycosyl acceptors, as well as efficient removal of protecting groups and appropriate purification protocols. Modern analytical tools of mass -spectrometry and sophisticated NMR techniques serve nowadays as checks of integrity and purity and validate the structure of synthetic glycans. A comprehensive overview of glycosylation reactions and glycan synthesis, however, is out of the scope of this chapter, and the reader is referred to recent relevant authoritative reviews (Crawford and Seeberger 2023; Oscarson and Cheallaigh 2021; Kinnaert et al. 2017; Boltje et al. 2009). This chapter will first describe the main features of glycosylation reactions and oligosaccharide synthesis, followed by selected recent representative examples of biomedically relevant glycans. Mainly donors that have found proven and general application in the solution synthesis of complex oligosaccharides will be described in more detail.

2 Principles of Glycoside Formation

Since the historic days of Emil Fischer, the glycosylation step has been in the main focus of carbohydrate chemistry. From a general perspective, a glycosyl donor equipped with a suitable leaving group is activated by catalytic or stoichiometric amounts of a promoter to undergo a series of transformations along two main pathways spanning a continuum of intermediates but essentially resembling S_N1 or S_N2-type reactions in the coupling step with the hydroxy-group of the glycosyl acceptor (Fig. 1). Formation of α- or β-configured contact ion pairs from the anomeric leaving groups followed by direct reaction with the hydroxy group of the glycosyl acceptor is envisaged to proceed in an S_N2-like scenario leading to glycosides with inverted anomeric configuration. The occurrence of a glycosyl oxocarbenium ion intermediate, however, has experimentally only been proven for highly polar systems in superacids (Franconetti et al. 2021). In case the contact ion pair equilibrates to the solvent-separated ion pair (SSIP), the stereoselectivity will be challenged and – depending on reaction conditions – the reaction will be more of an S_N1-character and will lead to anomeric product mixtures.

Computational studies indicate that oxocarbenium ion pairs adopt a variety of different ring conformations which strongly depend on the position of the counterion (Hosoya et al. 2014). S_N1-related glycosylation reactions sample an even larger

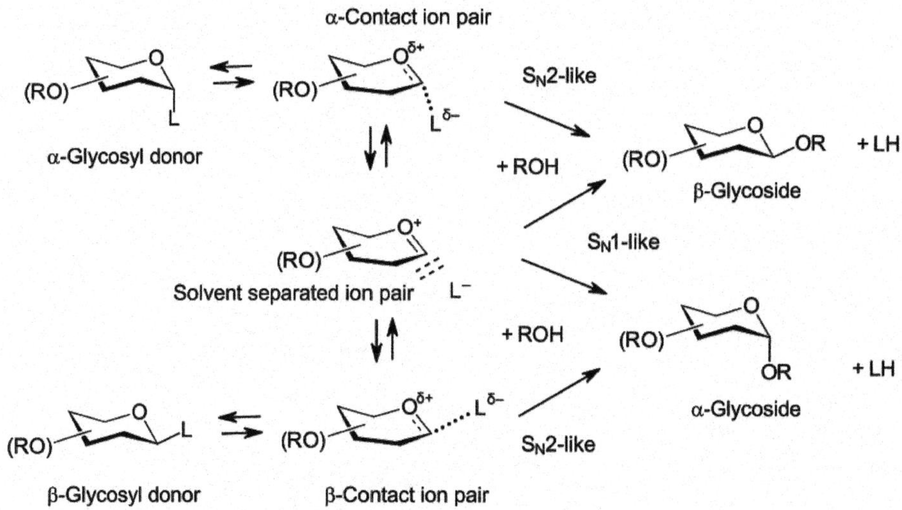

Fig. 1 General pathways of glycosylation reactions

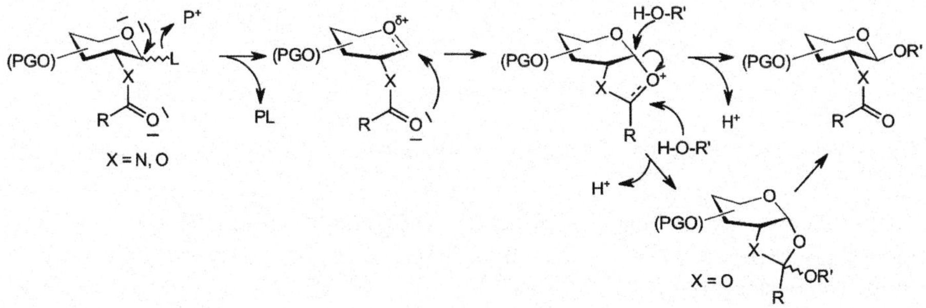

Fig. 2 Neighboring group participation leading to 1,2-*trans* glycosides and orthoester derivatives

space of intermediate conformations of glycosyl cations and are best represented by a conformational energy landscape that allows to rationalize the steric course of the most relevant glycosylation reactions (Hansen et al. 2019). Moreover, concentration and reactivity of donor and acceptor, temperature, solvent, promoter and additives all contribute to the complexity of glycosylation reactions (Andreana and Crich 2021; Chatterjee et al. 2018).

For the synthesis of 1,2-*trans* glycosides, usually neighboring group participation (anchimeric assistance) is exploited involving the presence of 2-*N*-acyl or 2-*O*-acyl groups (Fig. 2). Upon oxocarbenium ion formation, the acyl-oxygen intercepts the arising positive charge at the anomeric center, leading to a cyclic dioxolenium ion that effectively blocks the α-side and thus enables attack of the nucleophile from the β-face to eventually provide the 1,2-*trans* glycoside. Alternatively, the nucleophile may also react with the cyclic dioxolenium ion to generate *exo*/*endo* orthoester products. These often-undesired intermediates, however, may also act as glycosyl

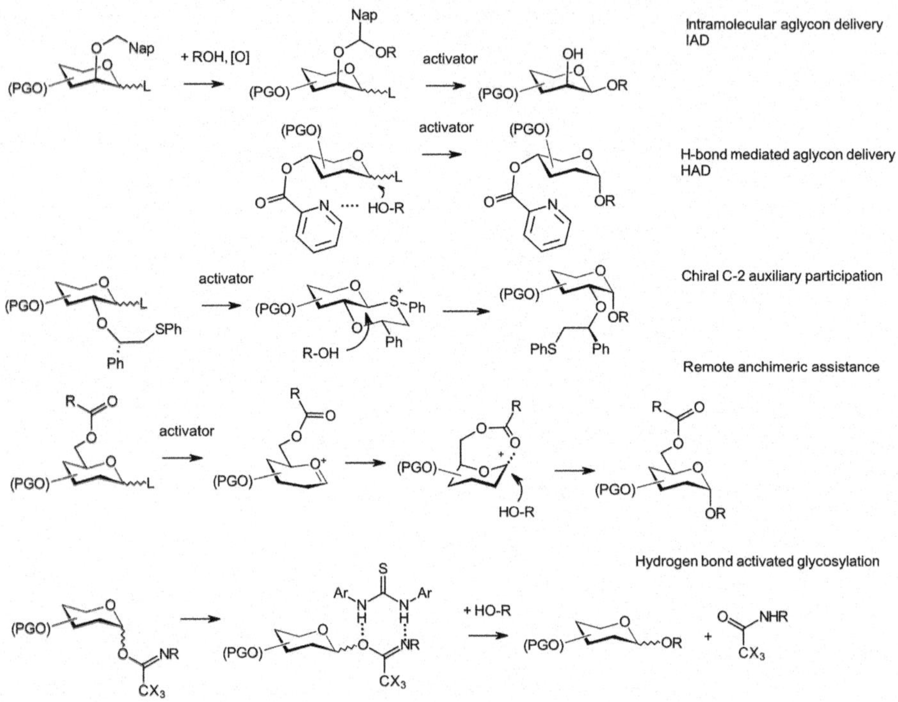

Fig. 3 Advanced methods for 1,2-*cis* glycoside formation

donors themselves, and under appropriate acidic conditions can be rearranged into the glycoside product, albeit frequently in lower yields due to competing hydrolysis of the orthoester group.

While neighboring group participation through acyl- and acylamido groups provides fairly robust access to 1,2-*trans* configured glycosides, the elaboration of 1,2-*cis* glycosides is significantly more challenging (Nigudkar and Demchenko 2015). Non-participating ether groups next to the anomeric center usually will lead to anomeric mixtures of glycosides, and more elaborate protecting schemes have been developed to improve the selectivity toward the formation of 1,2-*cis* connected products (Fig. 3).

Intramolecular aglycon delivery (IAD) has been elaborated to produce β-linked mannosides by the first oxidative coupling of the acceptor alcohol to a 2-*O*-ether-linked methylene carbon to give the corresponding acetal that upon activation transfers the aglycon from the top side to the anomeric center (Ishiwata et al. 2010). Chiral auxiliaries at C-2 exploit covalent neighboring participation (Kim et al. 2005; De Kleijne et al. 2020) leading to an intermediary sulfonium ion, but require additional steps for the introduction and subsequent cleavage of the auxiliary group. Hydrogen-bond mediated aglycon (HAD) delivery can be engaged by installing picolinyl or picoloyl substituents that position the incoming alcohol in a *syn* orientation to the activated donor (Yasomanee and Demchenko 2012).

Eventually, long-range anchimeric assistance can also be provided by the picolinyl, picoloyl or suitable ester groups that shield the top face of the activated donor to produce 1,2-*cis* configured glycosides (Lu et al. 2011). The contribution of distal ester participation in glycosylation reactions, however, has been challenged by recent NMR spectroscopic studies with bridged bicyclic glycosyl dioxacarbenium ions (Upadhyaya et al. 2021; Hettikankanamalage et al. 2020), and alternative explanations for the α-directing effects of 4-*O*-acyl groups in galactosyl donors have been presented (Ande and Crich 2023).

Hydrogen bonding to the leaving group has recently been exploited as a very mild activation procedure for various glycosyl imidate donors (Xiao et al. 2022). In this variant, thiourea derivatives facilitate the release of the leaving group by coordination to the oxygen and nitrogen of the imidate moiety, eventually leading to S_N1-type glycosylation reactions. For increased α-selectivity, however, additional measures, as conformational constraints in the imidate donor or solvent additives, are required.

2.1 Solvent Effects

Similar to the effects in nucleophilic substitution reactions, solvents provide an important contribution to steer the anomeric selectivity in glycosylation reactions by stabilizing the intermediate reaction transition states. Whereas dichloromethane or toluene are usually the solvents of choice for many glycosylation reactions, the addition of modifying cosolvents can be beneficial. Thus, the α-directing effect of ethereal solvents such as diethyl ether, THF, *tert*butyl-methyl or cyclopentyl-methyl ether has been described. Polar solvents may lead to the formation of additional reactive intermediates. Nitrile-based solvents such as propionitrile and acetonitrile are known to effect preferential formation of equatorial glycosides by intermediate conversion into a nitrilium donor of inverted configuration (Fig. 4).

In a recent example, the impact of solvent on the anomeric selectivity has been demonstrated for the assembly of an artificial construct harboring the amylose-specific α-(1→4)- as well as cellulose-based β-(1→4)-glucosidic linkages from

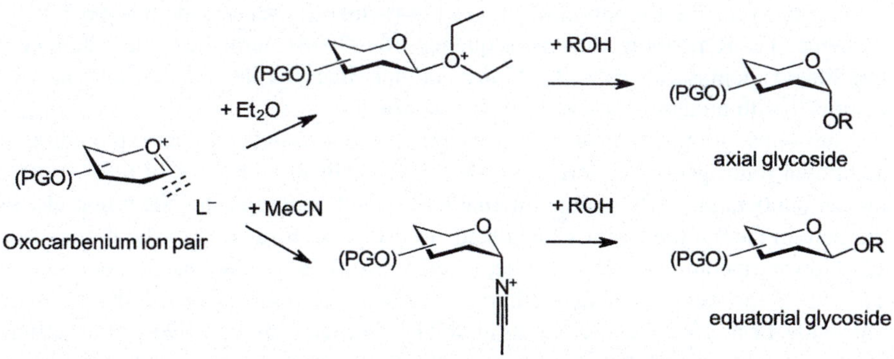

Fig. 4 Impact of solvents on the anomeric selectivity in glycoside formation

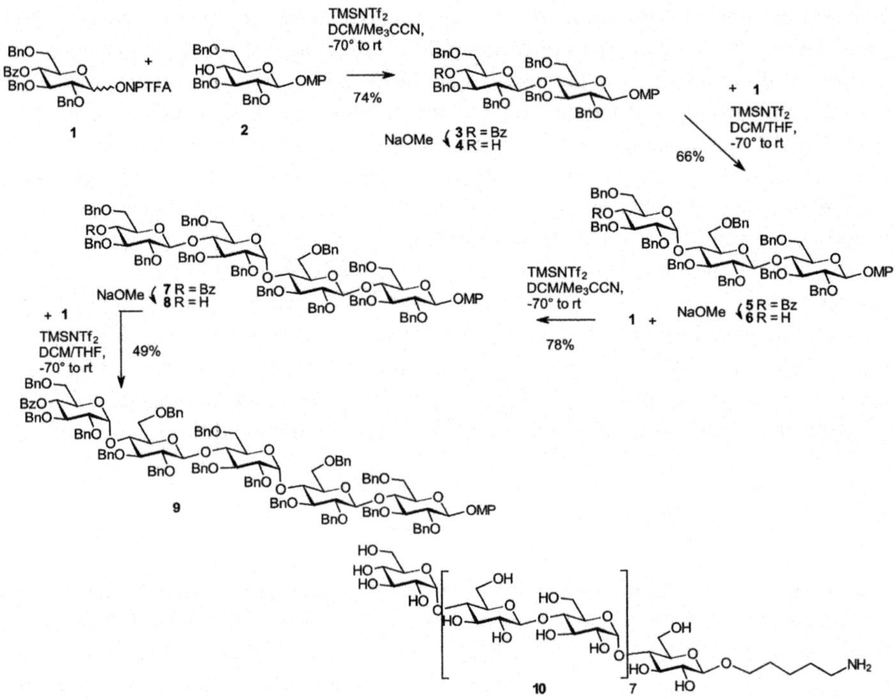

Scheme 1 Synthesis of oligoglucans with alternating α-(1→4) and β-(1→4) linkages

benzylated precursors (Yang et al. 2023). After screening numerous conditions, using a dichloromethane/pivalonitrile solvent mixture and $TMSNTf_2$ as promoter at low temperature, the disaccharide **3** was formed as a single β-anomer by coupling the *N*-phenyltrifluoroacetimidate donor **1** (without the presence of a β-directing 2-*O*-ester group) to the *p*-methoxyphenyl glycosyl acceptor **2**. Following debenzoylation of **3** with NaOMe, the resulting glycosyl acceptor **4** was then coupled with the same *N*-phenyltrifluoroacetimidate donor, albeit in a mixture of a dichloromethane/THF, leading to exclusive formation of the α-(1→4)-linked trisaccharide **5** in 66% yield (Scheme 1). Repetition of this sequence (**6→7→8**) furnished the alternately interlinked pentasaccharide **9**. Along similar lines, a hexadecasaccharide **10** equipped with an amino-spacer unit was obtained.

The contribution of solvent effects can be further expanded by the exploitation of additional hydrogen bond formation (Scheme 2). In the presence of a 2-*O*-*ortho*-tosylaminobenzyl (TAB) group, internal hydrogen bonding of the NH group blocks the bottom face of the donor **12** and leads to exclusive formation of the β-glucoside **13** at low temperature in propionitrile, whereas promotion of the reaction of acceptor **11** with triflic acid in diethyl ether at room temperature afforded the α-linked glucoside **14** in 75% yield, albeit as an 85:15 anomeric mixture (Ding et al. 2018). This bimodal approach has then successfully been applied to the synthesis of linear and branched α-glucans (Ding et al. 2020). Also, coordination of a Zn^{2+} ion to O-2

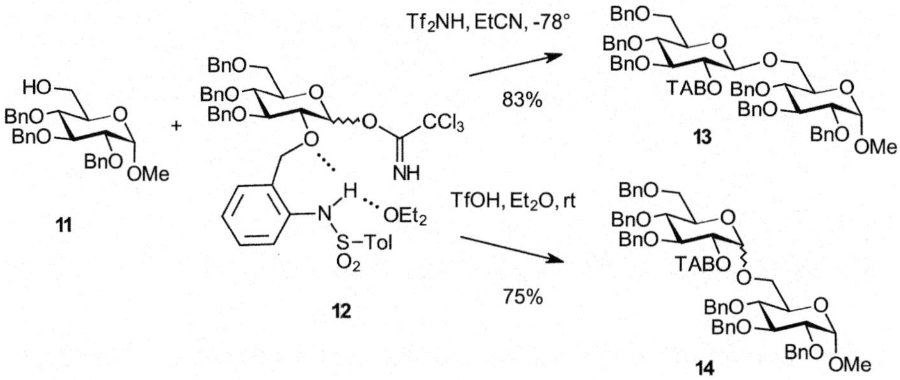

Scheme 2 Solvent participation in bimodal glycosylation

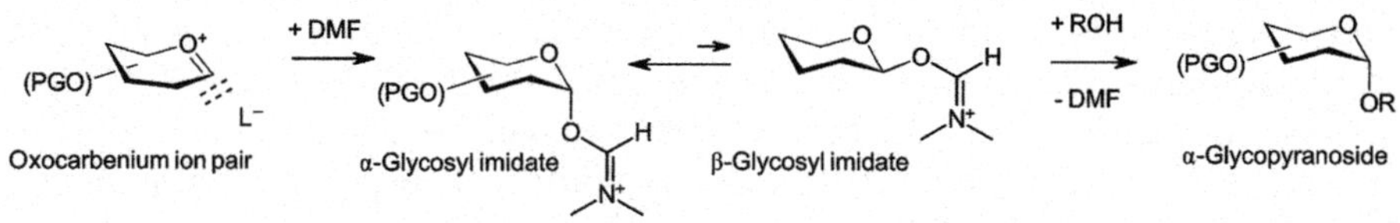

Fig. 5 DMF as a modifying solvent leading to glycosyl imidate intermediates

and O-3 by addition of Zn-iodide has been developed for α-glucosylation, α-galactosylation and β-mannosylation, respectively (Zhong et al. 2022).

Along similar lines, DMF has been shown to generate an equilibrium of glycopyranosyl imidate intermediates (Ingle et al. 2013). The more reactive equatorial imidate is invoked as the activated species leading to the 1,2-*cis*-configured glycoside in the presence of a glycosyl acceptor (Fig. 5).

A recent example was reported by Yang and coworkers for α-selective synthesis of Kdo glycosides (Lou et al. 2020). Assembly of ketosidic linkages of 3-deoxy-D-*manno*-oct-2-ulosonic acid (Kdo) residues is challenging due to the deactivating electron-withdrawing nature of the carboxyl group, the lack of participating groups at C-3 and the ready formation of elimination products (see also Sect. 3.2). The Yang group recently reported on an improved strategy using a per-*O*-acetylated Kdo *ortho*-hexynylbenzoate donor **15** with various glycosyl acceptors in the presence of DMF as modifying agent and reactive 2-dicyclohexylphosphino-2′,6′--dimethoxybiphenyl (Sphos) gold (I) triflylimide as catalyst.

As an example, a disaccharide synthesis is shown in Scheme 3. The reaction of Kdo donor **15** with primary alcohol **16** generated the disaccharide **17** in excellent yield. The presence of an imidate intermediate was supported by ^{1}H and ^{13}C NMR spectroscopic data recorded at −40°C. The spectral features indicated the presence of the α-configured imidate only; however, the authors reasoned that the α-anomer was in equilibrium with the more reactive β-anomeric imidate, thereby leading to the exclusive formation of the α-product via an S_N2-type pathway.

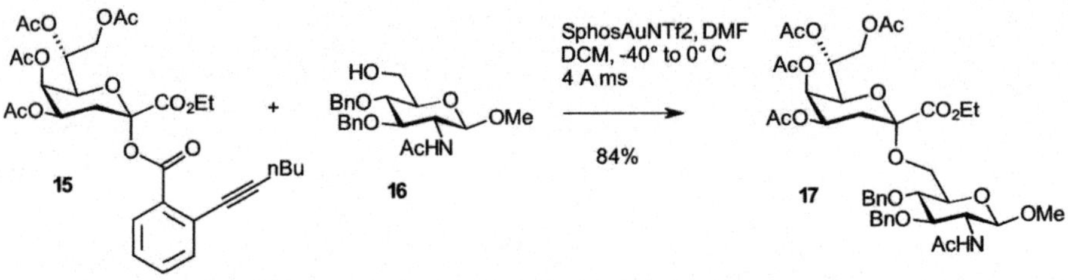

Scheme 3 Synthesis of an α-linked Kdo glycoside utilizing DMF as modulator

2.2 Impact of Configuration and Protecting Groups of Glycosyl Acceptors

While the main portion of the literature deals with the influence of leaving groups and the effect of protecting groups on the reactivity and anomeric selectivity of the glycosyl donors (Lian et al. 2015; Chatterjee et al. 2018; Zhang et al. 1999), much less systematic information has been accumulated regarding the role of the glycosyl acceptor (van der Vorm et al. 2018, van Van Vorm et al. 2019). In a recent seminal paper (van Hengst et al. 2023), the outcome of glycosylation reactions depending on the configuration of *gluco-*, *manno-*, *galacto-* D/L-*rhamno-*, D/L-*fuco-* and 2,6-dideoxy sugars containing varying benzyl or benzoyl protecting groups of the glycosyl acceptors has been thoroughly studied. As a standard protocol, the 4,6-*O*-benzylidene protected 2,3-di-*O*-benzyl donor **18** (and also a 2-azido-2-deoxy-3-benzyl-D-glucopyranosyl thiophenyl donor) was employed in glycosidation reactions with 67 different glycosyl acceptors. Pre-activation of the donors was carried out by the addition of diphenylsulfoxide/triflic anhydride at −40°C in dichloromethane, followed by the addition of the respective glycosyl acceptor **19**–**24** (Scheme 4). The main conclusions derived from these experiments were summarized as follows:

Equatorial hydroxy groups are more reactive than axial ones, as illustrated by the lower yields obtained with the galactoside acceptors **22**–**24** compared to the glucoside counterparts **19**–**21**.

This reactivity enhancement was also seen when a neighboring protected equatorial group was present.

Neighboring benzoyl esters led to decreased reactivity relative to benzyl ether groups (for **24** vs. **23**), and this disarming effect is more pronounced for benzoyl groups in equatorial positions than in axial ones. The deactivating effect of flanking benzoyl groups can be rationalized by the weaker nucleophilicity of the alcohol driving the glycosylation scenario towards an S_N1-type reaction of the intermediate oxocarbenium triflate ion pair rather than invoking the involvement of an intermediate covalent glycosyl triflate.

Thus, the influence of adjacent benzoyl participating groups was confirmed by the change from a non-selective formation of the 1→4 linkage in the tri-*O*-benzylated glucoside acceptor **19** towards a highly α-selective outcome in the presence of a

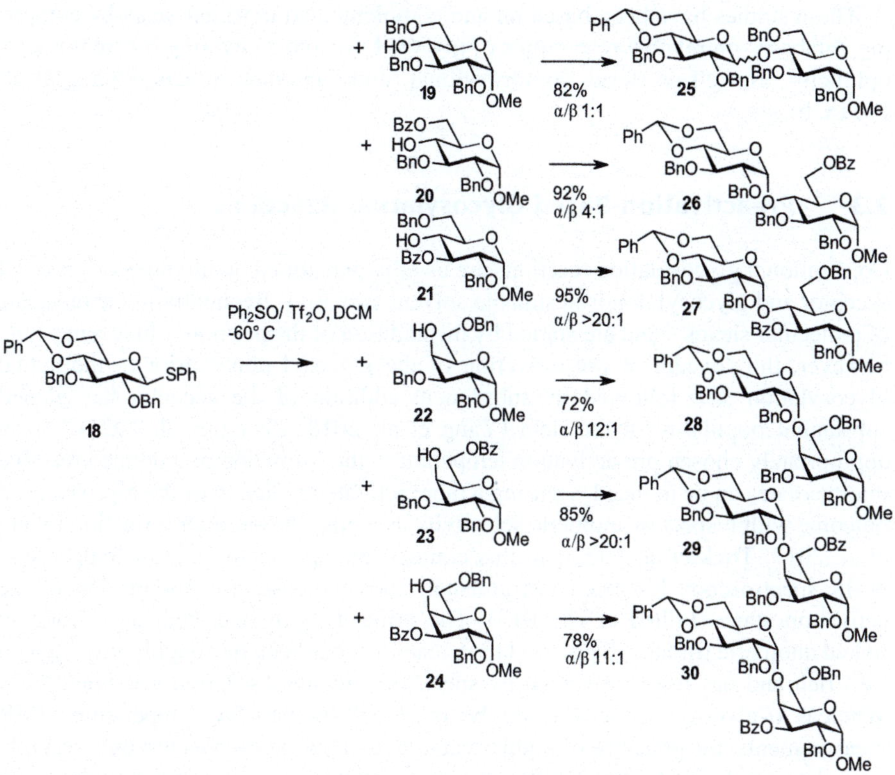

Scheme 4 Impact of acceptor configuration and protecting group pattern on anomeric selectivity

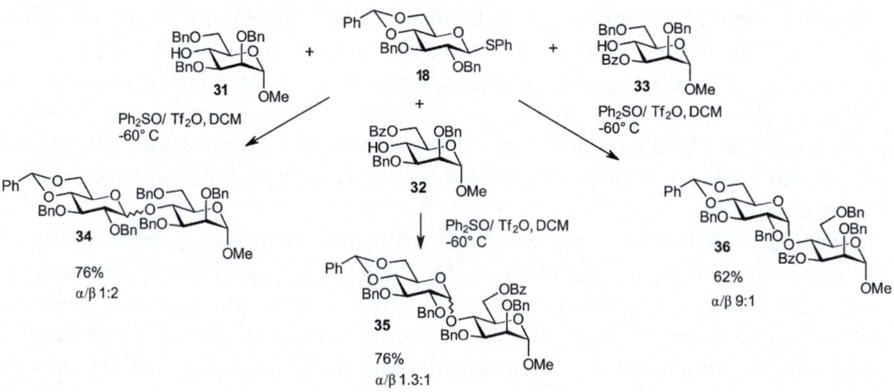

Scheme 5 Influence of protecting group pattern in glycosylation of mannoside acceptors **31**–**33**

deactivating 3-*O*-benzoyl group such as present in acceptor **21**. In case of the less reactive axial 4-OH group in the *galacto* series, the effect is less apparent, whereas for mannoside acceptors **31**–**33** again a switch from a β-selective to a highly α-selective outcome in the formation of disaccharide **34**–**36** was shown (Scheme 5).

These studies have been based on and complemented in recent years by compiling databases on relative reactivities of glycosyl acceptors covering inherent nucleophilicity, as well as steric, conformational, and electronic effects (Chang et al. 2021a, b).

2.3 Preactivation-Based Glycosylation Reactions

Conventional glycosylation reactions are usually performed on mixtures of glycosyl acceptor and glycosyl donor in suitable solvent systems – frequently in the presence of molecular sieves – and are started by the addition of the promoter. In recent years, however, the concept of pre-activation of the glycosyl donor prior to the actual glycosylation step followed by subsequent addition of the acceptor has gained substantial popularity in the field (Yang et al. 2018, Zhang et al. 2023). At an appropriately chosen pre-activation temperature, the formation of a defined reactive intermediate can be induced, securing a more reliable prediction of the glycosylation outcome with respect to anomeric selectivity as well as by-product formation (Yang et al. 2017, Tuck et al. 2022). In this context, the application of low-temperature NMR spectroscopy has been instrumental, allowing insightful monitoring of the activation stage (Frihed et al. 2015). The prime role of intermediate, covalently linked anomeric triflates in glycosylation reactions has been thoroughly investigated by Crich and has been confirmed by spectroscopic studies (Crich and Sun 1997). Whereas α-glycosyl triflates could be observed by variable temperature NMR measurements, the presence of highly reactive β-anomeric triflates has only recently been detected by dedicated NMR and kinetic studies in glycosylation studies with 2,3-di-*O*-methyl-4,6-*O*-benzylidene protected *gluco*-, *manno*- and *allo*-configured thiophenyl or sulfoxide donors (Santana et al. 2020). NMR data obtained at −65°C allowed to identify the formation of the respective β-glucosyl triflate in a range of 0.25–0.50% and in a substantially increased proportion for the β-allopyranosyl triflate in a 2.7:1 α/β ratio. A rapid equilibration of the anomeric triflates to the highly reactive β-triflate accounts for the high or even exclusive formation of α-glycosides. It should be noted that this equilibrium is furthermore sensitive to temperature, solvent polarity, nucleophilicity of the acceptor alcohol, and concentration of reactants.

The delicate balance of these factors governing the eventual outcome is illustrated in an example for the construction of the challenging β-selective 2-acetamido-2-deoxy-D-mannosaminuronic acid linkage that occurs in bacterial capsular polysaccharides, teichuronic acids and bacterial surface antigens. Irrespective of the anomeric thiophenyl or *N*-phenyltrifluoroacetimidate donor, activation with diphenyl sulfoxide and triflic anhydride at −80°C resulted in a 3:1 equilibrium in favor of the equatorial triflate **40** in the $^{1}C_{4}$ conformation despite sterically demanding axial substituents (Scheme 6). The reaction of the α-triflate mixture **39** and **40** with deuterated MeOH at −80°C then gave preferentially the β-mannoside (Walvoort et al. 2010). The anomeric sulfoxide donors **41** and **43** produced mixtures of anomeric sulfonium bistriflates **42** and **44** in addition to the anomeric triflates **39**

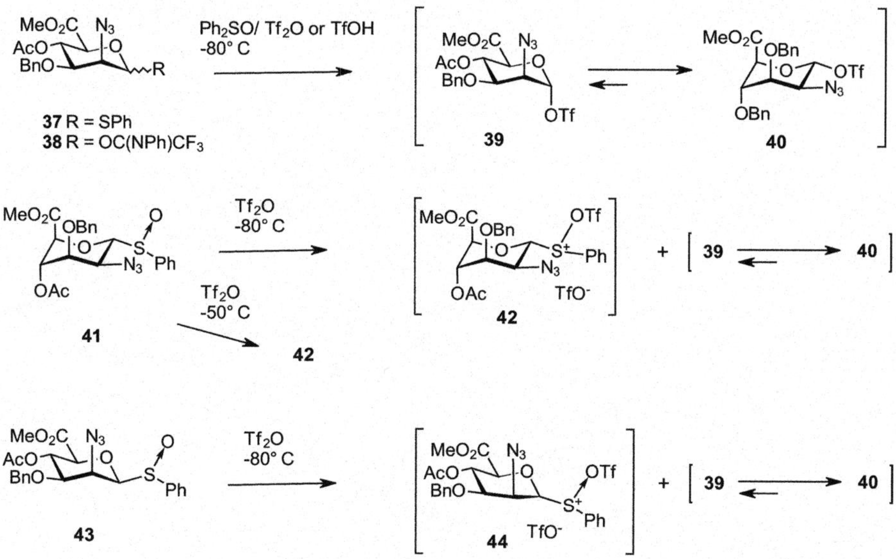

Scheme 6 Activated intermediates generated from donors **37**, **38** and **43** as identified by ^{1}H NMR

and **40**. It should be noted that the bistriflates **42** and **44** correspond to diastereoisomers arising from the respective diastereomeric sulfoxide donors.

Based on these mechanistic studies, a synthesis of tri-, penta- and heptasaccharide fragments of *Micrococcus luteus* teichuronic acid containing a challenging β-mannosaminyl linkage could be elaborated (Walvoort et al. 2010). Coupling of the 6-*O*-Fmoc protected glucosyl imidate donor **45** to O-4 of the 2-azido-2-deoxy-mannuronic thioglycoside acceptor **46**, promoted by triflic acid at low temperature, afforded disaccharide **47** in excellent yield (Scheme 7). Next, activation of the disaccharide thioglycoside donor with Ph_2SO-TfOH-TTBP followed by reaction with the primary alcohol **48** afforded the trisaccharide methyl glycoside **49** in good yield and high anomeric selectivity. The synthetic design allowed for the repetition of base-induced Fmoc deprotection and the coupling of the resulting alcohol **50** with donor **47** to lead to pentasaccharide **51** and eventually to heptasaccharide **53**. Deprotection of **53**, however, met with problems by β-elimination induced by the uronate, and furthermore, reduction of the azido groups and hydrogenolysis of the benzyl ethers could only be achieved by Birch reduction that was also accompanied by fragmentation of the product. Eventually, *N*-acetylation gave the deprotected heptasaccharide **54** in low yield over three steps. This example illustrates the fact that despite the main focus placed on the glycosylation reactions, similar efforts are needed to optimize and design reliable global deprotection protocols of protected intermediates to avoid loss of precious materials at late stages of the glycan assembly. Occasionally even deprotection of common protecting groups such as benzyl ethers may lead to complications due to partial saturation of the aromatic ring (Dhara et al. 2022; Crawford et al. 2020).

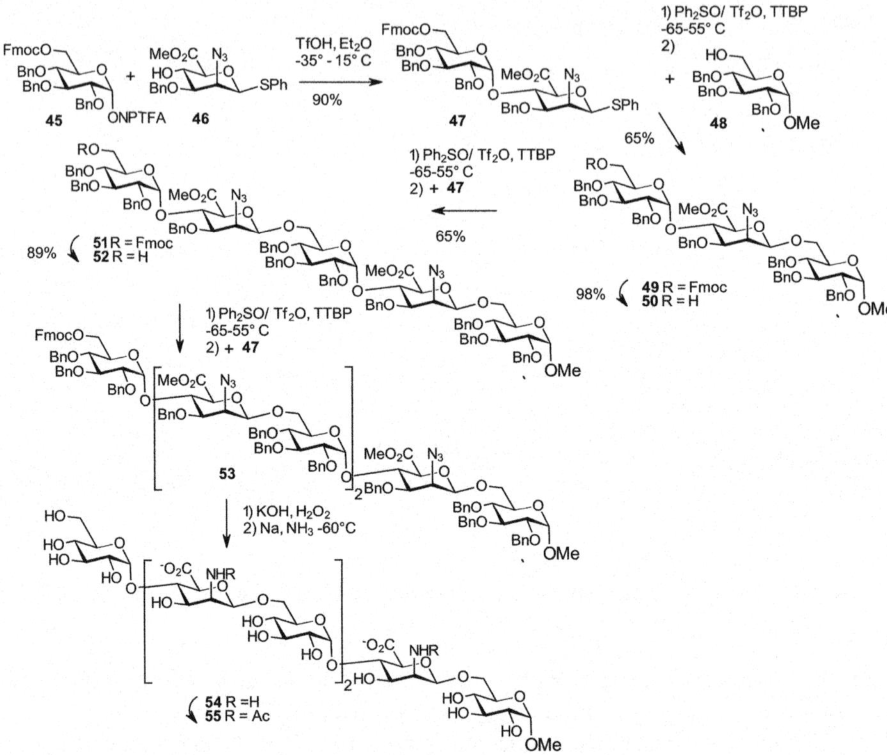

Scheme 7 Synthesis of oligosaccharide fragments of *Micrococcus luteus* teichuronic acid

2.4 Gold-Ion Catalyzed Glycosylation Reactions

The use of gold (I) and gold (III) catalysts has seen a dynamic increase in glycosylation chemistry within the past few years (Yu 2018; Li and Yu 2018; Hotha 2020). Whereas reactions promoted by gold (III) catalysts mainly operate through Lewis acidity and less via alkynophilic properties, the development of stable and reactive gold (I) catalysts allowed glycosylation reactions under mild and selective conditions. Basically, activation of the donor is initiated by the formation of a π-complex with an alkyne group followed by the formation of an isochromenylium intermediate that dissociates into an isochromen-4-yl gold(I) complex and an oxocarbenium type intermediate. Subsequent reaction with a glycosyl acceptor produces the glycoside and triflic acid that is necessary for the protodeauration step to give benzopyrenone and to close the catalytic cycle. Gratifyingly, the approach is suitable for acid-sensitive substrates, and the donors are shelf-stable (Scheme 8).

In one of the first studies, the *ortho*-hexynylbenzoyl donor **56** was coupled to the α-glucoside acceptor **57** in the presence of triphenylphosphine-gold (I) triflate at room temperature to give benzopyrenone **59** and the β-(1→4)-linked disaccharide **58**

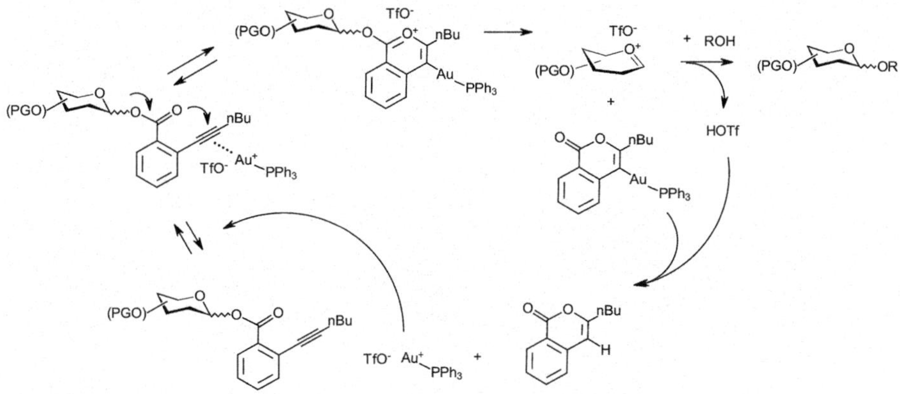

Scheme 8 Catalytic cycle of gold-catalyzed alkyne activation and glycosylation

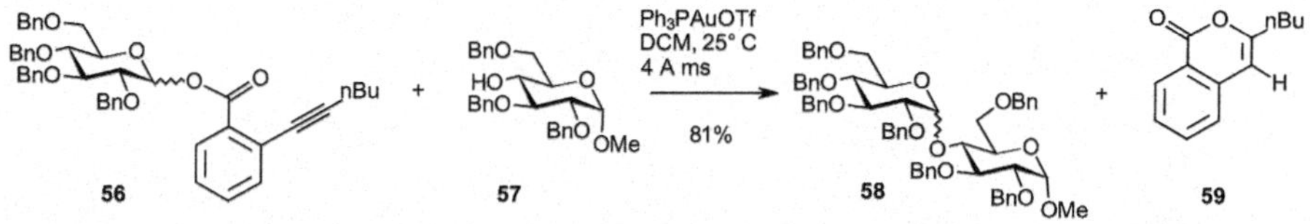

Scheme 9 Coupling of a glycosyl *ortho*-hexynylbenzoate **56** with alcohol **57**

in 71% yield as a 2:1 α/β mixture (Li et al. 2008). The reaction is triggered upon dissociation of the catalyst to release $[Ph_3PAu]^+$ as the actual activating agent (Tang et al. 2013) (Scheme 9).

In continuation of these earlier studies, the approach has been further developed and is now considered a valuable additional methodology in glycosylation chemistry. The main assets of gold (I)-catalyzed glycosylation beyond the stability of the ABz donors pertain to a broad scope of substrates, including complex triterpene saponins, orthogonality to conventional protecting and leaving groups and mild reaction conditions (Yu 2018). Anomeric selectivities have been improved by anchimeric assistance of protecting groups as well as by exploiting the impact of modifying solvents. As an example, illustrating the challenging assembly of *cis*-configured glycosidic bonds, the synthesis of a pentasaccharide fragment from the O-antigen of *Yersinia enterocolitica* O:3 is shown (Scheme 10). First, the anomeric *ortho*-hexynylbenzoyl donor **62** was prepared in excellent yield by carbodiimide promoted reaction of the 6-deoxy-*altro*-hemiacetal **60** with benzoic acid derivative **61**. Next, the donor was reacted with the glycosyl acceptor **63** in the presence of the gold bistriflylimide promoter to produce the β-(1 → 2)-linked disaccharide in excellent yield and high anomeric selectivity (Shen et al. 2017). Removal of the 2-*O*-benzyl group of **63** was sluggish and needed acidic conditions to eventually generate acceptor **64** in very high yield. Iteration of the reaction sequence and global deprotection finally gave the pentasaccharide **67** that adopts a left-handed helical structure.

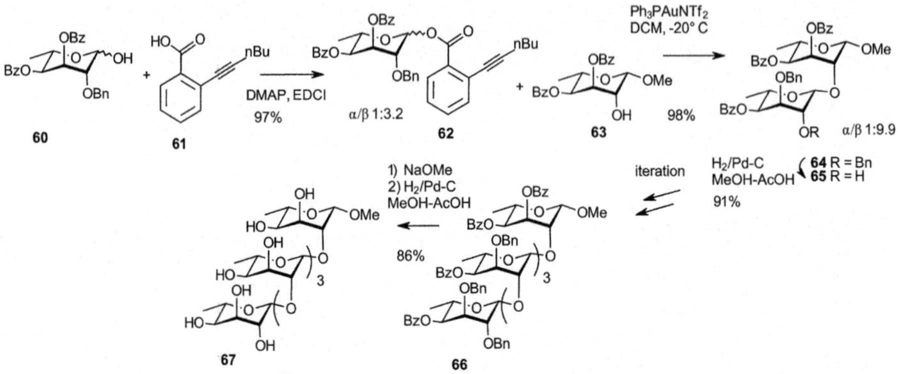

Scheme 10 Synthesis of an O-Antigen fragment of *Yersinia enterocolitica* O:3

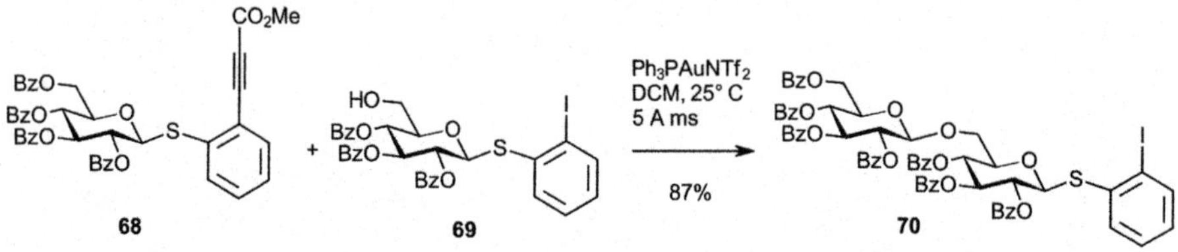

Scheme 11 Glycosidation of the latent 2-iodophenyl thioglycoside donor **68**

Besides the *ortho*-hexynylbenzoate esters, also alkynyl glycosyl carbonates and glycosyl 1,2-orthoesters have successfully been used (Mishra et al. 2016). The spectrum of related donors has recently been expanded by *ortho*-methoxycarbonylethynylphenyl (MCEPT) thioglycosides as complementary donors, allowing for divergent orthogonal one-pot protocols (Scheme 11). It combines the stability and versatility of thioglycoside donors with Yu donors and is suitable for active-latent donor activation strategies.

The reaction of the perbenzoylated MCEPT donor **68** with the *ortho*-iodophenyl thioglycoside **69** gave the β-(1→6)-linked disaccharide **70** in 87% yield (Liu et al. 2023). The thio-2-iodophenyl group can subsequently be converted into the MCEPT glycoside by Sonogashira coupling with propargylic acid methyl ester, allowing for iteration of glycosylation sequences. This approach has been applied to generate nonasaccharide units of the surface polysaccharide of *Acinetobacter baumannii* (O22), one of the notoriously resistant bacterial strains listed in the ESKAPE category (Scheme 12).

Coupling of the MCEPT acceptor **72** with the *N*-trichloroacetyl (TCA) protected lactosyl donor **71** in the presence of $Ph_3PAuOTf$ gave the β-(1→3)-linked tetrasaccharide MCEPT donor, which was not isolated but directly reacted with the disaccharide linker **74** in 75% overall yield in a one-pot set-up. Next, all three primary acetate esters were cleaved under acidic conditions to give acceptor **75**. Subsequent glycosylation with the DtBSi-protected galactosyl MCEPT donor **76**

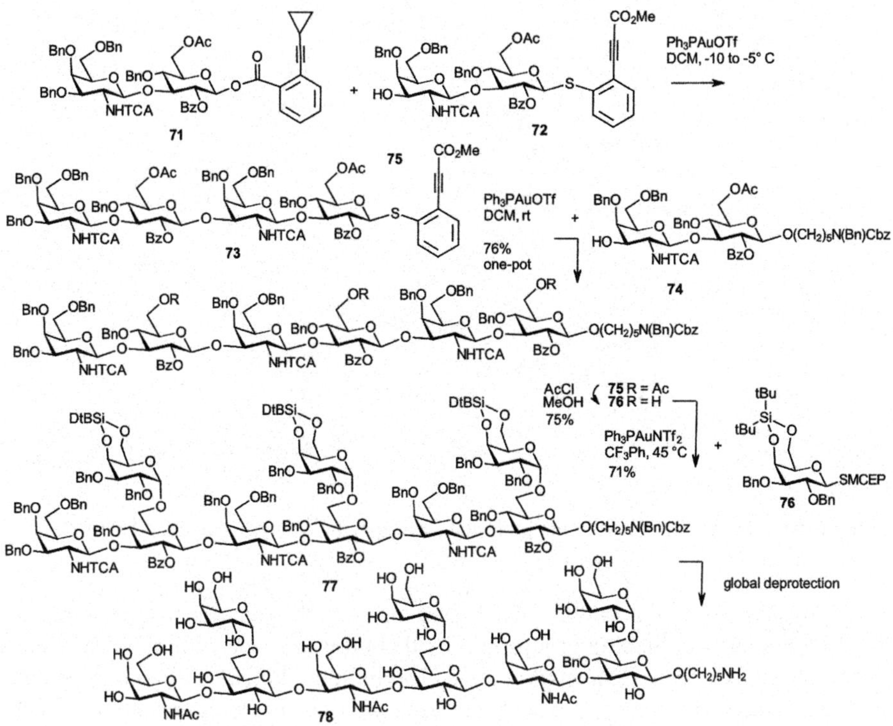

Scheme 12 Synthesis of a nonasaccharide ligand of *Acinetobacter baumannii* surface polysaccharide

promoted by $Ph_3PAuNTf_2$ produced the protected nonasaccharide **77** in excellent yield and α-selectivity. The bulky 4,6-*O*-silylene protecting group effectively blocks the β-face and secures perfect anomeric selectivity in this triple galactosylation reaction. Global deprotection and conversion of the trichloroacetamido into acetamido groups eventually afforded the linker nonasaccharide **78**, ready for conjugation to a protein carrier. The success of the ABz-based and related glycosylation methods is further documented by a hallmark synthesis of arabinofuranoside glycans (containing 23 arabinofuranosyl and 2 galactofuranosyl residues) from *Mycobacterium tuberculosis* cell wall polysaccharides, triterpene saponins, as well as the preparation of a 30-mer of branched α-linked glycans (Pasari et al. 2018; Li et al. 2010; Zhang et al. 2021).

2.5 One-Pot Glycosylation

The graded reactivity of glycosyl donors allows the stepwise elongation toward oligosaccharides to be carried out as one-pot protocols, thereby minimizing purification efforts. Thioglycoside donors in particular are suitable donors as their

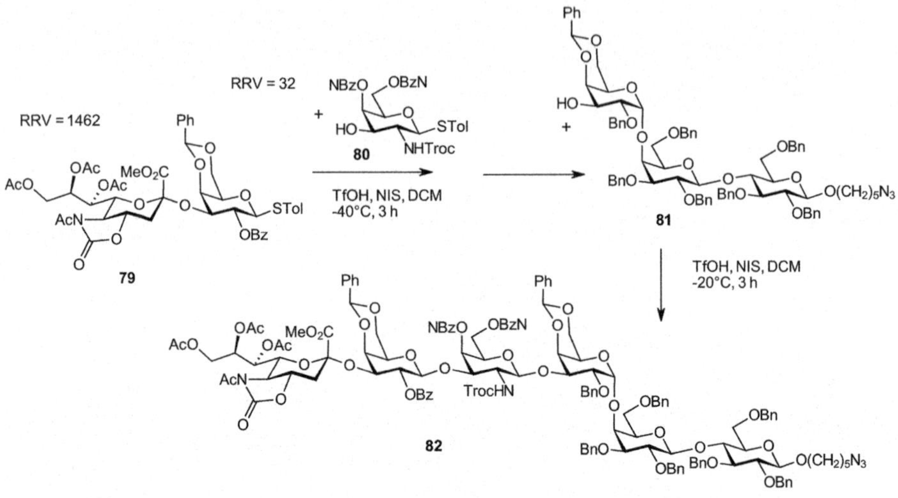

Scheme 13 One-pot synthesis of the protected stage-specific embryonic antigen SSEA-4

reactivity can be tuned by the installation of suitable electron-withdrawing and electron-donating protecting groups to provide building blocks with defined relative reactivity values (RRV). Meanwhile, over 150 building blocks with experimentally measured reactivity values have been accumulated, allowing a programmable one-pot synthesis of biologically relevant oligosaccharides (Cheng et al. 2018). The strategy is illustrated for the synthesis of the protected SSEA-4 hexasaccharide (Scheme 13). The highly reactive disaccharide donor **79** could be activated with TfOH/NIS at low temperature without affecting the leaving group of thioglycoside **80** to afford an intermediate trisaccharide donor, which was successfully activated at slightly higher temperature in the coupling reaction with the spacer trisaccharide acceptor **81** to eventually give the protected hexasaccharide **82** as a precursor to the stage-specific embryonic antigen 4 in 43% overall yield (Cheng et al. 2018).

As a second example, the synthesis of a hexasaccharide acceptor that has eventually been extended by enzyme-catalyzed galactosylation and sialylation to produce a lipooligosaccharide fragment of *Haemophilus ducrey* is shown in Scheme 14 (Luo et al. 2023). First, the trisaccharide **88** harboring an L-*glycero*-D-*manno*-heptopyranosyl unit was formed from a chain-elongated mannose precursor **84** that was coupled with STol glucosyl donor **83** prior to oxidative cleavage of the $PhMe_2Si$ group.

The resulting disaccharide **85** was then elaborated over several steps into the *ortho*-hexynylbenzoyl donor **86** via acetolysis of the methyl group and carbodiimide-promoted acylation with levulinic acid. Donor **86** was then coupled by a gold(I)-catalyzed reaction in good yield to the axial 5-OH group of a Kdo acceptor **87** to furnish the trisaccharide **88**. The exquisite α-anomeric selectivity was secured by the anchimeric assistance of the 2-*O*-benzoyl group in the donor **86**. Removal of the 6-*O*-levulinoyl group using hydrazine acetate afforded the trisaccharide acceptor **89**.

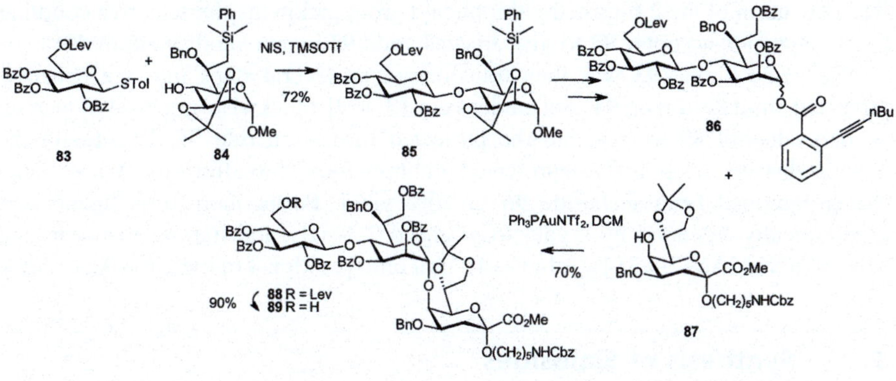

Scheme 14 Synthesis of LPS core trisaccharide acceptor of *Haemophilus ducrey*. (Adapted with permission from Luo et al. 2023, Copyright 2023 American Chemical Society)

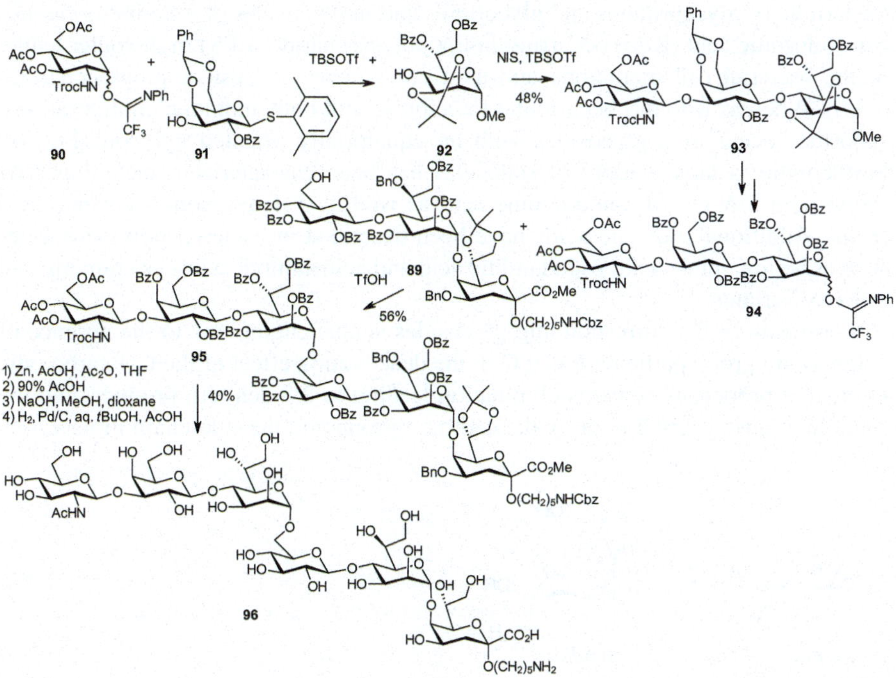

Scheme 15 Synthesis of a hexasaccharide ligand of a lipooligosaccharide from *Haemophilus ducrey*. (Adapted with permission from Luo et al. 2023, Copyright 2023 American Chemical Society)

Next, the trisaccharide donor comprising a 6-epimeric D-*glycero*-D-*manno*-heptopyranosyl moiety was assembled in a one-pot approach (Scheme 15). First the Troc-protected NPTFA-donor **90** was connected to O-3 of the benzylidene-protected galacto-thioglycoside **91** promoted by TBSOTf, followed by

NIS-activation of the 2,6-dimethylthiophenyl group, allowing the selective coupling to the heptoside acceptor **92** to give trisaccharide **93** in an overall yield of 48%.

The trisaccharide **93** was then transformed in several steps into the benzoyl-protected imidate donor **94** that underwent [3 + 3] blockwise glycosidation with primary alcohol **89** to generate the protected hexasaccharide **95**. Eventually the hexasaccharide was gobally deprotected and transformed to afford the *N*-acetylated spacer-equipped hexasaccharide **96** in 40% yield. Furthermore, this ligand was enzymatically extended by a galactosyl and α-2,3-sialyl transferase, incorporating a distal α-Neu5Ac-(2→3)-β-Gal-(1→4)- unit onto position 4 of the GlcNAc residue.

3 Synthesis of Sialosides

The synthesis of glycosides containing 2-ulosonic acid entities adds to the complexity of oligosaccharide assembly. These linkages are present in sialosides decorating the termini of glycoproteins and glycolipids and in glycosides of 3-deoxy-D-*manno*-oct-2-ulosonic acid (Kdo) occurring in lipopolysaccharide (LPS) in the outer leaflet of the bacterial cell membrane and as a component of capsular polysaccharides (CPS). Whereas Kdo has been found to be present in both anomeric configurations, sialosides occur as α-glycosides with an equatorially oriented aglycon (Fig. 6). Furthermore, a large variety of sialosides has been characterized, including 5-*N*-acetyl- and 5-*N*-glycolyl-neuraminic acid as well as 3-deoxy-non-2-ulosonic acid (Kdn). Additional ulosonic acids have been described in bacterial polysaccharides such as legionaminic acid, pseudaminic acid and acinetaminic acids as epimeric and 9-deoxy variants.

Formation of the corresponding glycosides is challenging due to the absence of neighboring group participation at C-3, the deactivating effect of the C-1 carboxylic group, the propensity towards elimination leading to 2,3-dehydro products and the steric hindrance exerted both by the carboxyl group and the side-chain moiety.

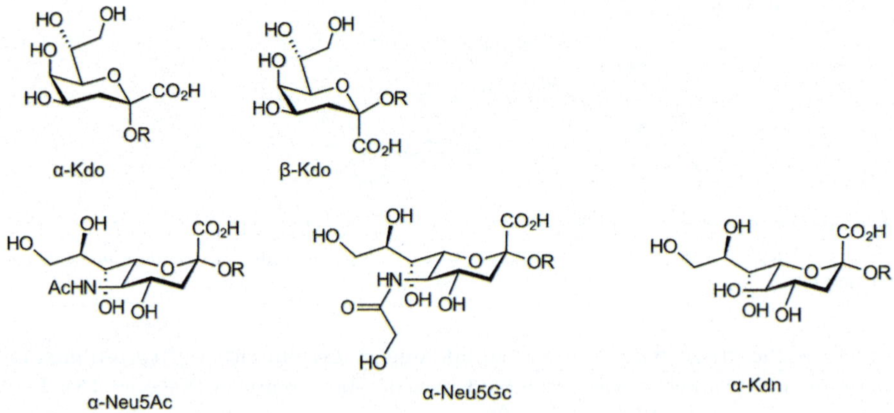

Fig. 6 Structure and anomeric configuration of naturally occurring 2-ulosonic glycosides

3.1 Synthesis of Glycosides of Sialic Acid

The synthesis of sialosides has been comprehensively reviewed (Vibhute et al. 2021; Panza et al. 2018; Krasnova and Wong 2019; De Meo and Gloeckner 2021) and of Kdo glycosides (Kosma 2016; Pradhan 2023), and thus only a few examples of recent developments will be covered herein.

Selectivity of α-anomeric Neu5Ac glycoside formation using a large variety of dedicated leaving and protecting groups has been prepared by nitrilium-ion intermediates in solvent mixtures containing acetonitrile as well as by neighboring group participation exerted by temporary protecting group at C-3. In a recent approach, immobilization of the carboxyl group by covalent tethering with a dedicated linker to the 5-position of neuraminic acid and position 5 of Kdo has been introduced, thereby effectively restricting unfavorable attack of the nucleophile to the anomeric center (Komura et al. 2019; Hamajima et al. 2022).

A major improvement has been the introduction of 4*O*,5*N*-oxazolidinone thioglycoside donors by Tanaka and its *N*-acetyl derivative by Crich under NIS-TfOH promotion (Tanaka et al. 2006; Crich and Li 2007). The potential of these donors is illustrated in the synthesis of α-(2→8)-linked oligosialosides (Koinuma et al. 2020) corresponding to subunits of biomedically important polysialic acids (Scheme 16). In order to increase the reactivity of the hindered 8-OH group, a 7,9-*O*-carbonyl group was introduced in the side chain of **97,** followed by *N*-acetylation **98** in excellent yield. This way, the detrimental effect of the intraresidue hydrogen bond from the 8-OH group to the 5-acetamido group could

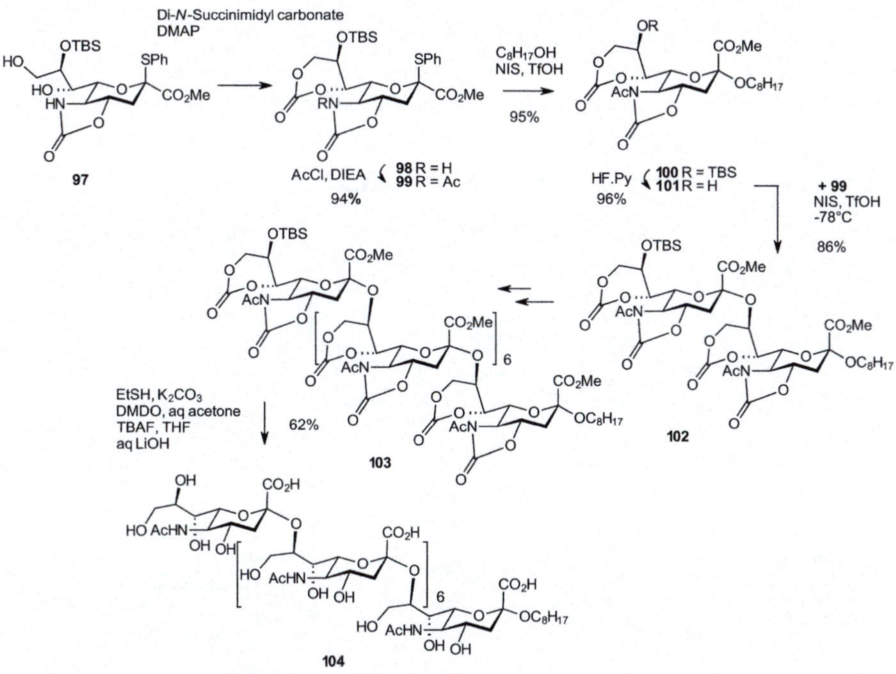

Scheme 16 Synthesis of an α-(2→8)-linked octasialoside

be eliminated. Donor **99** was then reacted with 1-octanol to give **100,** followed by the removal of the TBS ether to afford the glycosyl acceptor **101**. Subsequent coupling with donor **99** in the TfOH-NIS system at low temperature produced the trisaccharide **102** in excellent yield and with complete anomeric selectivity. Iteration of this sequence eventually gave access to a series of oligosialosides up to octasaccharide derivative **103**. Cleavage of the oxazolidinone units under basic conditions was not feasible but was eventually achieved by formation of the 4-*O*-ethylthiocarbamate derivative followed by its oxidative removal with DMDO. Removal of the TBS-ether by the action of TBAF and hydrolysis of the methyl ester groups with LiOH then gave the deprotected octasialoside **104**.

An alternative approach to avoid the impact of the hydrogen-bonding property of the 5-acetamido group was the introduction of a bis-*N*-acetyl moiety (NAc_2) or a 5-azido substituent (Demchenko and Boons 1998; Yu et al. 2001). In a recent variant of the latter version, a novel sialic acid donor harboring a 1-picolinyl ester group that exerts a neighboring group effect and enables hydrogen bonding to the acceptor alcohol has recently been elaborated (Chen et al. 2019). The main effect of the 1-picolinyl group, however, was rationalized on the basis that the picolinyl group intercepts the arising oxocarbenium ion, leading to a stabilization of the β-triflate that reacts with the nucleophile by inversion to give the α-anomeric product. The picolinyl group was introduced upon demethylation of the sialyl thioglycoside methyl ester **105,** followed by alkylation of the free acid with picolinyl bromide (Scheme 17). The resulting donor **106** was then tested in numerous model reactions, and its potency has been verified in the synthesis of type I and type II *N*-glycan antennae.

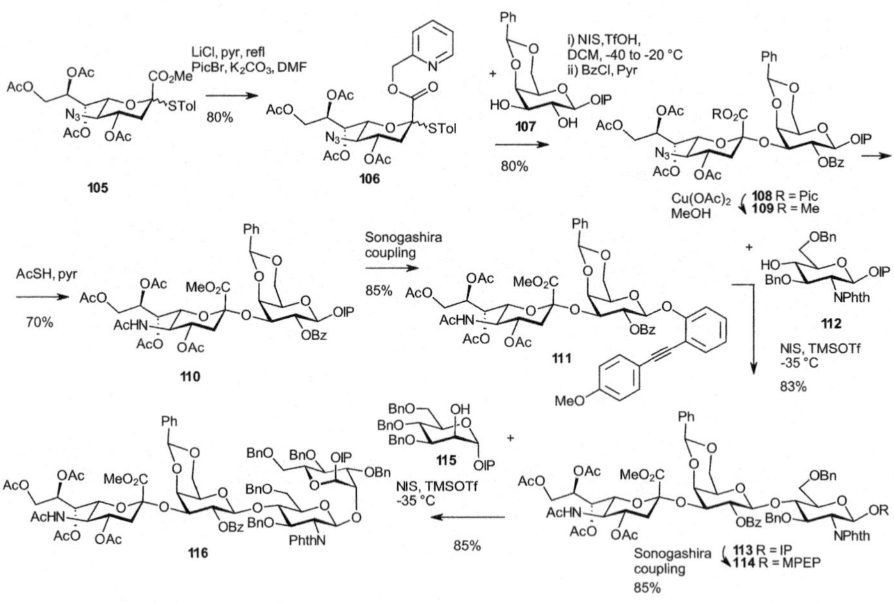

Scheme 17 Synthesis of a tetrasaccharide donor corresponding to type II antenna of *N*-glycans

Regioselective 3-*O*-glycosylation of the *ortho*-iodophenyl galactoside **107** followed by 2-*O*-benzoylation gave the α-(2→3)-linked disaccharide **108** as a latent glycosyl donor in a very good yield. Next, the picolinyl ester was reconverted into the methyl ester **109** by reaction with $Cu(OAc)_2$ in MeOH, followed by treatment of the 5-azido group with thioacetic acid to produce the acetamido derivative **110**. The latent iodophenyl group in **110** was then activated by Sonogashira coupling with *p*-methoxyphenylethyne (MPEP) to give disaccharide donor **111** that was subsequently coupled with the iodo-phenyl GlcN acceptor **112** promoted by NIS-TMSOTf to afford the β-(1→4)-linked trisaccharide **113** in 83% yield. Iteration of the activation step to generate the MPEP donor **114** was followed by coupling to the mannoside acceptor **115** and gave the latent tetrasaccharide donor **116** that is ready for extension of *N*-glycan core units.

A mixed strategy including the 5-azido and 5-NAc_2 approaches has recently been applied to the synthesis of asymmetrically deuterated undecasaccharides of sialylated *N*-glycans in order to assess recognition of the glycans by the viral neuraminidase H1N1 (Shirakawa et al. 2021). The 5-bis-*N*-acetylated sialyl donor provided sufficient α-selectivity for the assembly of the tetrasaccharide imidate donor **117**, which was coupled to OH-3 of the β-mannosyl unit in the trisaccharide acceptor **118** promoted by TMSO-triflate to afford the heptasaccharide **119** in 71% yield. The α-selectivity in the glycosylation step was achieved by using ether as solvent (Scheme 18).

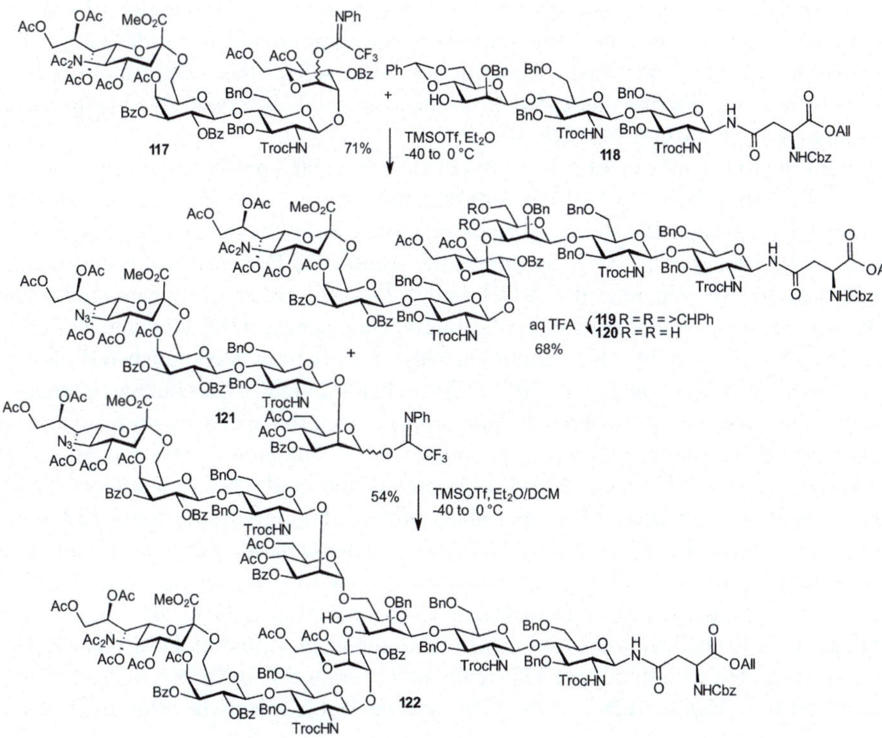

Scheme 18 Synthesis of an asymmetric sialylated undecasaccharide *N*-glycan

Hydrolysis of the 4,6-O-benzylidene group gave the diol **120** ready for a regioselective glycosylation of the primary alcohol. Next, the analogous 5-azido tetrasaccharide imidate donor **121** was prepared and reacted with the acceptor **120** under similar conditions, albeit in a 1:1 ether-dichloromethane mixture to improve solubility. This way, the protected undecasaccharide **122** was obtained, and – by inverting the sequence of sialylated donors in the glycosidation steps – the 5-azido sialyl moiety was additionally introduced in the *N*-glycan α-(1→3)-mannosyl arm. Selective reduction of the azido groups and reaction with deuterated acetyl chloride gave the deuterated acetamido analogues, which after global deprotection were subjected to H1N1 neuraminidase treatment. The assay showed that the sialic acid located at the α-(1→3) arm was preferentially cleaved by the viral enzyme. Eventually, NAc_2 donor **117** was also successfully employed for the construction of a tetrasialylated *N*-glycan.

3.2 Kdo-Glycoside Formation

The 8-carbon sugar 3-deoxy-D-*manno*-oct-2-ulosonic acid (Kdo) constitutes a highly conserved component in the inner LPS core of Gram-negative bacteria but is absent in mammalian organisms. Hence, there is significant pharmaceutical interest both concerning the development of inhibitors of the biosynthetic pathways of Kdo formation and its glycosyl transfer as novel antimicrobials and also to utilize synthetic Kdo ligands and antigens for diagnostics and for the generation of specific and cross-reactive antibodies. Whereas glycoside formation of sialosides targets the formation of the equatorial (α-anomeric) glycosides, the chemistry of Kdo glycosides has to address the selective synthesis of axial (α-anomeric) and equatorial (β-anomeric) glycosides (see Fig. 6).

For selective synthesis of α-Kdo glycosides, dedicated protecting groups such as 4,5-*O*-isopropylidene, 4,5-*O*-*tert*-butyldimethylsilyl, as well as 5,7-di-*tert*-butylsilylene groups that sterically block the β-face have been elaborated (Yoshizaki et al. 2001; Huang et al. 2015) as well as the addition of DMF as modifying solvent (see Scheme 3). Whereas the 4-OH and 8-OH groups of Kdo can readily be glycosylated to yield bacterial Kdo oligomers, the reaction at the axial and sterically hindered 5-OH group by a Kdo donor has only recently been accomplished (Pokorny and Kosma 2015a; Huang et al. 2015). The α-Kdo-(2→5)-α-Kdo linkage is present in the inner core of *Acinetobacter baumannii*, a bacterial species with pronounced antimicrobial resistance and which is commonly encountered in severe nosocomial infections. The synthesis of several fragments (Zhou et al. 2022) commenced with the α-selective coupling of the orthogonally protected thioglycoside donor **124** to the glycosyl acceptor **123** promoted by TBPA to give disaccharide **125** in excellent yield (Scheme 19). The 4-*O*-allyl group was then removed by isomerization of the allyl group into a propenyl moiety followed by hydrolysis to give the 4-OH acceptor **126** that partially formed an inter-residue lactone to the distal Kdo unit. Disaccharide **126** was next reacted with thiodonor **127** again under promotion by TBPA to produce the branched trisaccharide **128** in 57% yield and perfect α-selectivity. Removal of the

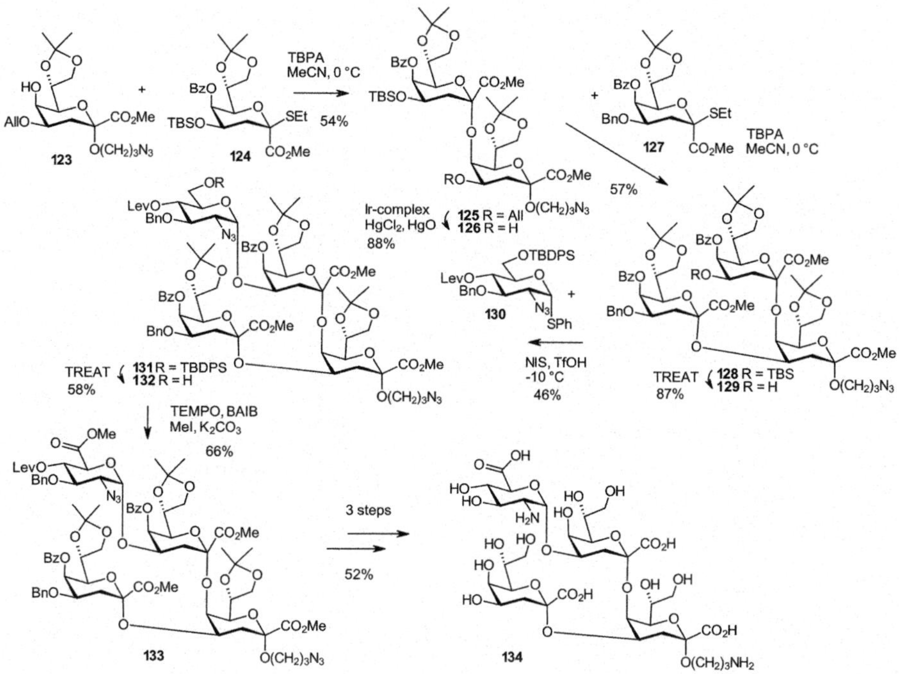

Scheme 19 Synthesis of a spacer derivative of the inner core LPS from *Acinetobacter baumannii*

4-*O*-silyl group by triethylamine hydrofluoride afforded the trisaccharide alcohol **129** with was condensed in the presence of NIS/triflic acid with the 2-azido-2-deoxy glucoside thiodonor **130** in fair yield but with excellent α-selectivity. The resulting tetrasaccharide **131** was then converted into the glucuronide derivative by cleavage to the TBDPS group followed by TEMPO and BAIB oxidation and esterification to generate the methyl ester **133**. Global deprotection involved acidic cleavage of the acetonides, ester saponification and hydrogenation to give the aminopropyl tetrasaccharide **134**.

This ligand and related fragments were then used to screen sera from infected patients, which showed that the branched 4,5-Kdo trisaccharide was a major epitope for IgG antibodies.

As an example, for the utilization of a temporary α-directing protecting group at C-3 of Kdo, the assembly of an inner core unit from *Campylobacter jejuni* is shown in Scheme 20 (Li et al. 2020).

The α-selectivity of the Kdo-fluoride donor **135** is based on the neighboring group effect of the axial 3-iodo group and moreover prevents formation of the 2,3-dehydro derivative that is frequently encountered in Kdo-glycosylations (Pokorny and Kosma 2015b). Regioselective glycosylation of the more reactive equatorial 4-hydroxy group of glycosyl acceptor **136** by the 3-iodo Kdo fluoride **135** in the presence of $BF_3{\cdot}Et_2O$ afforded the α-(2→4)-linked Kdo disaccharide **137** in

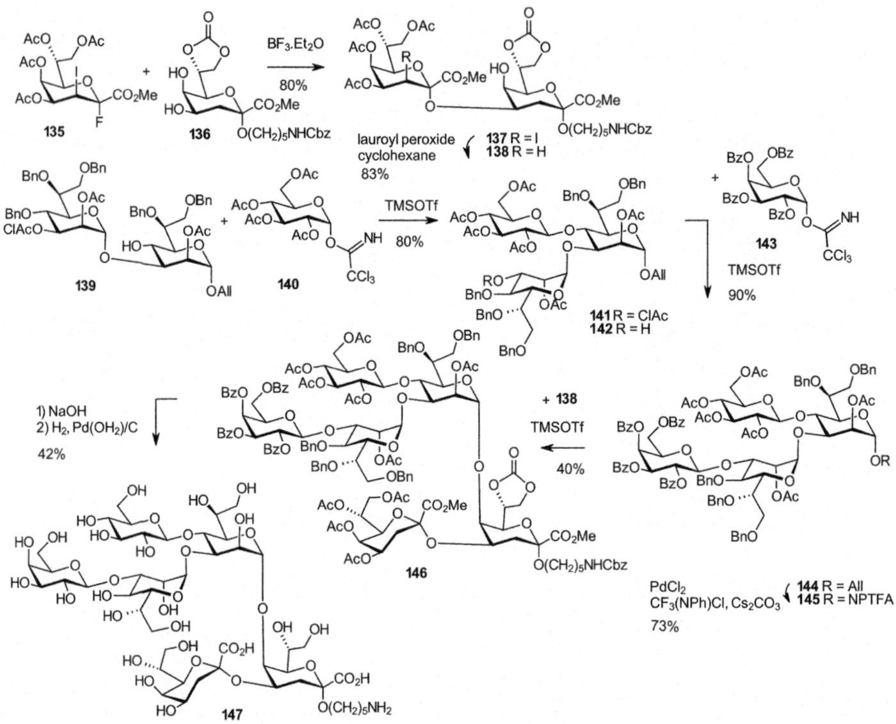

Scheme 20 Synthesis of the inner core hexasaccharide from *Campylobacter jejuni*

80% yield. Next, the iodo group was removed by catalytic hydrogen transfer to afford disaccharide **138**. For the second building block, reaction of the peracetylated glucosyl imidate donor **140** with the α-(1→3)-linked diheptoside **139** gave the β-(1→4)-connected trisaccharide allyl glycoside **141** in very good yield.

Selective cleavage of the distal 3-*O*-chloroacetyl group with hydrazine dithiocarbonate led to trisaccharide acceptor **142** that was subjected to galactosylation with the benzoylated trichloroacetimidate donor **143**. This way the tetrasaccharide **144** was obtained in excellent yield. The tetrasaccharide allyl glycoside **144** was next converted into the imidate donor **145** via intermediate formation of the hemiacetal and subsequent reaction with *N*-phenyl trifluoroacetimidoyl chloride in the presence of Cs_2CO_3.

The tetrasaccharide donor **145** was then reacted with the Kdo disaccharide acceptor in an excellent yield of 40% considering the steric demand in a [4 + 2] block synthesis and the low reactivity of the 5-OH group in the Kdo unit. Global deprotection of **146** eventually afforded the hexasaccharide **147**. The 5-aminohexyl spacer derivative was then subjected to enzymatic extension to introduce sialylated ganglioside epitopes and was printed as a glycan array. Binding studies with sera from patients inflicted with Guillain-Barré syndrome showed the presence of cross-reactive antibodies directed against the GM1a epitope and a heptosyl moiety.

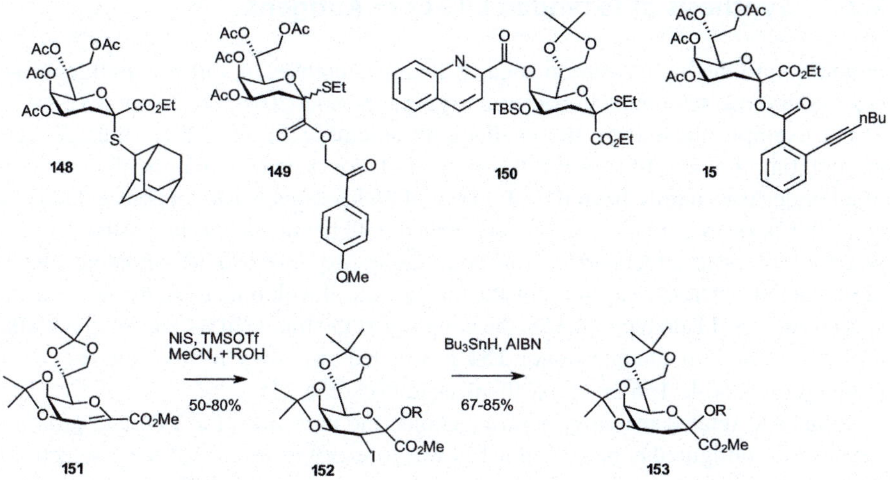

Scheme 21 Kdo donors used for β-ketoside formation

Synthesis of Kdo ß-Glycosides

Whereas the synthesis of α-ketosides of Kdo has been fairly well developed, the preparation of β-anomeric glycosides has only recently been targeted in more detail. This increased interest is due to the occurrence of β-Kdo units in bacterial exopolysaccharides and capsular polysaccharides. In particular, oligomeric β-(2→7)-Kdo-β-(2→4)-Kdo units serve as conserved anchors of capsular polysaccharides in the membrane (Willis and Whitfield 2013; Doyle et al. 2023). In a seminal paper, Crich described the major role of the side chain *trans-gauche* conformation that governs the outcome of glycosylation reactions of Kdo donors. Accordingly, S-adamantyl thioglycoside Kdo donor **148** at −78°C with NIS/TfOH as activators gave excellent β-selectivity (Ngoje and Crich 2020). Thioglycoside donors of Kdo 4′-*O*-phenacyl carboxylate derivatives **149** as well as 5-*O*-(2-quinolinecarbonyl) derivative **150** exhibited high β-anomeric selectivities upon activation with NIS/AgOTf in the former and NIS/TfOH in the latter case (Mazur et al. 2016; Huang et al. 2018). Gold-(I)-catalyzed reactions of ABz Kdo donor **15** in the absence of DMF as cosolvent (see also Scheme 3) with several glycosyl acceptors also proceeded with exclusive β-selectivity (Mi et al. 2017) (Scheme 21).

The bis-isopropylidene-protected glycal methyl ester **151** served as versatile precursor of β-Kdo glycosides. In the presence of NIS/TMSO-triflate, preferential formation of the 2,3-*trans* addition products **152** proceeded in good yield and high β-selectivity (Pradhan et al. 2014). Radical deiodination with AIBN and tributyl tinhydride eventually gave a series of protected β-Kdo derivatives **153** in good yields.

3.3 Synthesis of Extended LPS-Core Antigens

In addition to the higher-carbon sugar Kdo, also heptoses of various configurations are conserved constituents in bacterial glycans (Kosma 2008; Oscarson 2012). As a recent example illustrating the challenging assembly of six 1,2-*cis* linkages and highlighting the versatility and robustness of glycosyl imidate donors, the synthesis of an undecasaccharide from the LPS core of *Helicobacter pylori* is presented (Zou et al. 2022). *H. pylori* infections are associated with peptic ulcers and gastric cancers, and the molecular understanding of the immune response against antigenic sites at the bacterial surface is a prerequisite for vaccine development. Starting from the spacer-equipped Kdo alcohol **155**, the coupling at the hindered 5-OH group with the L-*glycero*-D-*manno* imidate donor **154** proceeded smoothly to give the α-(1→5)-linked disaccharide **156** in an excellent yield (Scheme 22).

Similar to related mannosylation reactions, the presence of a 2-*O*-acyl group in the *manno*-configured heptose donor **154** was foreseen to secure 1,2-*trans* selectivity due to neighboring group participation. Next, the heptose core was formed by attaching a second L-*glycero*-D-*manno* heptose unit in an α-(1→3)-linkage. Selective cleavage of the levulinoyl ester by hydrazine acetate furnished acceptor **157** to be condensed with the 2-*O*-levulinoyl protected imidate donor **158,** leading to trisaccharide **159** in good yield. Removal of the 2-*O*-levulinic ester gave acceptor **160** that

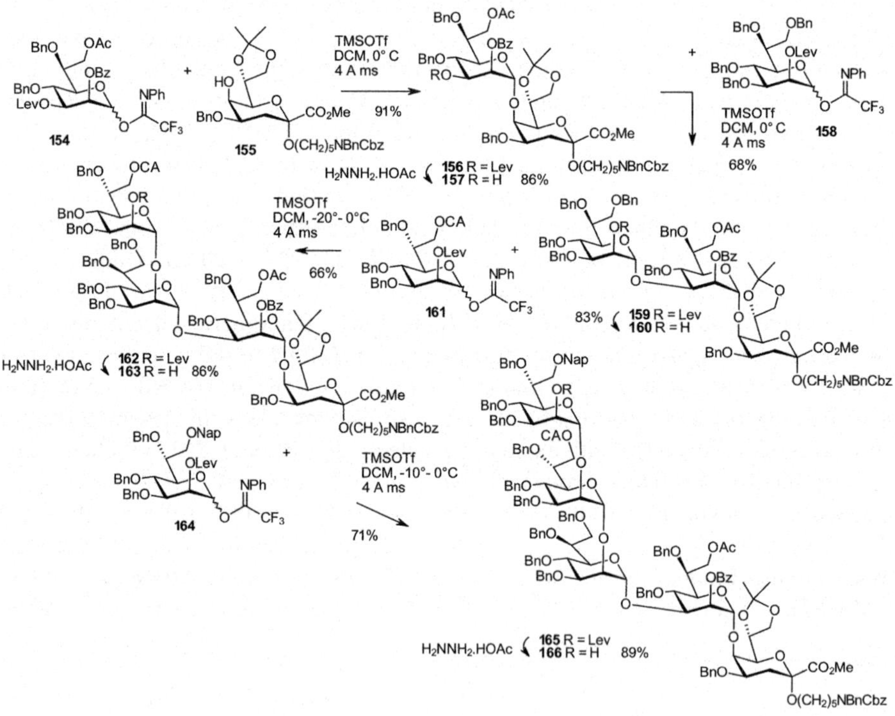

Scheme 22 Synthesis of the pentasaccharide acceptor **166**. (Adapted with permission from Zhou et al. 2022, Copyright 2022 American Chemical Society)

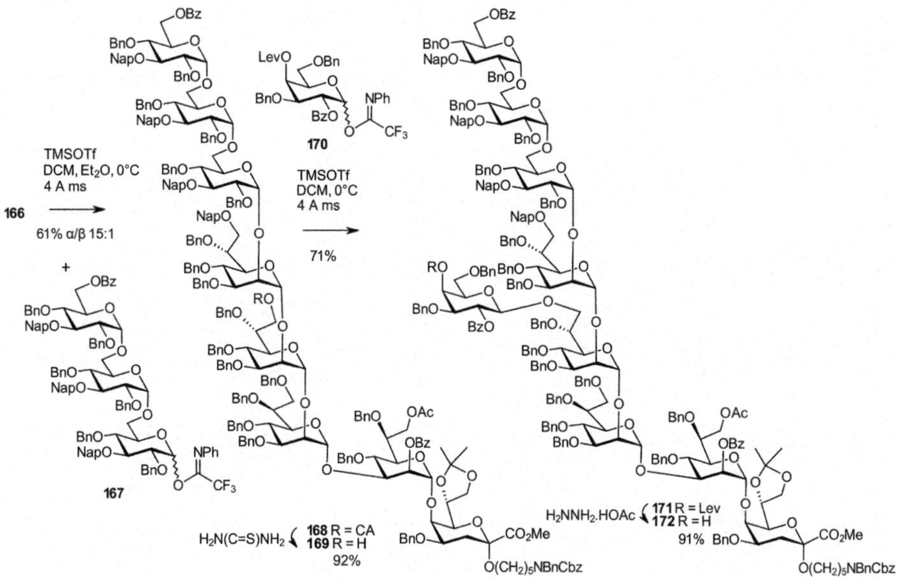

Scheme 23 Synthesis of nonasaccharide acceptor **172**. (Adapted with permission from Zhou et al. 2022, Copyright 2022 American Chemical Society)

was elongated by reaction with the 7-*O*-chloroacetyl protected the 6-epimeric D-*glycero*-D-*manno* imidate donor **161** in 66% yield. Repetition of the sequence (**162**→**163**) allowed the introduction of the fourth heptose unit at O2 of acceptor **163** using donor **164** containing an orthogonal 7-*O*-naphthylmethyl protection to afford the pentasaccharide core unit **165** in high yield. For the introduction of the α-(1→6)-linked glucotrioside fragment, the distal levulinoyl group was cleaved and gave the 2-OH acceptor **166**. Next, the building block **167** was separately elaborated and subjected to a block-wise [3 + 5] coupling reaction with acceptor **166**. The coupling reaction promoted by TMSOTf at 0°C was performed in an ethereal solvent mixture to support the formation of the α-anomeric product, and the octasaccharide **168** was produced in a high yield and remarkable 15:1 α/β ratio (Scheme 23). The lateral galactosyl unit was then introduced upon thiourea-promoted cleavage of the 7-*O*-chloroacetate to give **169** followed by reaction with the imidate donor **170**.

Due to the presence of a 2-*O*-benzoyl group, the glycosylation proceeded in a β-selective fashion and gave the branched nonasaccharide **171** in 71% yield. Cleavage of the 4-*O*-levulinoyl group at the galactosyl moiety eventually led to nonasaccharide acceptor **172**. The final [2 + 9] coupling step was carried out by the reaction of **172** with the α-(1→3)-linked glucobiosyl imidate donor **173** to produce the α-(1→4)-linked undecasaccharide **174** in very good yield considering the complexity and steric hindrance in the glycosylation reaction. Global deprotection eventually afforded the spacer ligand **175**, ready for testing of antibodies present in clinical samples. One has to note, however, that prior to the successful synthesis, other blockwise approaches had also been performed, which,

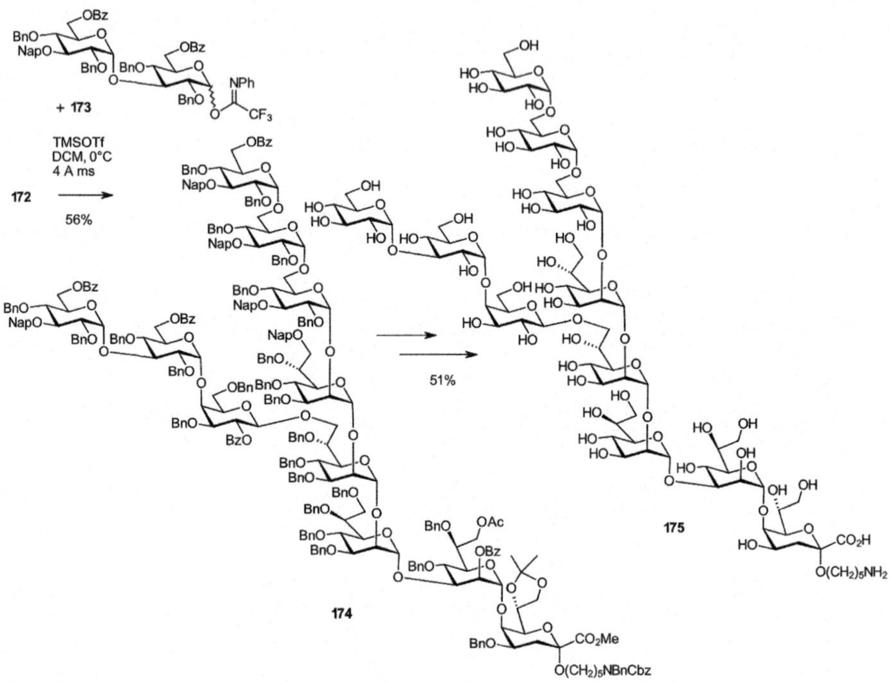

Scheme 24 Synthesis of the core undecasaccharide of *Helicobacter pylori* LPS. (Adapted with permission from Zhou et al. 2022, Copyright 2022 American Chemical Society)

however, met with difficulties. Furthermore, the 6-epimeric heptose donors and the oligoglucosyl donors had to be prepared separately, adding to the workload of this endeavor. The nonasaccharide and fragments thereof were then printed on glycan arrays via the terminal amino group, and antibody binding was assayed with several patient sera. The α-(1→6)-linked glucan epitope was readily recognized by IgG antibodies in the majority of samples, but in addition, phosphate substituents that are present in the native LPS species also play a major role in the immune recognition of the heterogeneous *H. pylori* core domain (Scheme 24).

4 Summary and Outlook

The elaboration of advanced methodologies to synthesize complex glycans has witnessed a remarkable dynamic development in the past few years. A main driving force has been a better understanding and mechanistic insight into the central glycosylation step, the design of novel leaving groups and monitoring of the variables governing selectivity and overall yield of glycoside formation. Despite the impressive construction of large, complex glycans, further improvements will still be needed to fully boost and harvest the impact of glycans in glycobiology, glycomedicine, diagnostics and vaccine technology.

References

Ande C, Crich C (2023) Stereodirecting effect of esters at the 4-position of galacto and glucopyranosyl donors: effect of 4-*C*-methylation on side-chain conformation and donor reactivity, and influence of concentration and stoichiometry on distal group participation. J Org Chem 88:13883–13893

Andreana PR, Crich C (2021) Guidelines for O-glycoside formation from first principles. ACS Cent Sci 7(9):1454–1462. https://doi.org/10.1021/acscentsci.1c0059

Boltje TJ, Buskas T, Boons G-J (2009) Opportunities and challenges in synthetic oligosaccharide and glycoconjugate research. Nat Chem 1:611. https://doi.org/10.1038/NCHEM.399

Chang C-W, Lin M-H, Chan C-K, Su K-Y, Wu C-H, Lo W-C, Lam S, Cheng Y-T, Liao P-H, Wong C-H, Wang C-C (2021a) Automated quantification of hydroxyl reactivities: prediction of glycosylation reactions. Angew Chem Int Ed 60:12413–12423

Chang C-W, Lin M-H, Wang C-C (2021b) Statistical analysis of glycosylation reactions. Chem Eur J 27:2556–2568

Chatterjee S, Moon S, Hentschel F, Gilmore K, Seeberger PH (2018) An empirical understanding of the glycosylation reaction. J Am Chem Soc 140:11942–11953

Chen J, Hansen T, Zhang Q-J, Liu D-Y, Sun Y, Yan H, Codée JDC, Schmidt RR, Sun J-S (2019) 1-Picolinyl-5-azido-thiosialosides: versatile donors for the stereoselective construction of sialyl linkages. Angew Chem Int Ed 58:17000–17008

Cheng C-W, Zhou Y, Pan W-H, Dey S, Wu C-Y, Hsu W-L, Wong C-H (2018) Hierarchical and programmable one-pot synthesis of oligosaccharides. Nat Commun 9:5202

Crawford C, Oscarson S (2020) Optimized conditions for the Palladium-catalyzed hydrogenolysis of benzyl and naphthylmethyl ethers: Preventing saturation of aromatic protecting group. Eur J Org Chem 2020: 3332–3337

Crawford CJ, Seeberger PH (2023) Advances in glycoside and oligosaccharide synthesis. Chem Soc Rev 52:7773. https://doi.org/10.1039/d3cs00321c

Crich D, Li W (2007) O-Sialylation with N-acetyl-5-N,4-O-carbonyl-protected thiosialoside donors in dichloromethane: facile and selective cleavage of the oxazolidinone ring. J Org Chem 72: 2387–2391

Crich D, Sun S (1997) Are glycosyl triflates intermediates in the sulfoxide glycosylation method? A chemical and ^{1}H, ^{13}C, and ^{19}F NMR spectroscopic investigation. J Am Chem Soc 119:11217–11223

de Kleijne FFJ, Moons SJ, White PB, Boltje TJ (2020) C-2 Auxiliaries for stereoselective glycosylation based on common additive functional groups. Org Biomol Chem 18:1165–1184

De Meo C, Gloeckner N (2021) Chapter 2: Synthesis of glycosides of sialic acid. In: Barchi JJ Jr, Vidal S (eds) Comprehensive glycoscience, vol 2, 2nd edn. Elsevier, Oxford, pp 228–266

Demchenko AV, Boons G-J (1998) *N*-diacetyl derivatives of neuraminic acid exhibits improved glycosyl donor properties and are very useful substrates for the preparation of glycosides of *N*-acetylneuraminic acid. Tetrahedron Lett 39:3065–3068

Dhara D, Dhara A, Murphy PV, Mulard LA (2022) Protecting group principles suited to late-stage functionalization and global deprotection in oligosaccharide synthesis. Carbohydr Res 521: 108644

Ding F, Ishiwata A, Ito Y (2018) Bimodal glycosyl donors protected by 2-O-(ortho-tosylamido) benzyl group. Org Lett 20:4384–4388

Ding F, Ishiwata A, Zhou S, Zhong X, Ito Y (2020) Unified strategy toward stereocontrolled assembly of various glucans based on bimodal glycosyl donors. J Org Chem 85:5536–5558

Doyle L, Ovchinnikova OG, Huang B-S, Forrester TJB, Lowary TL, Kimber MS, Whitfield C (2023) Mechanism and linkage specificities of the dual retaining β-Kdo glycosyltranserase modules of KpsC from bacterial capsule biosynthesis. J Biol Chem 299:104609

Franconetti A, Ardá A, Asensio JL et al (2021) Glycosyl oxocarbenium ions: structure, conformation, reactivity and interactions. Acc Chem Res 54:2552–2564

Frihed TG, Bols M, Pedersen CM (2015) Mechanisms of glycosylation reactions studied by low-temperature nuclear magnetic resonance. Chem Rev 115:4963–5013

Hamajima S, Komura N, Tanaka H-N, Imamura A, Ishida H, Noguchi H, Ichiyanagi T, Ando H (2022) Full stereocontrol in α-glycosidation of 3-deoxy-D-manno-2-octulosonic acid (Kdo) using macrobicyclic glycosyl donors. Org Lett 24:8672–8676

Hansen T, Lebedel L, Remmerswaal WA, van der Vorm S, Wander DPA, Somers M, Overkleeft HS, Filippov DV, Désiré J, Mingot A, Bleriot Y, van der Marel GA, Thibaudeau S, Codée JDC (2019) Defining the SN_1 side of glycosylation reactions: Stereoselectivity of glycopyranosyl cations. ACS Cent Sci 5:781–788

Hettikankanamalage AA, Lassfolk R, Ekholm FS, Leino R, Crich D (2020) Mechanisms of stereodirecting participation and ester migration from near and far in glycosylation and related reactions. Chem Rev 120:7104–7151

Hosoya T, Takano T, Kosma P, Rosenau T (2014) Theoretical foundation for the presence of oxacarbenium ions in chemical glycoside synthesis. J Org Chem 79:7889–7894

Hotha S (2020) In: Rauter AP, Christensen BE, Somsák L, Kosma P, Adamo R (eds) Recent trends in carbohydrate chemistry, vol 1. Elsevier, Amsterdam, pp 139–159

Huang J-S, Huang W, Meng X, Wang X, Gao P, Yang J-S (2015) Stereoselective synthesis of α-3-deoxy-D-manno-oct-2-ulosonic acid (α-Kdo) glycosides using 5,7-di-O-tert-butylsilylene-protected Kdo ethyl thioglycoside donors. Angew Chem Int Ed 54:10894–10898

Huang W, Zhou Y-Y, Pan X-L, Zhou X-Y, Lei J-C, Liu D-M, Chu Y, Yang J-S (2018) Stereodirecting effect of C5-carboxylate substituents on the glycosylation stereochemistry of 3-deoxy-D-*manno*-oct-2-ulosonic acid (Kdo) thioglycoside donors: Stereoselective synthesis of α- and β-Kdo glycosides. J Am Chem Soc 140:3574–3582

Ingle AB, Chao C-S, Hung W-C, Mong KKT (2013) Tuning reactivity of glycosyl imidinium intermediate for 2-azido-2-deoxyglycosyl donors in α-glycosidic bond formation. Org Lett 15: 5290–5293

Ishiwata A, Lee YJ, Ito Y (2010) Recent advances in stereoselective glycosylation through intramolecular aglycon delivery. Org Biomol Chem 8:3596–3608

Kim JH, Yang H, Boons GJ (2005) Stereoselective glycosylation reactions with chiral auxiliaries. Angew Chem Int Ed 44:947–949

Kinnaert C, Daugaard M, Nami F, Clausen MH (2017) Chemical synthesis of oligosaccharides related to the cell walls of plants and algae. Chem Rev 117:11337–11405

Koinuma R, Tohda K, Aoyagi T, Tanaka H (2020) Chemical synthesis of α(2,8) octasialosides, the minimum structure of polysialic acids. Chem Commun 56:12981–12984

Komura N, Kato K, Udagawa T, Asano S, Tanaka H-N, Imamura A, Ishida H, Kiso M, Ando H (2019) Constrained sialic acid donors enable selective synthesis of α-glycosides. Science 364: 677–680

Kosma P (2008) Occurrence, synthesis and biosynthesis of bacterial heptoses. Curr Org Chem 12: 1021–1039

Kosma P (2016) Progress in Kdo-glycoside synthesis. Tetrahedron Lett 57:2133–2142

Krasnova L, Wong C-H (2019) Oligosaccharide synthesis and translational innovation. J Am Chem Soc 141:3735–3654

Li W, Yu B (2018) Gold-catalyzed glycosylation in the synthesis of complex carbohydrate-containing natural products. Chem Soc Rev 47:7954–7984

Li Y, Yang Y, Yu B (2008) An efficient glycosylation protocol with glycosyl ortho-alkynylbenzoates as donors under the catalysis of $Ph_3PAuOTf$. Tetrahedron Lett 49:3604–3608

Li Y, Yang X, Liu Y, Zhu C, Yang Y, Yu B (2010) Gold(I)-catalyzed glycosylation with glycosyl *ortho*-alkynylbenzoates as donors: general scope and application in the synthesis of a cyclic triterpene saponin. Chem Eur J 16:1871–1882

Li T, Wolfert MA, Wei N, Huizinga R, Jacobs BC, Boons G-J (2020) Chemoenzymatic synthesis of Campylobacter jejuni lipopolysaccharide core domains to examine Guillain-Barré syndrome serum antibody specificities. J Am Chem Soc 142:19611–19621

Lian G, Zhang X, Yu B (2015) Thioglycosides in carbohydrate research. Carbohydr Res 403:13–22

Liu H, Liang Z-F, Liu H-J, Liao J-X, Zhong L-J, Tu Y-H, Zhang Q-J, Xiong B, Sun J-S (2023) Ortho-Methoxycarbonylethynylphenyl thioglycosides (MCEPTs): versatile glycosyl donors enabled by electron-withdrawing substituents and catalyzed by gold(I) or Cu(II) complexes. J Am Chem Soc 145:3682–3695

Lou Q, Hua Q, Zhang L, Yang Y (2020) Dimethylformamide-modulated Kdo glycosylation for stereoselective synthesis of α-Kdo glycosides. Org Lett 22:981–985

Lu SR, Lai YH, Chen JH, Liu CY, Mong KKT (2011) Dimethylformamide an unsual glycosylation modulator. Angew Chem Int Ed 50:7315–7320

Luo S, Liu Y, Hao T, Ma W, Luo Y, Wang S, Xu Z, Hu C, Wen L, Li T (2023) Chemoenzymatic total synthesis of Haemophilus ducrey lipooligosaccharide core octasaccharides containing natural and unnatural sialic acids. Org Lett 25:2312–2317

Mazur M, Barycza B, Andriamboavonjy H, Lavoie S, Tamigney Kenfack M, Laroussarie A, Blériot Y, Gauthier C (2016) 4′-Methoxyphenacyl-assisted synthesis of β-Kdo glycosides. J Org Chem 81:10585–10599

Meron-Sudai S, Asato V, Adler A, Bialik A, Goren S, Ariel-Cohen O, Reizis A, Mulard LA, Phalipon A, Cohen D (2023) A Shigella flexneri 2a synthetic glycan-based vaccine induces a long-lasting immune response in adults. NPJ Vaccines 8:35. https://doi.org/10.1038/s41541-023-00624-y

Mi X, Lou Q, Fan W, Zhuang L, Yang Y (2017) Gold(I)-catalyzed synthesis of β-Kdo glycosides using Kdo ortho-hexynylbenzoate as donor. Carbohydr Res 448:161–165

Mishra B, Neralka M, Hotha S (2016) Stable alkynyl glycosyl carbonates: catalytic anomeric activation and synthesis of a tridecasaccharide reminiscent of mycobacterium tuberculosis cell wall lipoarabinomannan. Angew Chem Int Ed Engl 55:7786–7791

Ngoje P, Crich D (2020) Stereocontrolled synthesisof the equatorial glycosides of 3-deoxy-D-manno-2-octulosonic acid: role of the side chain conformation. J Am Chem Soc 142:7760–7764

Nigudkar SS, Demchenko AV (2015) Stereocontrolled 1,2-cis glycosylation as the driving force of progress in synthetic carbohydrate chemistry. Chem Sci 6:2687–2704

Oscarson S (2012) Synthesis of bacterial surface structures containing Kdo and glycero-D-manno-heptose linkages. Carbohydr Chem 45:40–60

Oscarson S, Cheallaigh AN (2021) In: Vidal S (ed) Comprehensive glycoscience, vol 2, 2nd edn. Elsevier, pp 1–48

Panza M, Pistorio SG, Stine KJ, Demchenko AV (2018) Automated chemical oligosaccharide synthesis: novel approach to traditional challenges. Chem Rev 118:8105–8150

Pasari S, Manmode S, Walke G, Hotha S (2018) A versatile synthesis of pentacosafuranoside subunit reminiscent of mycobacterial arabinogalactan employing one strategic glycosidation protocol. Chem Eur J 24:1128–1139

Pokorny B, Kosma P (2015a) First and stereoselective synthesis of an α-(2→5)-linked disaccharide of 3-deoxy-D-*manno*-oct-2-ulosonic acid. Org Lett 17:110–113

Pokorny B, Kosma P (2015b) Synthesis of chlamydial haptens via 3-iodo Kdo fluoride donors. Chem Eur J 21:305–313

Pradhan TK (2023) Kdo glycosylations and their application in oligosaccharide synthesis. Eur J Org Chem 26:e202300146

Pradhan TK, Lin CC, Mong K-KT (2014) Preparation of a protected 3-deoxy-D-manno-oct-2-ulosonate glycal donor for the synthesis of β-KDO-containing oligosaccharides. Org Lett 16: 1474–1477

Reilly C, Stewart TJ, Renfrow M, Novak J (2019) Glycosylation in health and disease. Nat Rev Nephrol 15:346–366

Ruprecht C, Blaukopf M, Pfrengle F (2022) Synthetic fragments of plant polysaccharides as tools for cell wall biology. Curr Op Chem Biol 71:102208

Santana AG, Montalvillo-Jiménez L, Díaz-Casado L, Corzana F, Merino P, Cañada FJ, Jiménez-Osés G, Jiménez-Barbero J, Gómez AM, Asesnio JL (2020) Dissecting the essential role of anomeric β-triflates in glycosylation reactions. J Am Chem Soc 142:12501–12514

Shen Z, Mobarak H, Li W, Widmalm G, Yu B (2017) Synthesis of β-(1→2)-linked 6-deoxy-L-altropyranose oligosaccharides via gold (I)-catalyzed glycosylation of an *ortho*-hexynylbenzoate donor. J Org Chem 82:3062–3071

Shirakawa A, Manabe Y, Marchetti R, Yano K, Masui S, Silipo A, Molinaro A, Fukase K (2021) Chemical synthesis of sialyl *N*-glycans and analysis of their recognition by neuraminidase. Angew Chem Int Ed 60:24686–24693

Tanaka H, Nishiura Y, Takahashi T (2006) Stereoselective synthesis of oligo-α-(2,8)-sialic acids. J Am Chem Soc 128:7124–7125

Tang Y, Li J, Zhu Y, Li Y, Yu B (2013) Mechanistic insights into the gold (I)-catalyzed activation of glycosyl ortho-alkynylbenzoates for glycosidation. J Am Chem Soc 135:18396–18405

Tuck OT, Sletten ET, Danglad-Flores J, Seeberger PH (2022) Towards a systematic understanding of the influence of temperature on glycosylation reactions. Angew Chem Int Ed Engl 61: e2021115433

Turnbull JE, Field RA (2007) Emerging glycomics technologies. Nature Chem Biol 3: 74–77

Upadhyaya K, Subedi YP, Crich D (2021) Direct experimental characterization of a bridged bicyclic glycosyl dioxacarbenium ion by ^{1}H and ^{13}C NMR spectroscopy: importance of conformation on participation by distal esters. Angew Chem Int Ed 133:25601–25607

van Hengst JMA, Hellemons RJC, Remmerswaal WA, van de Vrande KNA, Hansen T, van der Vorm S, Overkleeft HS, van der Marel GA, Codée JDC (2023) Mapping the effect of configuration and protecting group pattern on glycosyl acceptor reactivity. Chem Sci 14:1532–1542

Van der Vorm, van Hengst JMA, Bakker M, Overkleeft HS, van der Marel GA, Codeé JDC (2018) Mapping the relationship between glycosyl acceptor reactivity and glycosylation stereoselectivity. Angew Chem Int Ed 57:8240–8244

Van der Vorm, Hansen T, van Hengst JMA, Overkleeft HS, van der Marel GA, Codeé JDC (2019) Acceptor reactivity in glycosylation reactions. Chem Soc Rev 48:4688–4706

Varki A, Cummings RD, Esko JD, Stanley P, Hart GW, Aebi M, Mohnen D, Kinoshita T, Packer NH, Prestegard JJ, Schnaar RL, Seeberger PH (eds) (2022) Essentials of glycobiology, 4th edn. Cold spring Harbor Laboratory Press. https://doi.org/10.1101/glycobiology.4e.53

Vibhute AM, Komura N, Tanaka H-N, Imamura A, Ando H (2021) Advanced chemical methods for stereoselective sialylation and their application in sialoglycan syntheses. Chem Rec 21: 3194–3223

Walvoort MTC, Lodder G, Overkleeft HS, Codée JDC, van der Marel GA (2010) Mannosazide methyl uronate donors. Glycosylating properties and use in the construction of β-ManNAcA-containing oligosaccharides. J Org Chem 75:7990–8002

Werz DB, Ranzinger R, Herget S, Adibekian A, van der Lieth C-W, Seeberger PH (2007) Exploring the structural diversity of mammalian carbohydrates ("glycospace") by statistical databank analysis ACS. Chem Biol 2:685–691

Willis LM, Whitfield C (2013) KpsC and KpsS are retaining 3-deoxy-D-manno-oct-2-ulosonic acid (Kdo) transferases involved in synthesis of bacterial capsules. Proc Natl Acad Sci USA 110: 20753–20758

Xiao K, Hu Y, Wan Y, Li X, Nie Q, Yan H, Wang L, Liao J, Liu D, Tu Y, Sun J, Codée JDC, Zhang Q (2022) Hydrogen bond activated glycosylation under mild conditions. Chem Sci 13:1600–1607

Yang W, Yang B, Ramadan S, Huang X (2017) Preactivation-based chemoselective glycosylations: a powerful strategy for oligosaccharide assembly. Beilstein J Org Chem 13:2094–2114

Yang B, Yang W, Ramadan S, Huang X (2018) Pre-activation-based stereoselective glycosylations. Eur J Org Chem 2018:1075–1096

Yang F, Sun Y, Xu P, Molinaro A, Silipo A, Yu B (2023) Synthesis of unprecedented α/β (1→4)-glucans via stereoselective iterative glycosylation. Chem Eur J 29:e202300659

Yasomanee JP, Demchenko AV (2012) Effect of remote picolinyl and picoloyl substituents on the stereoselectivity of chemical glycosylation. J Am Chem Soc 134:20097–20102

Yoshizaki H, Fukuda N, Sato K, Oikawa M, Fukase K, Suda Y, Kusumoto K (2001) First total synthesis of the Re-type lipopolysaccharide. Angew Chem Int Ed 40:14755–11480

Yu B (2018) Gold (I) catalyzed glycosylation with glycosyl *o*-alkynylbenzoates as donors. Acc Chem Res 51:507–516

Yu C-S, Niikura K, Lin C-C, Wong C-H (2001) The thioglycoside and glycosyl phosphite of 5-azido sialic acid: excellent donors for the *α*-glycosylation of primary hydroxy groups. Angew Chem Int Ed 40:2900–2903

Zhang Z, Ollmann IR, Ye X-S, Wischnat R, Baasov T, Wong C-H (1999) Programmable one-pot synthesis. J Am Chem Soc 121:734–753

Zhang Y, He H, Chen Z, Huang Y, Xiang G, Li P, Yang X, Lu G, Xiao G (2021) Merging reagent modulation and remote anchimeric assistance for glycosylation: highly stereoselective synthesis of α-glycans up to 30-mer. Angew Chem Int Ed 60:12597–12606

Zhang J-D, Gao X-J, Liu S-C, Geng Z-F, Chang L, Liu Y-J, Ma Q-Y, Xing G-SW, Liu G-J, Fang D-C (2023) α-Stereoselective 3-deoxy-D-manno-oct-2-ulosonic acid (Kdo) O-glycosylation with a *p*-toluenethioglycoside donor by the $(p\text{-Tol})_2SO/Tf_2O$ preactivation strategy. Org Lett 25:4150–4155

Zhong X, Zhou S, Ao J, Guo A, Xiao Q, Huang Y, Zhu W, Cai H, Ishiwata A, Ito Y, Liu X-W, Ding F (2022) ZnI(II) Iodide directed β-mannosylation: reaction selectivity, mode and application. J Org Chem 86:16901–16915

Zhou X-Y, Li L-X, Zhang Z, Duan S-C, Huang Y-W, Luo Y-Y, Mu X-D, Chen Z-W, Qin Y, Hu J, Yin J, Yang J-S (2022) Chemical synthesis and antigenic evaluation of Inner Core oligosaccharides from Acinetobacter baumannii lipopolysaccharide. Angew Chem Int Ed 61: e202204420

Zou X, Hu J, Zhao M, Qin C, Zhu Y, Tian G, Cai J, Seeberger PH, Yin J (2022) Chemical synthesis of the highly sterically hindered core undecasaccharide of Helicobacter pylori lipopolysaccharide for antigenicity evaluation with human serum. J Am Chem Soc 144: 14535–14547

Synthesis of Sulfated Carbohydrates – Glycosaminoglycans

Rakesh Raigawali, Sharath S. Vishweshwara, Saurabh Anand, and Raghavendra Kikkeri

Contents

Abstract

Glycosaminoglycans (GAG) are polysaccharides that are ubiquitous on the surface of all mammalian cells, interacting with a multitude of proteins and orchestrating essential physiological and pathological processes. Among various GAG structures, heparan sulfate (HS) stands out for its intricate structure, positioning it as a significant cell-surface molecule capable of regulating wide range of cellular functions. Consequently, investigating the structure-activity relationships (SARs) with well-defined HS ligands emerges as an attractive avenue advancing drug discovery and biosensors. This chapter outlines a modular divergent strategy for synthesizing HS oligosaccharides to elucidate SARs. Here, we provide a literature overview on the synthesis of disaccharide building blocks, employing different orthogonal protecting groups, promoters, and optimization

R. Raigawali · S. S. Vishweshwara · S. Anand · R. Kikkeri (✉)
Indian Institute of Science Education and Research, Pune, India
e-mail: rkikkeri@iiserpune.ac.in

F. Pfrengle et al. (eds.), *Complex Carbohydrates in Health and Disease*, Handbook of Experimental Pharmacology 288, https://doi.org/10.1007/164_2025_742

conditions to improve their suitability for subsequent oligosaccharide synthesis. Further, we highlight the synthesis of universal disaccharide building blocks derived from natural polysaccharides. We also provide insights of one-pot method and automated solid-phase synthesis of HS oligosaccharides. Finally, we review the status of SARs of popular heparan sulfate binding proteins (HSBPs).

Keywords

Glycosaminoglycan · Glycosylation · Heparan sulfate · Orthogonal protection · Sulfation

1 Introduction

Mammalian cell surfaces display a variety of carbohydrate structures, ranging from monosaccharides to polysaccharides. Glycosaminoglycans (GAGs) stand out as complex polyanionic polysaccharides, comprising hexosamine-uronic acid disaccharide repeating units (Taylor and Gallo 2006). GAGs are ubiquitous in cell surfaces and extracellular matrices, with each tissue having a unique composition of GAGs that interact with various proteins (Bulow and Hobert 2006). GAGs play crucial roles in both physiological and pathological processes by acting as repositories for proteins, shielding them from proteases, increasing local protein concentrations, facilitating interactions with receptors and other proteins, and even acting as receptors themselves (Sasisekharan et al. 2006; Kamhi et al. 2013). Thus, comprehending the structure-activity relationship of GAGs could yield advancements in biosensors and drug molecules for the next generation.

GAGs can be categorized into four primary groups according to their structural features: hyaluronan or hyaluronic acid (HA), chondroitin sulfate (CS) and dermatan sulfate (DS), heparin and heparan sulfate (HS), and keratan sulfate (KS) (Scheme 1) (Kjellén and Lindahl 1991; Couchman and Pataki 2012). Further classification of these polysaccharides includes sulfated and non-sulfated variants. HA is a non-sulfated GAG composed of *N*-acetyl β-D-glucosamine (β-GlcNAc) and β-D-glucuronic acid (β-GlcA). Whereas sulfated GAGs, like CS and DS composed of *N*-acetyl-β-D-galactosamine (β-GalNAc), and heparin/HS, contain α-GlcNAc as hexosamine unit. Furthermore, tissue-specific modifications contribute to the structural diversity observed in GAGs. Particularly, DS, heparin, and HS chains are converted by changing β-GlcA to α-L-iduronic acid (α-IdoA) by the action of C5-epimerases (Xu and Esko 2014). Sulfation can occur at specific sites, such as O2 of the uronic acid and O4 and O6 of β-GalNAc in CS and DS, whereas heparin and HS may be sulfated at O2 of the uronic acid and O3 and O6 of the glucosamine. Additionally, the glucosamine units may be *N*-acetylated or *N*-sulfated, while the occurrence of free amine groups (*N*-unsubstituted) is rare (Couchman and Pataki 2012; Bhoge et al. 2023). Unlike other GAGs, KS lacks a uronic acid and consists of

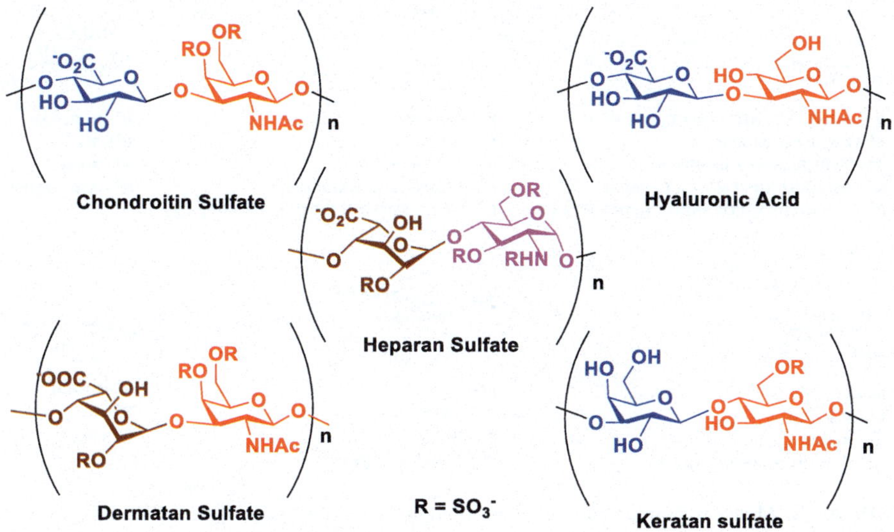

Scheme 1 Structures of GAG disaccharides

β-GlcNAc and β-Gal residues. Sulfations and fucosylations can occur at specific positions in KS. Due to the complexity and heterogeneity of GAG structures, the chemical synthesis of pure GAGs of different chain lengths is highly challenging. Emerging chemoenzymatic and chemical methods, including automated synthesis, offer potential for obtaining specific GAG structures for research purposes (Panza et al. 2018). This chapter provides a summary of the chemical and automated synthesis strategies employed for synthesizing GAGs, with specific emphasis on heparan sulfate oligosaccharide synthesis.

2 Synthesis of Heparan Sulfate Oligosaccharides

2.1 Solution-Phase Synthesis

Over the past two decades, extensive investigations have delved into the synthesis of HS oligosaccharides (Panza et al. 2018; Mende et al. 2016). However, the chemical synthesis of well-defined HS ligands remains a formidable challenge due to its extensive structural diversity. Moreover, the absence of a commercially available source for the L-iduronic acid (Hansen et al. 2015) and the complexities surrounding stereocontrolled glycosylation between glucosamine and uronic acid residues further compound the difficulty. As a result, chemists are driven to investigate various synthetic methods to generate a broad spectrum of HS oligosaccharides. Within these methods, the modular convergent synthetic approach stands out as a promising strategy (Codee et al. 2005). Here, common disaccharide building blocks are synthesized, which can be effortlessly transformed into either glycosyl donors or

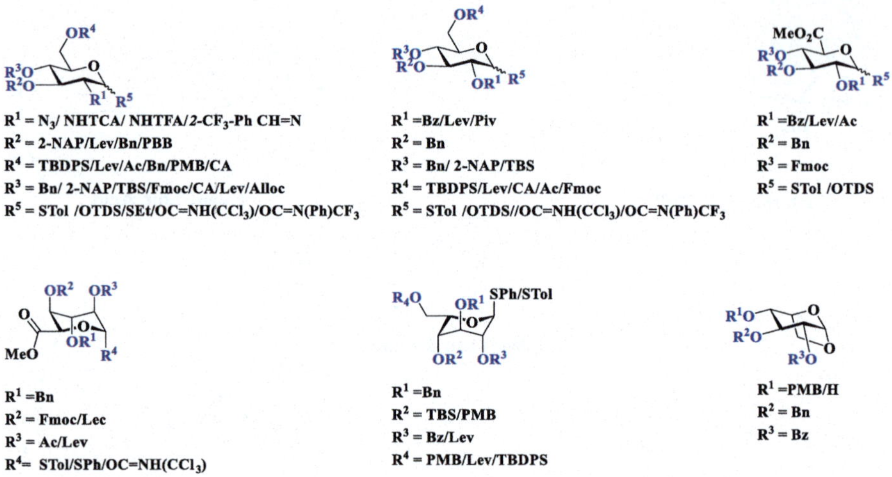

Scheme 2 Monosaccharide building blocks used in the assemble HS oligosaccharides

acceptors, facilitating the synthesis of HS oligosaccharides (Zong et al. 2016, 2017a, b; Zulueta et al. 2012; Noti et al. 2006; Hu et al. 2011; Wang et al. 2010; Jain et al. 2021a, b; Anand et al. 2020; Dey and Wong 2018). Here, the anomeric hydroxyl of the glucosamine building block has been shielded either with a stable thiotoluene donor or with *t*-hexyldimethylsilyl (TDS) ether, which can be easily removed and converted into a donor prior to glycosylation. The C4 hydroxyls of the glucosamine and uronic acid building blocks have been safeguarded with temporary protective groups such as acetate (Ac), levulinoyl esters (Lev), 9-fluorenylmethyl carbonate (Fmoc), or 2-naphthylmethyl (NAP) groups, which can be selectively cleaved to generate glycosyl acceptors. In contrast, permanent protection groups like benzyl (Bn), *tert*-butyldimethylsilyl (TBS), or 2-NAP have been incorporated to the C4 hydroxyl of the non-reducing end glucosamine and the C3 hydroxyl of uronic acid to prevent sulfation. These protections are subsequently removed during the final global deprotection step. A suitable combination of temporary protective groups like *tert*-butyldiphenylsilyl (TBDPS), 2-NAP, Lev, *p*-bromobenzyl (PBB), and allyl ether (Alloc) at the C6 and C3 hydroxyl groups of glucosamine is expected to provide remote and steric participation in α-glycosylations followed by site-selective deprotection and sulfation, resulting in the desired sulfated HS sequences (Scheme 2). The amine of GlcN has been masked as azide or imide groups to promote α-stereoselective glycosylation via the anomeric effect. Likewise, the C2 hydroxyl of uronic acid building blocks has been temporarily protected with benzoyl (Bz), Lev, or 4-acetoxy-2,2-dimethyl butanoyl (PivOAc) ester groups to take advantage of neighboring group participation and accomplish 1,2-trans glycosylation, followed by subsequent selective deprotection and sulfation. Notably, the L-iduronic acid building block was synthesized from 1,2:5,6-di-O-isopropylidene-α-D-glucofuranose through six straightforward steps to yield anhydrous L-idose, which was glycosylated with glucosamine, followed by anhydrous

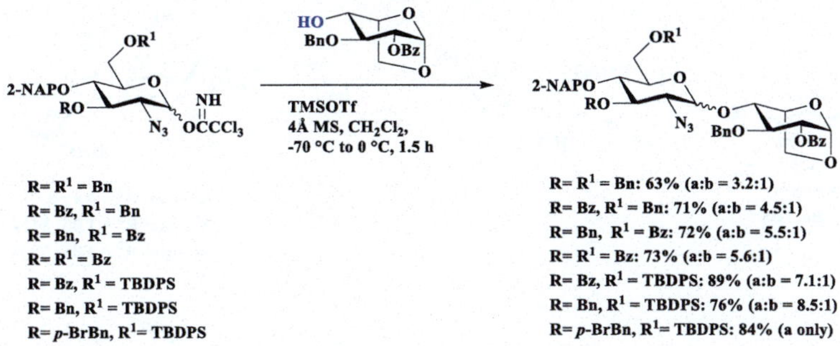

Scheme 3 Synthesis of HS disaccharides in the presence of different orthogonal protections

ring opening and oxidized to synthesize GlcN-IdoA disaccharide building block. Ultimately, [2 + 2] glycosylation resulted in HS oligosaccharides.

Trihaloacetimidate and thioglycosyl donors have been extensively used to enhance the stereoselectivity in the formation of the 1,2-cis α-D-glucosamine glycosidic bond with meticulous selection of orthogonal protecting groups and glycosylation promoters (Scheme 3) (Zulueta et al. 2012; Hung et al. 2017). Alternatively, HS disaccharides were synthesized through the activation of glucosamine thiosyl donor in the presence of *p*-toluene sulfenyl chloride (*p*-TolSCl) and silver triflate (AgOTf), followed by glycosylation of thiosyl idose or glucose thioglycosyl acceptor forming exclusively the α-anomer (Scheme 4) (Yang et al. 2012, 2016; Dulaney et al. 2015; Wang et al. 2021; Yoshida et al. 2014; Dhamale et al. 2014; Baryal et al. 2023). Subsequent adjustment of protecting groups, disaccharide donor/acceptors, post-glycosylation oxidation, sulfation, and global deprotection resulted in the synthesis of HS oligosaccharides. However, all these methods are primarily aimed to synthesize HS oligosaccharides with homologous sulfation patterns. For heterogeneous sulfation patterns, two distinct disaccharide building blocks were synthesized, where the C2 position of glucosamine was either protected with trichloroacetyl group (TCA) or ortho-trichloro benzylimine and another disaccharide building composed 2-azido 2-deoxy D-glucose building block to introduce selective *N*-sulfation and *N*-acetylation on HS oligosaccharides (Chopra et al. 2021).

2.2 One-Pot Synthesis

The one-pot assembly of HS oligosaccharides has been carried out by glycosylating separate building blocks, leading to the formation of oligosaccharides within a few hours, in contrast to the step-by-step assembly, which can extend over several days to weeks. Such approaches demand building blocks where the donors exhibit relative reactivity differences, and each glycosylation must be 100% efficient to prevent the formation of side products that could complicate the synthesis process. Through this strategy, fondaparinux was synthesized by activating a more reactive donor in the presence of the acceptor glucuronic acid disaccharide, followed by the

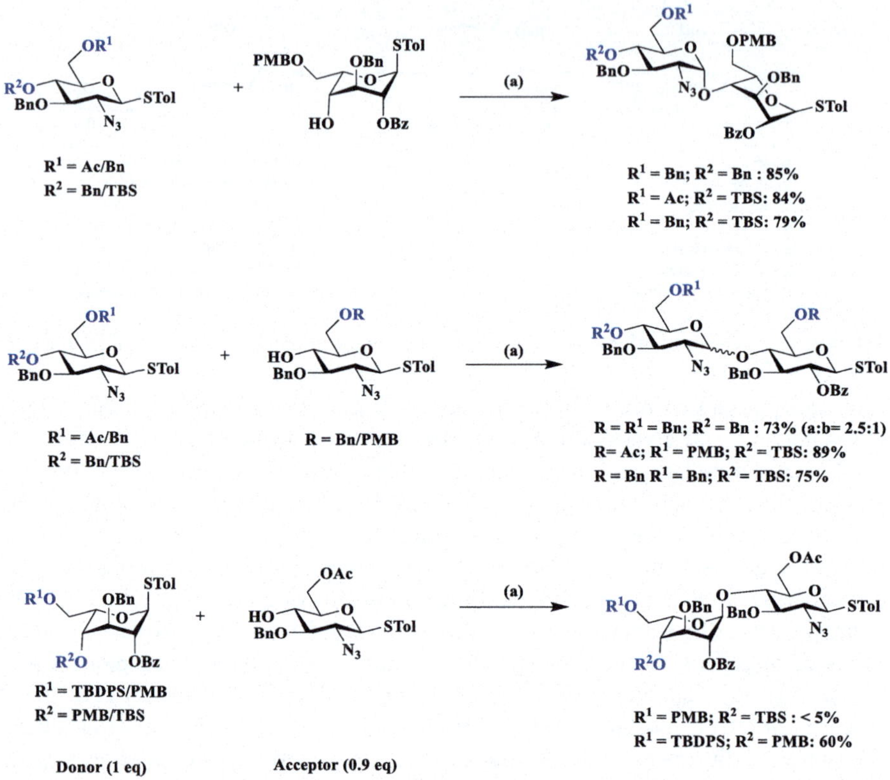

Scheme 4 Synthesis of HS disaccharides: (a) AgOTf, *p*-TolSCl, −78 °C, acceptor, TTBP

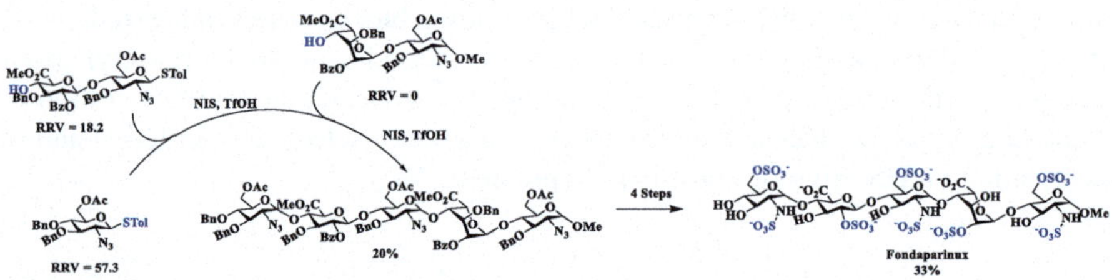

Scheme 5 One-pot synthesis of fondaparinux

activation of the second disaccharide acceptor with differing reactive reactivity values (RRVs) (Scheme 5). Subsequent selective deprotection, *O*- and *N*-sulfation, and global deprotection resulted in the required pentasaccharides (Scheme 5) (Polat and Wong 2007). Despite its rapidity, the implementation of this method with identical repeating units and similar RRVs poses a challenge. Thus, the one-pot assembly by preactivation proves to be a more effective strategy. A HS hexasaccharide was assembled by preactivating the non-reducing end disaccharide

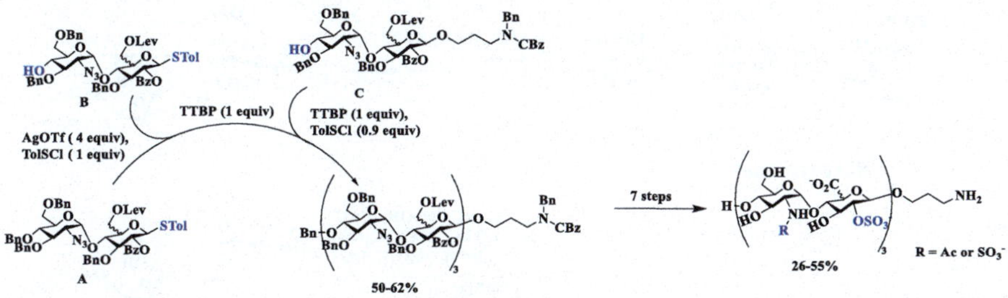

Scheme 6 One-pot synthesis of HS oligosaccharides

with TolSOTf, followed by the addition of the acceptor and 2,4,6-tri-*tert*-butylpyrimidine (TTBP) (Scheme 6) (Wang et al. 2010). Subsequently, after glycosylation was completed, the tetrasaccharide was activated and glycosylated with the reducing end HS disaccharide unit to assemble the oligosaccharide.

2.3 Universal Building Blocks

To date, all 48 combinations of HS disaccharides have been synthesized and characterized (Hu et al. 2012; Lu et al. 2006). However, for biological investigations, a tetrasaccharide is the minimum required length of HS oligosaccharide (Pawar et al. 2019). Hence, synthesizing all 2,304 combinations of HS tetrasaccharides necessitated a novel and robust approach to synthesizing disaccharide building blocks. To enhance the prospect of synthesizing disaccharide building blocks on gram scales, heparin/heparan sulfate polysaccharides were hydrolyzed and chemically modified to synthesize disaccharide intermediates in a minimal number of steps for the production of universal HS disaccharide ligands to construct an oligosaccharide library (Pawar et al. 2019; Wang et al. 2023). Controlled hydrolysis of *N*-acetyl heparin was carried out in the presence of 1 M TfOH at 100 °C, followed by esterification and acetylation, resulting in the GlcN–IdoA disaccharide. Similarly, the hydrolysis of sodium heparinate with triflic acid (TfOH), followed by esterification and azidation in the presence of triflic azide (TfN_3), potassium carbonate (K_2CO_3), and zinc chloride ($ZnCl_2$), and subsequent anomeric demethylation and peracetylation, yielded the $GlcN_3$–IdoA disaccharide in a 20% yield over four steps. Alternatively, the use of benzyl chloroformate introduced an *N*-Cbz protecting group in place of the *N*-acetyl group in HS disaccharide. Through the hydrolysis of heparosan and subsequent esterification and peracetylation, the GlcA–GlcN disaccharide was obtained with an overall yield of 16% over four steps (Scheme 7). Utilizing these uronic acid HS disaccharide building blocks, other uronic acid-based HS disaccharides were synthesized through epimerization. Here, the GlcA-containing building block was subjected to bromination using NBS in the presence of UV light, resulting in the C-5 bromo derivatives with a yield of 75%. Subsequently, treatment with Et_3B and Bu_3SnH at 20 °C yielded the highest amount of

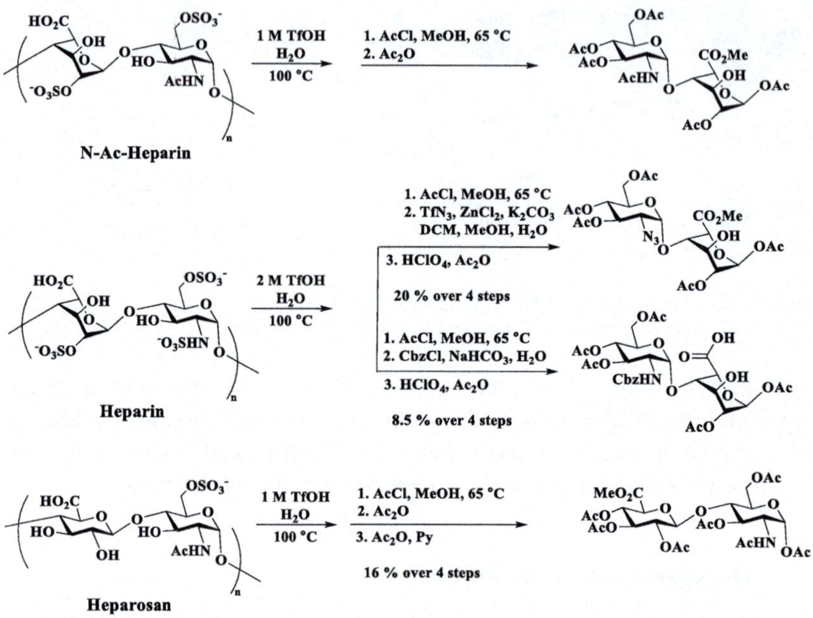

Scheme 7 Synthesis of HS disaccharide building blocks from natural polysaccharides

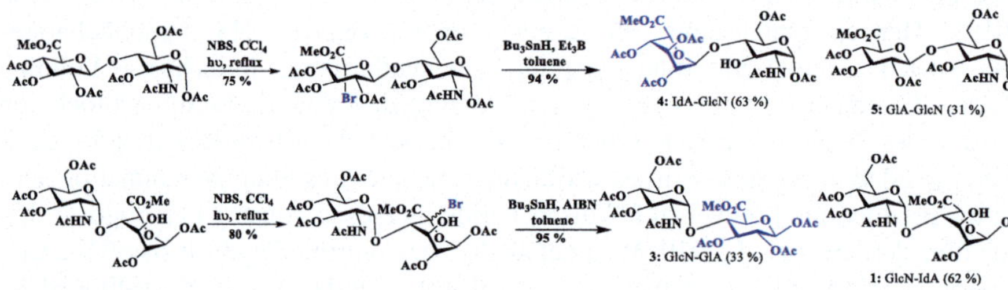

Scheme 8 Synthesis of epimerized HS disaccharide building blocks

epimerized IdoA–GlcN product (63% yield), alongside the recovery of the valuable starting material with a yield of 31%. On the other hand, the NBS-mediated bromination of GlcN-IdoA resulted in an epimeric mixture of the C-5 bromo HS disaccharides. Subsequent α-dehalogenation using AIBN and Bu_3SnH at 110 °C yielded only 33% epimerization (Scheme 8). In summary, these streamlined routes offer a novel approach to synthesize core HS disaccharide modules for oligosaccharide assembly. Subsequently, utilizing these disaccharide building blocks, 64 potential HS tetrasaccharides were synthesized and characterized.

2.4 Solid-Phase Synthesis

In order to rationalize and simplify the assembly of disaccharides into oligosaccharides, it is necessary to automate the synthesis using a solid-phase oligosaccharide synthesizer, which can improve the assembly of defined oligosaccharides in days rather than years. The solid-phase synthesis approach has been explored to assemble HS oligosaccharides. To this end, a semi-automated glycan assay was performed on Merrifield resin, which was functionalized with a photolabile nitrobenzyl ether-based linker and glycosylated with IdoA and glucosamine building blocks to synthesize hexasaccharide analogs with overall yields of 11% (Guedes et al. 2015). In another development, a HS disaccharide intermediate was immobilized onto synphase lanterns, and divergent deprotection and sulfation strategy was employed to synthesize 16 different HS disaccharides in high yield (Ramadan et al. 2022). The first fully automated HS tetrasaccharide assembly was reported by using a D-GlcN-α-D-GlcA disaccharide building block. The chloroacetate group of the tetrasaccharide precursor was selectively cleaved on resin, sulfonated, and finally photocleaved from the resin to obtain 3% overall

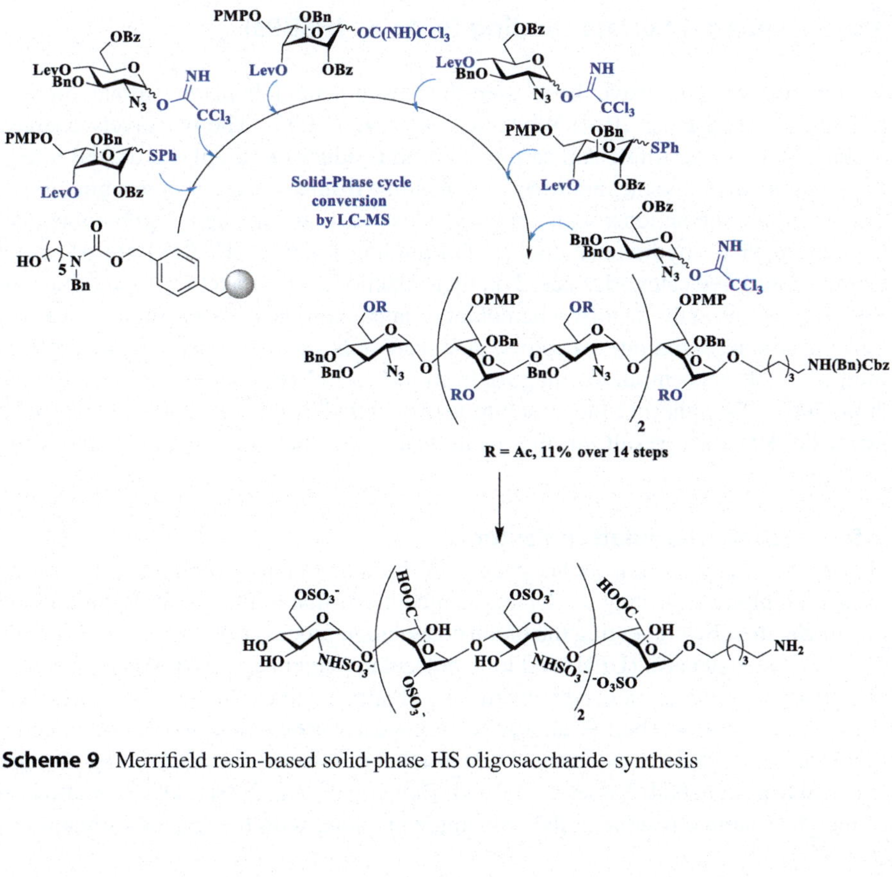

Scheme 9 Merrifield resin-based solid-phase HS oligosaccharide synthesis

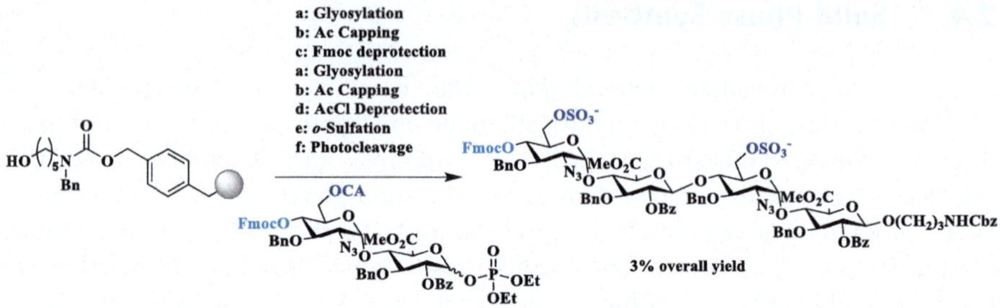

Scheme 10 Automated glycan assembly of HS tetrasaccharide

yield of the product (Pongener et al. 2023). In conclusion, automated solid-phase HS oligosaccharide synthesis offers promising method to assemble the HS oligosaccharides in quick time. Nevertheless, additional optimization of glycosylation, sulfation, and final deprotection steps are necessary to obtain final HS oligosaccharides (Schemes 9 and 10).

2.5 Heparan Sulfate Binding Proteins (HSBPs)

Nearly 500 proteins have been identified as potential heparan sulfate binding proteins (Xu and Esko 2014; Whitelock and Iozzo 2005). The negatively charged groups, such as the sulfate and carboxylic acid residues of HS, play a crucial role in these interactions. Lysine, arginine, and histidine residues in the proteins bind to HS through ionic interactions, while polar residues such as asparagine, glutamine, and histidine mediate hydrogen bonding. HS-binding proteins (HSBPs) are primarily classified into five categories based on their functions. These include blood coagulation factors; growth factors, chemokines, and cytokines; extracellular proteins; transmembrane signaling receptors; and cell adhesion proteins. These proteins bind to HS at concentrations ranging from μM to nM (Raman et al. 2005; Bishop et al. 2007). Deciphering the structure-activity relationships of these proteins with synthetic HS may result in the identification of potential markers and drug molecules.

2.5.1 Blood Coagulation Factors

Heparin is widely used as an anticoagulant. It binds to antithrombin (AT), a serine protease inhibitor, forming an AT-heparin complex that inhibits both thrombin and Factor Xa, thereby preventing the blood coagulation process (Rosenberg et al. 1973). Unfractionated heparin (UFH) and low molecular weight heparin (LMWH), derived from porcine mucosa, have been mainstays of anticoagulant therapy for many years (Bray et al. 1989; Herbert et al. 1997). In the early 1980s, researchers identified a unique pentasaccharide sequence in heparin that is essential for AT binding (Petitou and Van Boeckel 2004; Visser et al. 1994; Dementiev et al. 2004). The 3-*O*-sulfation of the glucosamine residue at the reducing end, along with the $^2S_0$ conformation of

L-iduronic acid in this sequence, was found to be critical for AT activity (Demeter et al. 2018). This breakthrough paved way for the development of fondaparinux (Arixtra), which remains the only heparin-based drug approved by the FDA. HS regulates serine protease inhibitors such as antithrombin III (ATIII) and heparin cofactor II (HCII), both of which play roles in thrombin inhibition. Recently, Desai and coworkers used computational methods and a synthetic HS hexasaccharide library to identify the rare role of 2-O-sulfated glucuronic acid in HCII activation (Sankarayanarayanan et al. 2017). Overall, serine proteases such as thrombin, HCII, Factor Xa, and Factor IXa are critical enzymes in the blood coagulation cascade. Despite successful studies on the SARs of HS-serine proteases interaction, challenges remain, such as selectively targeting specific serine proteases without affecting others to minimize side effects.

2.5.2 Growth Factors

Growth factors (GFs) are a broad class of proteins that play a crucial role in regulating various cellular processes, including cell proliferation, migration, differentiation, and survival (Beenken and Mohammadi 2009; Turner and Grose 2010; Yun et al. 2010; Pellegrini 2001; Maccarana et al. 1993). Dysregulated GF signaling can lead to a range of pathological conditions, including cancer. Many growth factors bind to heparan sulfates (HS) on heparan sulfate proteoglycans (HSPGs), thereby modulating cellular signaling. For example, fibroblast growth factors (FGF-1 and FGF-2) bind to HS, forming a ternary complex with FGF receptors (FGFRs), which is essential for their activation and signaling pathways. HS also plays a protective role by preventing FGFs from proteolytic degradation and inactivation (Allen et al. 2001; Ashikari-Hada et al. 2004; Turnbull et al. 1992). Hung and coworkers reported the synthesis of 48 HS-based disaccharide analogues and systematically explored their binding to FGF-1 and FGF-2 (Hu et al. 2012; Li et al. 2014). ITC binding studies and X-ray crystallography confirmed that 3-O-sulfated *N*-sulfated glucosamine and L-iduronic acid residues are essential for FGF-1 and FGF-2. These findings were confirmed at tetrasaccharide and hexasaccharide level with limited synthetic HS ligands (Zong et al. 2017a; Horton et al. 2021). In addition to FGFs, HS oligosaccharides play a critical role in regulating other growth factors, such as bone morphogenetic protein-2 (BMP-2), heparin-binding epidermal growth factor (HB-EGF), and VEGFs. Linhardt and coworkers demonstrated that highly sulfated *N*-sulfated HS ligands with a minimum hexasaccharide length are essential for VEGF activity (Zhao et al. 2012). While BMP-2 and HB-EGF binds to 6-O-sulfated HS oligosaccharides (Jain et al. 2021c). However, due to the lack of comprehensive tetrasaccharide and hexasaccharide HS libraries, identifying an ideal ligand for these growth factors remains an area of active research.

2.5.3 Chemokines

A major biological role of HS is to modulate the activity of cytokines and chemokines (Hughes and Nibbs 2018; Farrugia et al. 2018; Proudfoot et al. 2017). HS serves as a co-receptor linking approximately 50 distinct chemokines to its cell surface receptors to regulate leukocyte migration, angiogenesis, breast cancer metastasis, and leukocyte degranulation (Proudfoot et al. 2017). Upon external stimuli such as bacterial or viral infections, various cells release chemokines to draw leukocytes to the site of damage, promoting immune responses to eradicate the pathogen (Raman et al. 2011; Parish 2005; Belperio et al. 2000; Handel et al. 2005). Under normal conditions, chemokines exist in a monomeric form. When inflammation occurs, HS binds to chemokines, leading to their polymerization and increased activity. Gaining insight into the SARs between HS and chemokines can aid in the creation of new anti-inflammatory scaffolds with reduced side effects. By utilizing a synthetic HS library, researchers examined the SARs of chemokines. Lindahl and coworkers reported that interleukin-8 (CXCL8 or IL-8) prefers the 2-*O*-sulfated L-iduronic acid and 6-O sulfated *N*-sulfated glucosamine unit to activate neutrophil trafficking (Spillmann et al. 1998). Gallagher and coworkers identified that 2-*O*-sulfated iduronic acid (IdoA) in HS is essential for CXCL4 chemokine tetramerization and cell surface receptor binding (Stringer and Gallagher 1997). Hsieh-Wilson and coworkers revealed that *N*-sulfated HS oligosaccharides displayed strong binding to CCL5 chemokine and mitigate cell migration (Sheng et al. 2013). Seeberger and coworkers observed that a simple 2,4-*O*- disulfate iduronic acid had strong binding to CCL20 and inhibit asthamatic response (Nonaka et al. 2014). Boons and coworkers discovered that CCL2 binds to highly sulfated HS compounds without preference for the uronic acid component (Zong et al. 2017a). Despite uncovering essential features for targeting chemokines with HS ligands, limitations in the selectivity and specificity of current libraries have restricted their promise as drug candidates.

3 Conclusions

Efficient strategies for synthesizing building blocks and precise stereo-selective glycosylation techniques are crucial for the chemical synthesis of HS oligosaccharides. Using a wide range of orthogonal protecting groups and promotors, α-glycosylation between glucosamine and uronic acid residues has been optimized. Additionally, advancements such as the solution-based one-pot approach and the activation of glucosamine thiosyl donor in the presence of p-toluene sulfenyl chloride (p-TolSCl) and silver triflate (AgOTf) have bolstered the potential for synthesizing higher HS oligosaccharides. Nonetheless, the synthesis of HS disaccharide building blocks, involving 18–24 steps, remains a significant challenge. A recent method involving obtaining disaccharide universal building blocks from natural heparin polysaccharides with 4–5 steps has improved the overall outlook for HS synthesis. Moreover, automated HS synthesis is anticipated to offer rapid access to diverse and challenging sequences to meet biological requirements. However, HS synthetic libraries showed unprecedented binding capacity to blood coagulants, growth factors, and chemokines. Further systematic exploration with an expanded HS library could yield novel therapeutic molecules for various diseases and infections.

References

Allen BL, Filla MS, Rapraeger AC (2001) Role of heparan sulfate as a tissue specific regulator of FGF-4 and FGF receptor recognition. J Cell Biol 155:845–857

Anand S, Mardhekar S, Raigawali R, Mohanta N, Jain P, Shanthurthy CD, Gnanaprakasam B, Kikkeri R (2020) Continuous-flow accelerate sulfation of heparan sulfate intermediates. Org Lett 22:3402–3406

Ashikari-Hada S, Habuchi H, Kariya Y, Itoh N, Reddi AH, Kimata K (2004) Characterization of growth factor-binding structures in heparin/heparan sulfate using an octasaccharide library. J Biol Chem 279:12346–12354

Baryal KN, Ramadan S, Su G, Huo C, Zhao Y, Liu J, Hsieh-Wilson LC, Huang X (2023) Synthesis of a systematic 64-membered heparan sulfate tetrasaccharide library. Angew Chem Int Ed 63: E202211985

Beenken A, Mohammadi M (2009) The FGF family: biology, pathophysiology and therapy. Nat Rev Drug Discov 8:235–253

Belperio JA, Keane MP, Arenberg DA, Addison CL, Ehlert JE, Burdick MD, Strieter RM (2000) CXC chemokines in angiogenesis. J Leukoc Biol 68:1–8

Bhoge PR, Raigawali R, Mardhekar S, Anand S, Kikkeri R (2023) Synergestic interplay of uronic acid and sulfation composition of heparan sulfate on molecular recognition to activity. Carbohydr Res 532:108919

Bishop JR, Schuksz M, Esko JD (2007) Heparan sulphate proteoglycans fine tune mammalian physiology. Nature 446:1030–1037

Bray B, Lane DA, Freyssinet JM, Pejler G, Lindahl U (1989) Anti-thrombin activities of heparin. Effect of saccharide chain length on thrombin inhibition by heparin cofactor II and by anti-thrombin. Biochem J 262:225–232

Bulow HE, Hobert O (2006) The molecular diversity of glycosaminoglycans shapes animal development. Annu Rev Cell Dev Biol 22:375–407

Chopra P, Joshi A, Wu J, Lu W, Yadavalli T, Wolfert MA, Shukla D, Zaia J, Boons GJ (2021) The 3-O-sulftion of heparan sulfate modulates protein binding and lyase degradation. PNAS 118: e2012935118

Codee JD, Stubba B, Schiattarella M, Overkleeft HS, van Boeckel CAA, van Boom JH, van der Marel GA (2005) A modular strategy toward the synthesis of heparin-like oligosaccharides using monomeric building blocks in a sequential glycosylation strategy. J Am Chem Soc 127: 3767–3773

Couchman JR, Pataki CA (2012) An introduction to proteoglycans and their localization. J Histochem Cytochem 60:885–897

Dementiev A, Petitou M, Herbert JM, Gettins PGW (2004) The ternary complex of antithrombin-anhydrothrombin-heparin reveals the basis of inhibitor specificity. Nat Struct Mol Biol 11:863–867

Demeter F, Gyöngyösi T, Bereczky Z, Kövér KE, Herczeg M, Borbás A (2018) Replacement of the L-iduronic acid unit of the anticoagulant pentasaccharide idraparinux by a 6-deoxy-L-talopyranose – synthesis and conformational analysis. Sci Rep 8:2–11

Dey S, Wong CH (2018) Programmable one-pot synthesis of heparin pentasaccharides enabling access to regiodefined sulfate derivatives. Chem Sci 9:6685–6691

Dhamale OP, Zong C, Al-Mafraji K, Boons GJ (2014) New glucuronic acid donors for the modular synthesis of heparan sulfate oligosaccharides. Org Biomol Chem 12:2087–2096

Dulaney SB, Xu Y, Wang P, Tiruchinapally G, Wang Z, Kathawa J, El-dakdouki MH, Yang B, Liu J, Huang X (2015) Divergent synthesis of heparan sulfate oligosaccharides. J Org Chem 80: 12265–12279

Farrugia BL, Lord MS, Melrose J, Whitelock JM (2018) The role of heparan sulfate in inflammation, and the development of biomimetics as anti-inflammatory strategies. J Histochem Cytochem 66(4):321–336

Guedes N, Kopitzki S, Echeverria B, Pazos R, Elosegui E, Calvo J, Reichardt NC (2015) Solid-phase assembly of glycosaminoglycan oligosaccharide precursors. RSC Adv 5:9325–9327
Handel TM, Johnson Z, Crown SE, Lau EK, Proudfoot AE (2005) Regulation of protein function by glycosaminoglycans – as exemplified by chemokines. Annu Rev Biochem 74:385–410
Hansen SU, Dalton CE, Barath M, Kwan G, Raftery J, Jayson GC, Miller GJ, Gardiner JM (2015) Synthesis of l-iduronic acid derivatives via [3.2.1] and [2.2.2] l-iduronic lactones from bulk glucose-derived cyanohydrin hydrolysis: a reversible conformationally switched superdisarmed/rearmed lactone route to heparin disaccharides. J Org Chem 80:3777–3789
Herbert JM, Petitou M, Lormeau JC, Cariou R, Necciari J, Magnani HN, Zandberg P, van Amstredam RGM, van Boeckel CAA, Meuleman DG (1997) SR 90107A/Org 31540, a novel anti-factor Xa antithrombotic agent. Cardiovasc Drug Rev 15:1–26
Horton M, Su G, Yi L, Wang Z, Xu Y, Pagadala V, Zhang F, Zaharoff DA, Pearce K, Linhardt RJ, Liu J (2021) Construction of heparan sulfate microarray for investigating the binding of specific saccharide sequences to proteins. Glycobiology 31:188–199
Hu YP, Lin SY, Huang CY, Zulueta MML, Liu JY, Chang W, Hung SC (2011) Synthesis of 3-O-sulfonated heparan sulfate octasaccharides that inhibit the herpes simplex virus type 1 host–cell interaction. Nat Chem 3:557–563
Hu YP, Zhong YQ, Chen ZG, Chen CY, Shi Z, Zulueta MML, Ku CC, Lee PY, Wang CC, Hung SC (2012) Divergent synthesis of 48 heparan sulfate-based disaccharides and probing the specific sugar-fibroblast growth factor-1 interaction. J Am Chem Soc 134:20722–20727
Hughes CE, Nibbs RJ (2018) A guide to chemokines and their receptors. FEBS J 285(16): 2944–2971
Hung TY, Irena D, Zulueta MML, Tai TJ, Lain SH, Cheng CP, Tsai PX, Lin SY, Chen ZG, Ku CC, Hsiao CD, Chyan CL, Hung SC (2017) Structure of the complex between a heparan sulfate octasaccharide and mycobacterial heparin binding hemoaglutinin. Angew Chem Int Ed 56: 4192–4196
Jain P, Shanthamurthy CD, Leviatan Ben-Arye S, Woods RJ, Kikkeri R, Padler-Karavani V (2021a) Discovery of rare sulfated *N*-unsubstituted glucosamine based heparan sulfate analogs selectively activating chemokines. Chem Sci 12:3674–3681
Jain P, Shanthamurthy CD, Leviatan Ben-Arye S, Yeshuda S, Nandikol SS, Thulasiram HV, Palder-Karavani V, Kikkeri R (2021b) Synthetic heparan sulfate ligands for vascular endothelial growth factor to modulate angiogenesis. Chem Commun 57:3516–3519
Jain P, Shanthamurthy CD, Leviatan Ben-Arye S, Yehuda S, Nandikol SS, Thulasiram HV, Padler-Karavani V, Kikkeri R (2021c) Synthetic heparan sulfate ligands for vascular endothelial growth factor to modulate angiogenesis. Chem Commun 57:3516–3519
Kamhi E, Joo EJ, Dordick JS, Linhardt R (2013) Glycosaminoglycans in infectious disease. J Biol Rev 88:928–943
Kjellén L, Lindahl U (1991) Proteoglycans: structures and interactions. Annu Rev Biochem 60: 443–475
Li YC, Ho IH, Ku CC, Zhong YQ, Hu YP, Chen ZG, Chen CY, Lin WC, Zulueta MM, Hung SC, Lin MG, Wang CC, Hsiao CD (2014) Interactions that influence the binding of synthetic heparan sulfate based disaccahrides to fibroblast growth factor-2. ACS Chem Biol 8:1712–1717
Lu LD, Shie CR, Kulkarni SS, Pan GR, Lu XA, Hung SC (2006) Synthesis of 48 disaccharide building blocks for the assembly of a heparin and heparna sulfate oligosaccharide library. Org Lett 8:5995–5998
Maccarana M, Casu B, Lindahl U (1993) Minimal sequence in heparin/heparan sulfate required for binding of basic fibroblast growth factor. J Biol Chem 268(32):23898–23905
Mende M, Bednarek C, Wawryszyn M, Sauter P, Biskup MB, Schepers U, Bräse S (2016) Chemical synthesis of glycosaminoglycans. Chem Rev 116(14):8193–9255
Nonaka M, Bao X, Matsumura F, Götze S, Kandasamy J, Kononov A, Broide DH, Nakayama J, Seeberger PH, Fukuda M (2014) Synthetic di-sulfated iduronic acid attenuates asthmatic response by blocking T-cell recruitment to inflammatory sites. Proc Natl Acad Sci USA 111: 8173–8178

Noti C, de Paz JL, Polito L, Seeberger PH (2006) Preparation and use of microarrays containing synthetic heparin oligosaccharides for the rapid analysis of heparin-protein interactions. Chem Eur J 12:8664–8686

Panza M, Pistorio SG, Stine KJ, Demchenko AV (2018) Automated chemical oligosaccharide synthesis: novel approach to traditional challenges. Chem Rev 118:8105–8150

Parish CR (2005) Heparan sulfate and inflammation. Nat Immunol 6:861–862

Pawar NJ, Wang L, Higo T, Bhattacharya C, Kancharla PK, Zhang F, Baryal K, Huo CX, Liu J, Linhardt RJ, Huang X, Hsieh-Wilson LC (2019) Expedient synthesis of core disaccharide building blocks from natural polysaccharides for heparan sulfate oligosaccharide assembly. Angew Chem Int Ed 58:18577–18583

Pellegrini L (2001) Role of heparan sulfate in fibroblast growth factor signalling: a structural view. Curr Opin Struct Biol 11:629–634

Petitou M, Van Boeckel CAA (2004) A synthetic antithrombin III binding pentasaccharide is now a drug! What comes next? Angew Chem Int Ed 43:3118–3133

Polat T, Wong C-H (2007) Anomeric reactivity based one-pot synthesis of heparin-like oligosaccharides. J Am Chem Soc 129:12795–12800

Pongener I, Sletten ET, Danglad-Flores J, Seeberger PH, Miller GJ (2023) Synthesis of a heparan sulfate tetrasaccharide using automated glycan assembly. Org Biomol Chem 22:1395–1399

Proudfoot AE, Johnson Z, Bonvin P, Handel TM (2017) Glycosaminoglycan interactions with chemokines add complexity to a complex system. Pharmaceuticals 10(3):70

Ramadan S, Su G, Baryal K, Hsieh-Wilson LC, Liu J, Huang X (2022) Automated solid phase assisted synthesis of heparan sulfate disaccharide library. Org Chem Front 9:2910–2920

Raman R, Sasisekharan V, Sasisekharan R (2005) Structural insights into biological roles of protein-glycosaminoglycan interactions. Chem Biol 12(267):277

Raman D, Sobolik-Delmaire T, Richmond A (2011) Chemokines in health and disease. Exp Cell Res 317(5):575–589

Rosenberg DR, Hichs M, Damus SP (1973) Anticoagulant action of heparin. Nature 246:355–357

Sankarayanarayanan NV, Strebel TR, Boothello RS, Sheerin K, Raghuraman A, Sallas F, Mosier PD, Watermeyer ND, Oscarson S, Desai UR (2017) A hexasaccharide containing rare 2-O-sulfate-glucuronic acid residues selectively activates heparin cofactor II. Angew Chem Int Ed 56:2312–2317

Sasisekharan R, Raman R, Prabhakar V (2006) Glycomics approach to structure-function relationships of glycosaminoglycans. Annu Rev Biomed Eng 8:181–231

Sheng GJ, Oh YI, Chang S-K, Hsieh-Wilson LC (2013) Tunable heparan sulfate mimetics for modulating chemokine activity. J Am Chem Soc 135(30):10898–10901

Spillmann D, Witt D, Lindahl U (1998) Defining the interleukin-8-binding domain of heparan sulfate. J Biol Chem 273(25):15487–15493

Stringer SE, Gallagher JT (1997) Specific binding of the chemokine platelet factor 4 to heparan sulfate. J Biol Chem 272(33):20508–20514

Taylor KR, Gallo RL (2006) Glycosaminoglycans and their proteoglycans: host-associated molecular patterns for initiation and modulation of inflammation. FASEB J 20:9–22

Turnbull JE, Fernig DG, Ke Y, Wilkinson MC, Gallagher JT (1992) Identification of the basic fibroblast growth factor binding sequence in fibroblast heparan sulfate. J Biol Chem 267:10337–10341

Turner N, Grose R (2010) Fibroblast growth factor signalling: from development to cancer. Nat Rev Cancer 10:116–129

Visser A, Grootenhius PDJ, Van Boeckel CAA (1994) A mechanism of haparin indused potentiation of antithrombin III. Nat Struct Biol 1:423–425

Wang Z, Xu Y, Yong B, Tiruchinapally G, Sun B, Liu R, Dulaney S, Liu J, Huang X (2010) Preactivation-based, one-pot combinatorial synthesis of heparin-like hexasaccharides for the analysis of heparin-protein interactions. Chem Eur J 16:8365–8375

Wang P, Zhao J, Nasr SH, Otieno SA, Zhang F, Qiang W, Linhardt RJ, Huang X (2021) Probing amyloid β interactions with synthetic heparan sulfate oligosaccharides. ACS Chem Biol 16: 1894–1899

Wang L, Sorum AW, Huang BS, Kern MK, Guowei S, Pawr N, Huang X, Liu J, Pohl NLB, Hsieh-Wilson LC (2023) Efficient platform for synthesizing comprehensive heparan sulfate oligosaccharide libraries for decoding glycosaminoglycan-protein interactions. Nat Chem 15:1108–1117

Whitelock JM, Iozzo RV (2005) Heparan sulfate: a complex polymer charged with biological activity. Chem Rev 105:2745–2764

Xu D, Esko JD (2014) Demystifying heparan sulfate-protein interactions. Annu Rev Biochem 83: 129–157

Yang B, Yoshida K, Yin Z, Dai H, Kavunja H, El-Dakdouki MH, Sungsuwan S, Dulaney SB, Huang X (2012) Chemical synthesis of a heparan sulfate glycopeptide: syndecan-1. Angew Chem Int Ed 51:10185–10189

Yang W, Yoshida K, Yang B, Huang X (2016) Obstacles and solutions for chemical synthesis of syndecan-3 (53-62) glycopeptides with two heparan sulfate chains. Carbohydr Res 435:180–194

Yoshida K, Yang B, Yang W, Zhang Z, Zhang J, Huang X (2014) Chemical synthesis of syndecan-3 glycopeptides bearing two heparan sulfate glycan chains. Angew Chem Int Ed 53:9051–9058

Yun YR, Won JE, Jeon E, Lee S, Kang W, Jo H, Jang JH, Shin US, Kim HW (2010) Fibroblast growth factors: biology, function, and application for tissue regeneration. J Tissue Eng 1:1–18

Zhao W, McCallum SA, Xiao Z, Zhang F, Linhardt RJ (2012) Binding affinities of vascular endothelial growth factor (VEGF) for heparin-derived oligosaccharides. Biosci Rep 32:71–81

Zong C, Huang R, Condac E, Chiu Y, Xiao W, Li X, Lu W, Ishihara M, Wang S, Ramiah A, Stickney M, Azadi P, Amster IJ, Moremen KW, Wang L, Sharp JS, Boons GJ (2016) Integrated approach to identify heparan sulfate ligand requirements of Robo1. J Am Chem Soc 138:13059–13067

Zong C, Venot A, Li X, Lu W, Xiao W, Wilkes J-SL, Salanga CL, Handel TM, Wang L, Wolfert MA, Boons G-J (2017a) Heparan sulfate microarray reveals that heparan sulfate–protein binding exhibits different ligand requirements. J Am Chem Soc 139:9534–9543

Zong C, Venot A, Li X, Lu W, Xiao W, Wilkes JSL, Salanga CL, Handel TM, Wang L, Wolfert MA, Boons GJ (2017b) Heparan sulfate microarray reveals that heparan sulfate-protein binding exhibits different ligand requirements. J Am Chem Soc 139:9534–9543

Zulueta MM, Liu SY, Lin YT, Huang CJ, Wang CC, Ku CC, Shi Z, Chyan CL, Irene D, Lim LH, Tsai TI, Hu YP, Arco SD, Wong CH, Hung SC (2012) α-Glycosylation by D-glucosamine-derived donors: synthesis of heparosan and heparin analogues that interact with mycobacterial heparin-binding hemagglutinin. J Am Chem Soc 134:8988–8995

Activity-Based Profiling of Retaining Glycosidases in Disease Diagnosis and Their Application in Drug Discovery

Yevhenii Radchenko, Johannes M. F. G. Aerts, Gideon J. Davies, Jeroen D. C. Codée, and Herman S. Overkleeft

Contents

Abstract

Retaining glycosidases employ a two-step double displacement mechanism to hydrolyze their substrate glycosides. This mechanism involves a covalent enzyme-substrate adduct, and irreversible retaining glycosidase inhibitors have been designed based on this mechanism. Tagging such inhibitors with a reported moiety (biotin, fluorophore, bioorthogonal tag) provides activity-based retaining glycosidase probes. This chapter describes research on such activity-based probes that are inspired by the natural product retaining β-glucosidase inhibitor, cyclophellitol. Modulation of the configuration and substitution pattern yielded a suite of probes with which a host of retaining glycosidases are inhibited, and reported on, including enzymes involved in human pathologies (cancer, inherited lysosomal storage disorders). This chapter provides insights into their design and

Y. Radchenko · J. M. F. G. Aerts · J. D. C. Codée · H. S. Overkleeft (✉)
Leiden Institute of Chemistry, Leiden University, Leiden, The Netherlands
e-mail: h.s.overkleeft@chem.leidenuniv.nl

G. J. Davies
York Structural Biology Laboratory, Department of Chemistry, The University of York, York, United Kingdom

F. Pfrengle et al. (eds.), *Complex Carbohydrates in Health and Disease*, Handbook of Experimental Pharmacology 288, https://doi.org/10.1007/164_2025_743

synthesis, their application in disease diagnosis, and their application in drug discovery, both as tools to uncover competitive inhibitors and as starting point for the design of covalent inhibitors.

Keywords

Activity-based protein profiling · Cyclophellitol · Glycomimetics · Glycosidases · Lysosomal storage disorders · Mechanism-based inhibitors

1 Introduction

Activity-based protein profiling (ABPP) is one of the main driving forces in chemical biology that impacts fundamental as well as applied biological and biomedical research (Cravatt et al. 2008; Fonovíc and Bogyo 2008; Willems et al. 2011). In ABPP, an activity-based probe (ABP) irreversibly inhibits an enzyme through the formation of a stable covalent enzyme-substrate adduct. A reporter, which is an integral part of the ABP structure, and which may be a fluorophore, an enrichment/capture agent such as biotin or a bioorthogonal group (azide, alkyne, strained alkene), enables specific detection and annotation of ABP-selected enzymes in various assay formats. Comparative ABPP allows for the discovery of wholly new enzyme activities. Competitive ABPP enables inhibitor discovery and the design and exploitation of high-throughput drug discovery platforms, including ones based on fluorescence polarisation (Dillon et al. 2012). ABPP is complementary to contemporary methodology in biochemistry, cell biology, and molecular genetics and enables studies on enzymes in their physiological context that cannot be done by any other means. Whereas, for instance, fluorogenic substrates allow for quantification of bulk enzyme activities, ABPP discriminates the individual enzymes contributing to such activities. Moreover, and unlike any other contemporary assay, ABPP in all circumstances allows backtracking of the activities detected, either by gel electrophoresis or by proteomics.

Key to the success of ABPP is access to potent and (in-class) selective mechanism-based enzyme inactivators. With notable exceptions (as for example in the kinase field) ABPP has therefore met with most success in areas where enzymes process their substrates through mechanisms involving a transient, covalent enzyme-substrate adduct. This is well illustrated by the first successful activity-based probe (ABP) designs – those for serine hydrolases (Liu et al. 1999) and those for cysteine proteases (Greenbaum et al. 2000). Mechanism-based, covalent, and irreversible inhibitors for these large families of enzymes have been known for decades, and at the turn of the century, the groups of Cravatt and Bogyo independently realized that tagging known inhibitor scaffolds would yield tools that, instead of studying isolated enzymes, enable monitoring whole enzyme families at once and in complex biological samples. Cravatt and coworkers in 1999 introduced the concept of ABPP by grafting a biotin onto a fluorophosphonate (a broad-spectrum serine hydrolase inhibitor) (Liu et al. 1999). The resulting FP-biotin proved able to report at once on tens to hundreds of serine hydrolases in one single experiment. Bogyo and coworkers at around the same time realized that extending the natural product cysteine protease inactivator, E-64, with a biotin yielded effective ABPs with which the majority of the cathepsin family of lysosomal cysteine proteases could be studied (isolated, identified, and the selectivity of inhibitors on individual enzyme members monitored) in one single experiment (Greenbaum et al. 2000). A wealth of ABPs targeting various enzyme families have followed these pioneering studies, including threonine proteases (proteasomes) and kinases (Fang et al. 2021). One such family and the subject of this chapter comprises the broad family of retaining glycosidases. As with the serine hydrolase ABPs and cysteine protease ABPs, the mechanism of action followed by most retaining glycosidases allows for mechanism-based inactivation, and from there, for ABP designs. In contrast to serine hydrolases and cysteine proteases, however, glycosidases are often very particular to the nature (configuration, substitution pattern) of their substrate, which is reflected by retaining glycosidase ABP reactivity. As a result, and as discussed in this chapter, a suite of ABPs geared to emulate the configuration of the glycosides processed by glycosidases have been reported in the past 15 years (Willems et al. 2014a; Wu et al. 2019).

Glycosidases, also termed glycoside hydrolases (GHs) and classified in the CAZy database (Henrissat and Davies 1997; Lombard et al. 2014) into over a hundred families based primarily on sequence and structure/fold, in general follow one of two distinct mechanisms (exemplified for β-glucosidases in Fig. 1). As can be seen (Fig. 1a) and as first described by Koshland (1953), retaining β-glucosidases employ two S_N2 displacement steps. In the first, the aglycon is protonated by the catalytic acid-base residue and expelled by the active site nucleophile, yielding the transient, covalent enzyme substrate acylal adduct, with inversion of configuration at the substrate glucoside anomeric carbon. The leaving group is then expelled from the

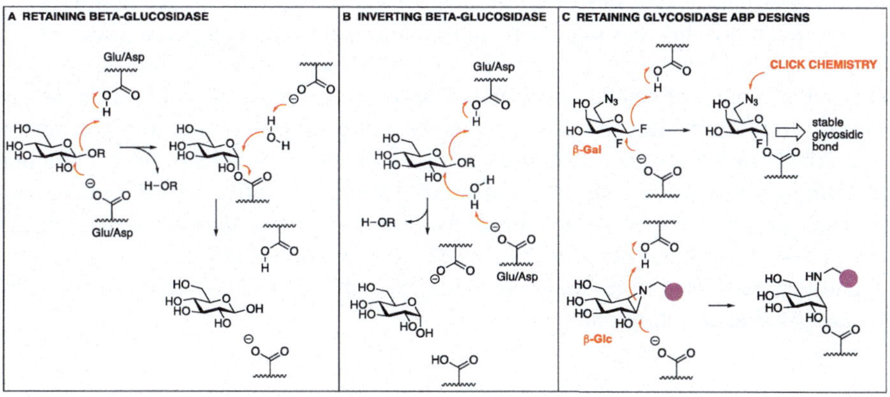

Fig. 1 Mechanism of action of retaining β-glucosidases (**a**) inverting β-glycosidases (**b**), and mode of action (**c**) of the two most employed retaining glycosidase ABP designs: deoxyfluorosugars (top) and cyclitol aziridines (bottom – the purple bulb denotes a reporter group: a fluorophore, biotin, or a bioorthogonal handle)

enzyme active site, then water enters, and in a reversal of steps, again with inversion (so net retention) of configuration, glucose is formed and released from the active site. Inverting β-glucosidases (Fig. 1b panel) in contrast employ a one-step mechanism which does not involve a covalent enzyme-substrate intermediate adduct.

Inverting glycosidases, which comprise roughly half of the glycoside hydrolases known, and because of the lack of a covalent enzyme-substrate intermediate, have to date been rather resistant to ABPP, some isolated examples such as those utilizing photoaffinity labeling aside (Gandy et al. 2011; van Scherpenzeel et al. 2010). Retaining glycosidases in contrast have met with considerably more success in the past decades (Stubbs and Vocadlo 2009). Several glycomimetic scaffolds have been put forward over the years, the two most prominent of which are given in Fig. 1c and comprise the deoxyfluoroglycosides (top) and cyclophellitols/cyclophellitol aziridines (bottom). Substitution of a hydroxyl group by fluorine (or installation of a fluorine at C5) renders a glycoside less reactive compared to the native sugar (Withers et al. 1987). This holds true both for the initial reaction within a retaining glycosidase active site (which can be compensated for by tuning the reactivity of the leaving group – here shown is a fluoride) and the enzyme substrate acylal adduct. Depending on the configuration of the sugar and the nature of the enzyme, the covalent adduct may be stabilized to such an extent that the enzyme is effectively irreversibly inhibited and can be detected by the reporter group (here an azide) attached (Vocadlo and Bertozzi 2004; Williams et al. 2006; Stubbs et al. 2008; Chauvigné-Hines et al. 2012; Tsai et al. 2013; Jain et al. 2018). The second scaffold design comprises a sugar-configured cyclitol epoxide or (as shown in Fig. 1c) aziridine (Willems et al. 2014a; Wu et al. 2019). Protonation of the epoxide/aziridine and concommitant nucleophilic displacement leads to a stable enzyme-inhibitor ester adduct. The aziridine nitrogen allows attachment of a reporter group, yielding one-step ABPs. Alternatively, and when working with epoxides, the reporter moiety must replace one of the hydroxyls, but (and this is one of the disadvantages of the fluoroglycosides as well) this may be detrimental for enzyme recognition. Cyclitol aziridines are for this reason perhaps most suitable for retaining glycosidase ABPP. This chapter will highlight the status on such ABPs with a focus on human enzymes and in the context of health and disease. Following a section on synthesis strategies toward cyclitols/cyclitol aziridines, examples will be given on their use in comparative ABPP in tissues of diseased individuals in which deficiency in a particular retaining glycosidase is causative of disease. Then, competitive ABPP is introduced to discover new competitive retaining glycosidase ligands, after which the chapter concludes by detailing how mechanism-based glycosidase inhibitors based on the cyclitol epoxides/aziridines themselves may become clinical candidate mechanism-based glycosidase inhibitors.

2 Synthesis of Cyclitol Epoxides and Cyclitol Aziridines

All the work on cyclitol epoxide–/aziridine-based retaining glycosidase ABPs focus on the mode of action of the natural product cyclophellitol (Fig. 2a). This molecule was first isolated from the mushroom *Phellinus* sp. in 1990 at which time also its mode of action was unveiled (Atsumi et al. 1990; Withers and Umezawa 1991). Cyclophellitol inhibits retaining β-glucosidases in a mechanism like that depicted for its aziridine analogue in Fig. 1c (protonation and nucleophilic displacement of the epoxide). As such, cyclophellitol emulates the mode of action of the retaining glucosidase (α and β) inactivator, CBE (Fig. 2a inset), which at that time had been known for decades (Legler 1966, 1968). Cyclophellitol however proved at once more potent and more selective (β over α) than CBE. This feature, and likely also its structure (a small cyclohexane but with all ring carbons chiral and featuring an epoxide), attracted the interest of synthetic organic chemists, with several total syntheses in the decades following its discovery (Marco-Contelles 2001; Jiang et al. 2016a).

Arguably the most effective route of synthesis and the strategy from which most of today's cyclophellitol-based ABPs are derived is the one from Madsen and coworkers (Fig. 2a) (Hansen et al. 2005). The Madsen synthesis has as key step the indium-catalyzed Barbier addition of bromocrotonate **4** to chiral, protected 2,3-dihydroxy-4-pentenal **3**, which is accessible in a few standard transformations from D-xylose **1**. The ester in the resulting 1,7-diene, featuring four of the six chiral centers inherent to cyclophellitol and obtained in good diastereomeric ratio, is then reduced to the primary alcohol after which ring-closing metathesis give partially

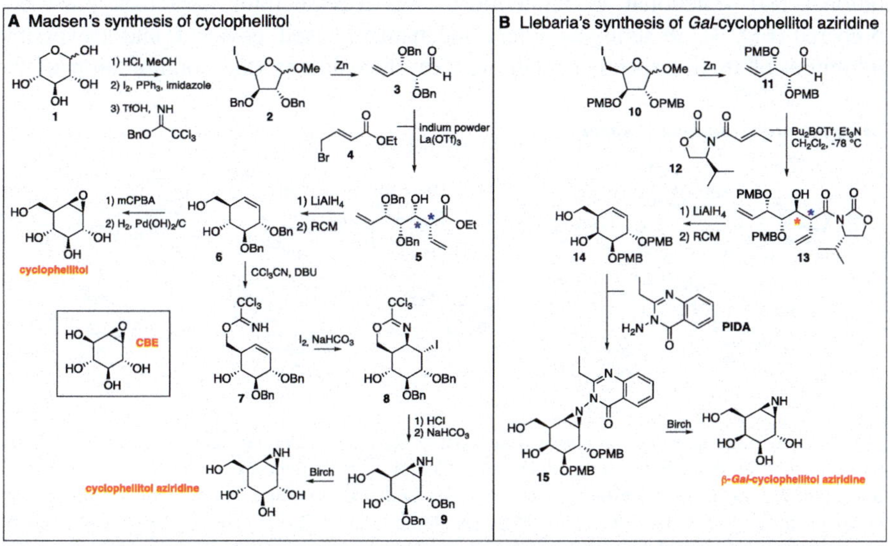

Fig. 2 Representative synthesis strategies toward cyclitol epoxides and cyclitol aziridines: synthesis of (**a**) cyclophellitol and cyclophellitol aziridine and (**b**) β-D-*Gal*-cyclophellitol aziridine

protected cyclohexene **6**. Madsen's total synthesis then capitalizes on the homoallylic alcohol in **6** to do a directed *m*CPBA-mediated epoxidation, after which palladium-catalyzed hydrogenolysis yields cyclophellitol. Cyclohexene **6** can also be used for a stereoselective synthesis of cyclophellitol aziridine and to this end the primary alcohol in cyclohexene **6** is transformed into trichloroacetimidate **7** (Kallemeijn et al. 2012; Li et al. 2014). Iodocyclisation then yields in a stereospecific manner iodo-iminal **8**, which after acid treatment (removal of the iminal) and then base treatment (yielding the free amine) gives in another stereospecific process, and after intramolecular iodine displacement aziridine **9**. Birch reduction of the two benzyl ethers then gives cyclophellitol aziridine. Another adaptation of the Madsen route is depicted in Fig. 2b and comprises the route toward galactose-configured cyclophellitol aziridine as published by Llebaria and coworkers (Harrak et al. 2011). Here, instead of an enantioselective Barbier, an enantioselective aldol condensation features as the key step. Xylose-derived pentenal **11** (differing from **3** only in the two secondary alcohol protective groups) is condensed with Evans-templated crotonate **12** to give 1,7-diene **13** again with good stereoselectivity. The stereochemical outcome in the Llebaria route differs from that in the Madsen route. Both routes create two additional chiral centers in the condensation step. One (the blue asterisk) is the same in both; however, the other (blue asterisk in Scheme 1A, red asterisk in Fig. 2b) is opposite. This opposite outcome allows for the synthesis of the galactose-configured cyclophellitol aziridine by first reducing off the Evans template in **13**, then ring-closing metathesis, followed by direct aziridination and final global deprotection. Both routes are scalable and adaptable (varying in starting material and order of reactions). In this way, and through independent routes of synthesis (Ofman et al. 2021), a series of configurational and functional cyclophellitols and cyclophellitol aziridines has been prepared (Fig. 3), all acting as potent, mechanism-based, covalent, and irreversible inhibitors of their respective retaining exo-glycosidases (the configuration of the

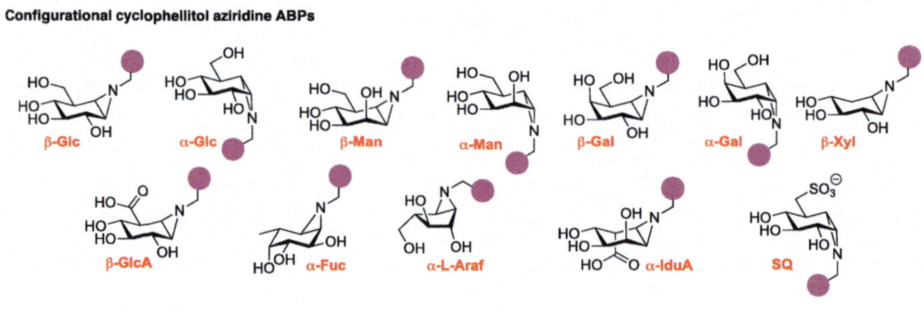

Fig. 3 Structure of retaining glycosidase ABPs. Red color denotes the configuration of the monosaccharide the ABPs are modeled on. The pink bulb denotes a reporter group (biotin, fluorophore, bioorthogonal group). The syntheses and full structures of the depicted probes can be found in the following references: β-Glc (Kallemeijn et al. 2012; Li et al. 2014), α-Glc (Jiang et al. 2016b), β-Man (McGregor et al. 2022), α-Man (Armstrong et al. 2020), β-Gal (Kuo et al. 2023), α-Gal (Willems et al. 2014b), β-Xyl (Schröder et al. 2019), β-GlcA (Wu et al. 2017), α-Fuc (Jiang et al. 2015), α-L-Araf (McGregor et al. 2020), α-IduA (Artola et al. 2018), SQ (Li et al. 2024)

parent sugar to which the compounds have been modeled is given in red). Grafting a reporter group to the aziridine (and in some cases the primary alcohol) gives access to the panel of retaining exoglycosidase ABPs that are subject of this chapter (reporter groups are shown as pink bulbs in Fig. 3; some selected examples of specific structures will be given in the next sections).

3 Retaining Glycosidase ABPP in Monitoring Glycosidases in Lysosomal Storage Disorders

Lysosomal storage disorders (LSDs) comprise a family of inherited genetic disorders in which either a lysosomal hydrolase or a factor assisting a lysosomal hydrolase is genetically impaired. Depending on the nature of the disease and that of the genetic mutation, one specific activity is either partially or completely impaired, with a less to more severe disease state as the result. Many of these inherited metabolic disorders are related to (partial) impairment of lysosomal glycosidases and characterized by accumulation of their substrate as the primary storage material. These include Gaucher disease (deficiency in lysosomal glucosylceramidase (GBA1), storage of glucosylceramide), Fabry disease (deficiency in lysosomal α-galactosidase (αGalA), storage of globotriaosylceramide), and Pompe disease (deficiency in lysosomal α-glucosidase (GAA), glycogen storage). These three lysosomal glycosidases all employ a Koshland two-step retaining mechanism, and all are amenable to cyclophellitol-type mechanism-based inhibition and, therefore, to activity-based glycosidase profiling. ABPP counts active enzyme molecules, rather than reporting on combined enzyme activity as is done by for instance fluorogenic substrate assays. This feature allows for the development of diagnostics ABPP assays, as is illustrated in Fig. 4 for Gaucher, Fabry, and Pompe disease. Figure 4a depicts general ABPP workflows in which retaining glycosidases are detected, identified, and modulated. In comparative ABPP, cells (extracts, tissues, whole species) are treated with an ABP, then lysed, protein contents denatured, and applied to either sodium dodecylsulfate polyacrylamide gel electrophoresis (SDS-PAGE) or biotin-streptavidin pulldown. The former allows for in-gel detection (fluorescence scanning in case of usage of a fluorescent ABP, Western blotting in case of a biotin-carrying ABP) of ABP-labeled proteins, whereas the latter allows for LC-MSMS identification of ABP-modified proteins. In competitive ABPP, samples are preincubated with putative inhibitors (which may be competitive inhibitors as indicated, or mechanism-based ones) prior to SDS-PAGE or biotin-streptavidin pull-down.

The first cyclophellitol-based ABP was developed with the purpose to detect relative numbers of active acid glucosylceramidase (GBA1) molecules in the context of Gaucher disease (Fig. 4b) (Witte et al. 2010). In contrast to most exoglycosidases (enzymes that process glycans by removing one monosaccharide at a time from the non-reducing end), GBA1 appeared to accept substituted glucose(like) ligands, and installing a fluorophore at C6 of cyclophellitol (glucose numbering, position indicated in Fig. 4b, see for the full structure the original paper) yielded a highly

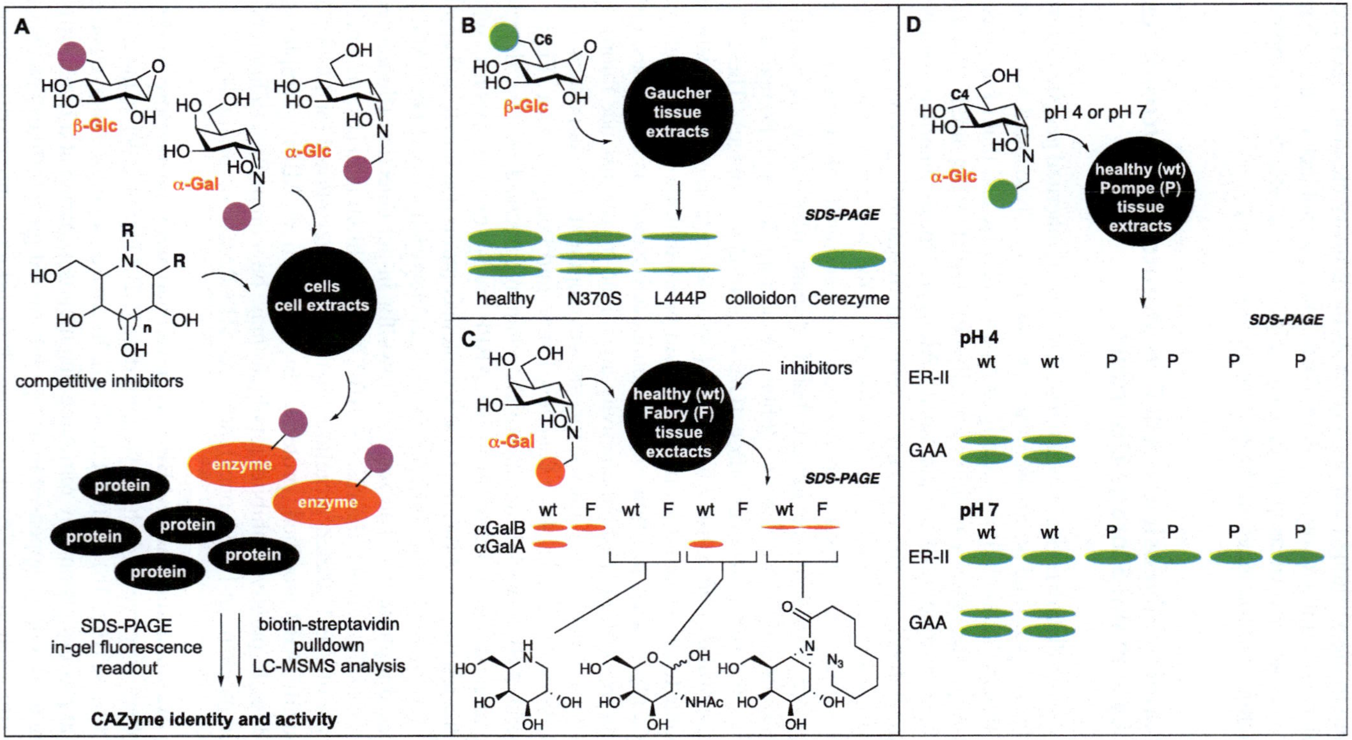

Fig. 4 (**a**) General comparative and competitive ABPP workflow for detecting and annotating retaining glycosidases. (**b**) Diagnostics assay for determining active glucosylceramidase (GBA1) molecules in Gaucher patient tissue. (**c**) Diagnostics assay for αGalA; the enzyme deficient in Fabry disease. (**d**) Diagnostics assay for GAA; the enzyme deficient in Pompe disease

potent and selective ABP that did not cross-react with the two other two human retaining β-glycosidases, GBA2 and GBA3. Severity of Gaucher disease correlates with the nature of the mutation, which is reflected by the fluorescence intensity of protein-normalized cell extracts from healthy individuals compared to those obtained from various GBA1 gene-mutant individuals (here and onward cartoons of the gel images are shown; see the corresponding papers for original gel images, and in this experiment a BODIPY-green-fluorescent ABP was used). The N370S mutation comprises a more common, and relatively mild, mutation, which is reflected in the significant, but only partial, decrease in labeling intensity in the extract derived from an individual having this mutation (GBA1 appears as various isoforms on SDS-PAGE due to the heterogenicity in its four N-glycans). The L444P mutation is more severe, and the various mutations that lead to the lethal Collodion phenotype result in complete absence of detectable active GBA1 molecules. The right-most lane depicts labeling of recombinant, homogeneous GBA1 (Cerezyme – all N-glycans are trimmed down to the core pentasaccharide) that is used in the clinic in what is termed enzyme replacement therapy (ERT) for the treatment of non-neuronopathic (type 1) Gaucher disease. In contrast to Gaucher disease, where some GBA1 activity remains in most genetic mutations, Fabry disease is characterized by the complete absence of the enzyme underlying this disease, αGalA. This is readily detected by comparative ABPP (Fig. 4c). Healthy individuals express two highly homologous genes, αGalA and αGalB, both of which react with α-galactose-configured cyclophellitol aziridine probes (the two bands in the left-most lane; here a red fluorescent ABP is used) (Willems et al. 2014b). In contrast, extracts from tissue from a Fabry diseased individual reveal the upper (αGalB) band only. Both αGalA and αGalB are inhibited by the competitive inhibitor, deoxygalactonojirimycin (lanes three and four), and αGalB, which is an α-*N*-acetyl-galactosaminidase activity, is inhibited by its product *N*-acetylgalactosamine. Labeling of both αGalA and (though to a lesser extend) αGalB can be blocked by inclusion of the non-fluorescent αGal-cyclophellitol aziridine in an experiment to confirm enzyme specificity of the ABP. The power of the technology finally is revealed by the experiment depicted in Fig. 4d. Changing only the configuration at C4, compared to the αGalA/B probe, yields a probe (here endowed with a green fluorophore) providing a completely different labeling pattern (Jiang et al. 2016b). Treatment of tissue extracts from healthy individuals at pH 4 yields a double-digit band which corresponds (as was established by biotin-pulldown-ABPP-proteomics as shown in Fig. 4a) to two isoforms of the lysosomal glycogen-degrading enzyme α-glucosidase A (GAA). Treatment of the same tissue extract at pH 7 yields these two bands as well and on top of that also a higher running band that corresponds to the ER-quality control partaking enzyme ER-α-glucosidase II (GANAB, ER-II). ER-II is a retaining α-glucosidase, this in contrast to the inverting enzyme also involved in ER quality control; ER-I. This experiment shows that retaining glycosidase ABPP can be pH dependent, and report on true enzymatic activities: GAA is indeed active at pH 4 and 7 whereas ER-II activity is lost at pH 4. Comparative ABPP on samples from four Pompe disease individuals shows the absence of GAA labeling at pH 4, but not that of ER-II labeling at pH 7, which serves as a nice internal control: the protein samples are bona fide, and the absence of GAA labeling is caused by the absence of (active) enzyme.

4 Retaining Glycosidase ABPP in Identifying Competitive Inhibitors

Competitive glycosidase inhibitors are sought-after modalities in various disease areas. The competitive intestinal α-glucosidase inhibitors, miglitol (*N*-hydroxyethyl-deoxynojirimycin) and acarbose, are used in the clinic for the treatment of type 2 diabetes, whereas the sialic acid analogues, oseltamivir (Tamiflu) and zanamivir (Relenza) are used as anti-influenza agents based on their influenza neuraminidase inhibitor activity. In the context of LSDs, competitive inhibitors of mutant enzymes are currently pursued in what is termed pharmacological chaperone therapy (PCT) to stabilize the fold of nascent mutant lysosomal glycosidases. In this view (and which is also evidenced from the studies described in Fig. 4b–d), it is not so much the activity of mutant enzymes that is impaired as their ability to survive ER quality control and/or ability to traffic to lysosomes. Exposing mutant enzyme to a high affinity active site ligand (that is, an effective competitive inhibitor) is thought to stabilize its proper fold and thus the chance it survives the secretory pathway. In an alternative therapeutic strategy to stabilize it in circulation, an ERT-enzyme is administered jointly with a competitive ligand, and in both scenarios, the competitive ligand is expected to be competed out by the actual substrate once within lysosomes. PCT, either as standalone therapy or in combination with ERT, is currently pursued for a range of LSDs including Gaucher, Fabry, and Pompe. For this, potent and selective inhibitors for the respective enzymes are sought for. Retaining glycosidase ABPP can be of assistance in the search for these as is demonstrated in Fig. 5 in a series of experiments in which competitive ABPP is merged with the often-applied high-throughput technique of fluorescence polarization. In fluorescence polarization (FP) assays, a fluorophore is excited at a specific

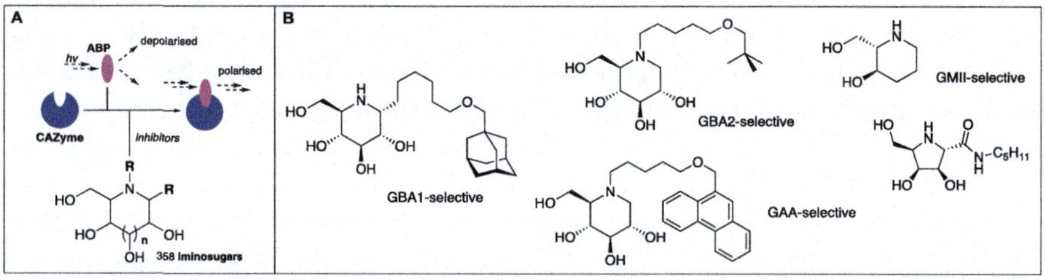

Fig. 5 Fluorescence polarization activity-based protein profiling (FP-ABPP) in the discovery of competitive retaining glycosidase inhibitors. (**a**) General principle. (**b**) Selection of competitive inhibitors out of the in-house 358 entry iminosugar library discovered for the retaining β-glucosidases GBA1, GBA2, the retaining α-glucosidase GAA, and the retaining α-mannosidase GMII

wavelength with linear polarized light. The orientation of emitted light and thereby the degree of polarized light depend on the movement (Brownian motion) of the excited fluorophore has undergone, which depends on its size: large molecules, such as an enzyme) bound ABP, move and rotate slower than small molecules, such as unbound ABPs. The degree of loss of polarized emitted light therefore presents the degree of inhibition of the target enzyme. Mechanism-based, covalent, and irreversible inhibitors are ideally suited for FP, which relies on tight binding of a fluorescent reporter molecule to a target enzyme (Dillon et al. 2012). This technique, termed fluorescence polarization activity-based protein profiling (FP-ABPP), has been used to identify candidate inhibitors for a range of hydrolytic enzymes (Dillon et al. 2012; Bachovchin et al. 2009) including serine hydrolases, cysteine proteases, and, as illustrated in Fig. 5, retaining glycosidases.

The general workflow of FP-ABPP is depicted in Fig. 5a and comprises treatment of the retaining glycosidase of interest with a pyrrolidine/piperidine iminosugar from focused libraries (the here-presented studies are done with 358 configurational and functional pyrrolidine and piperidine analogues) and at various concentrations, prior to treatment with a fluorescent ABP. For this purpose, purified recombinant enzymes can be used, but also lysates from cells in which particular glycosidases have been brought to overexpression. Follow-up studies on the determination of activity and selectivity (with respect to off-target glycosidases) have yielded for instance α-C-deoxynojirimycins as GBA1-selective competitive inhibitors (van der Gracht et al. 2023). In a similar vein, branched *N*-alkyldeoxynojirimycins were identified from the same library as GAA-selective competitive inhibitors (van der Gracht et al. 2023). The GAA selectivity of these compounds, which may serve as starting points for PCT development targeting Pompe disease, is thought to be due to the branched nature of the hydrophobic group, as linear hydrophobic moieties at this position have in the past been shown to lead to potent inhibition of GBA1, neutral glucosylceramidase (GBA2), and glucosylceramide synthase (GCS). FP-ABPP can be utilized to identify competitive inhibitors for any retaining glycosidase for which fluorescent ABPs are available, as was demonstrated for the identification of GBA2-selective deoxynojirimycins (Lahav et al. 2017) as well as some fragment pyrrolidines-piperidines for the anticancer target, Golgi-residing α-mannosidase, GMII (Armstrong et al. 2020).

5 Functionalized Cyclophellitols as Starting Point in Drug Discovery

Besides their use in ABPP, covalent and irreversible inhibitors are increasingly seen as promising starting points for drug discovery. Covalent inhibitors have historically been avoided in medicine for fears of off-target toxicity. Recent studies have however shown that tuning inhibitor reactivity (electrophilicity) together with optimizing initial target binding efficiency can lead to very efficient and very selective enzyme inhibitors. These include several blockbuster anticancer agents that covalently and irreversibly inhibit kinases and proteasomes. Cyclophellitols

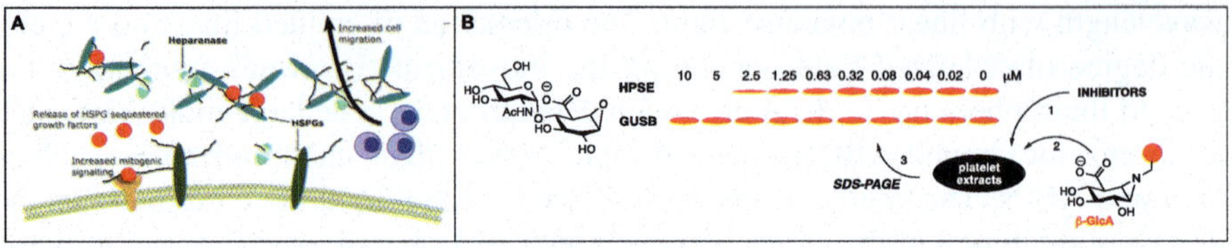

Fig. 6 (**a**) Extracellular heparanase (HPSE) promotes tumor metastasis by remodeling the extracellular matrix, allowing for tumor cells to migrate and tumor-promoting signal molecules to escape. (**b**) GlcNAc-glucuronic cyclophellitol is an efficient inhibitor of the retaining endoglucuronidase, HPSE, without touching the retaining exoglucuronidase, GUSB

may serve as good starting points for drug discovery by targeting disease-causative retaining glycosidases, one example of which comprises human heparanase (HPSE). HPSE is the main human enzyme responsible for extracellular heparan sulfate breakdown and strongly drives metastasis when overexpressed and excreted to the extracellular matrix (ECM). Cancer-associated extracellular matrix changes promote metastatic spread by the release of tumor proliferation and angiogenesis-promoting growth factors and cytokines, as well as the release of tumor cells from primary tumor sites (Fig. 6a). HPSE has therefore been considered as potential anti-metastatic cancer target, yet the most effective HPSE inhibitors to date suffer from multiple off-target effects, and no HPSE inhibitors have reached the clinic to date. These contemporary HPSE inhibitors are often large, substrate (heparan sulfate)-derived, heterogeneous, and highly anionic species that act as competitive inhibitors. HPSE is a retaining glycosidase which was found (Fig. 6b) to react with a β-glucuronic-type cyclophellitol aziridine ABP, just like the initial target of the studies: the exoglucuronidase, GUSB (Wu et al. 2017). Follow-up studies revealed that functionalization of C4 of the glucuronic acid cyclophellitol scaffold, for instance, with an α-configured GlcNAc, yielded compounds both more potent and more selective than the non-substituted uronic cyclophellitols, as was demonstrated (Fig. 6b) in a competitive ABPP experiment (de Boer et al. 2022).

The disaccharidic cyclophellitol shown in Fig. 6b has been investigated for its ability to suppress tumor growth in in vivo murine models for breast cancer, lung cancer, and bone marrow invading multiple myeloma, where it is shown to be equally active to the current best-in class compound, SST0001, and to work (in the bone marrow model) synergistically with the established clinical multiple myeloma drug and proteasome inhibitor, bortezomib (de Boer et al. 2022). Some initial structure-activity relationship studies have indicated that sulfation of the primary OH in the disaccharidic cyclophellitol, as well as deletion of the ACHN moiety, yields compounds with comparable in vitro HPSE inhibition potency. Though far away from actual clinical use, these studies hold promise for the development of a clinical HPSE-targeting drug: the lead is small, well defined, and amenable to medicinal chemistry optimization with respect to pharmacokinetic and pharmacodynamic behavior. This last example also highlights the potential of covalent and irreversible inhibitors where competitive ones may fail, such as appears the case for HPSE. This enzyme, like endoglycosidases in general, features a large substrate binding site with many interactions with its substrate required for effective substrate binding and turnover, but also for effective binding of competitive inhibitors. The initial binding efficiency is much less a determining factor for covalent (and irreversible) inhibitor activity, for which slow initial binding is offset by long (and for irreversible inhibitors, indefinite) K_{off} rates. This example finally also shows how ABPP and covalent inhibitor design/discovery/development go hand in hand. A suitable, partially selective ABP (the βGlcA ABP) served as starting point for the design of more potent and more selective inhibitors, the efficacy of which in in situ and in vivo studies could then be established assisted by the self-same ABP.

6 Conclusion and Outlook

This review summarized research from the past 15 years on the use of cyclophellitol-based ABPs in the study of human retaining glycosidases in health and disease. Retaining glycosidase ABPs can be used to monitor retaining glycosidase activities in relation to disease, to screen for new competitive inhibitors for these, and to be used as starting point for the development of more potent and more selective inhibitors by removing the reporter moiety and optimizing the structure of the cyclitol epoxide/aziridine scaffold. Besides the examples shown, there are more human retaining glycosidases of biomedical relevance, in the context of inherited metabolic disorders (for instance, galactosylceramidase – Krabbe disease; acid neuraminidase (NEU1) – sialidosis). All these enzymes may be targeted by similar approaches, both for their imaging and pharmacological inhibition. The same holds true for ABPP of gut microbiome retaining glycosidases, for instance in the context of individual-specific glucuronidase-mediated drug reactivation (Jariwala et al. 2020), whereas beyond the arena of biomedicine, cyclophellitol-ABPs have found use in dissecting retaining exoglycosidase and retaining endoglycosidase pools in secretomes of microorganisms grown on selected biomass polysaccharide sources (Schröder et al. 2019; McGregor et al. 2020, 2023). Finally, ABPP of inverting glycosidases has not yet been achieved to the same extent, and suitable designs to trap their catalytic machineries covalently and irreversibly are lacking. The discovery of such scaffolds would also open this family, likely as large as that of the retaining glycosidases, to the field of ABPP.

References

Armstrong Z, Kuo C-L, Lahav D, Liu B, Johnson R, Beenakker TJM, de Boer C, Wong C-S, van Rijssel ER, Debets MF, Florea BI, Hissink C, Boot RG, Geurink PP, Ovaa H, van der Stelt M, van der Marel GA, Codée JDC, Aerts JMFG, Wu L, Overkleeft HS, Davies GJ (2020) Manno-epi-cyclophellitols enable activity-based protein profiling of human α-mannosidases and discovery of new Golgi mannosidase II inhibitors. J Am Chem Soc 142:13021–13029

Artola M, Kuo C-L, McMahon SA, Hansen T, van der Lienden M, He X, van den Elst H, Florea BI, Kermode AR, van der Marel GA, Gloster T, Codée JDC, Overkleeft HS, Aerts JMFG (2018)

New irreversible α-L-iduronidase inhibitors and activity-based probes. Chem Eur J 24:19081–19088

Atsumi S, Umezawa K, Iinuma H, Naganawa H, Nakamura H, Iitaka Y, T. (1990) Takeuchi Production, isolation and structure determination of a novel beta-glucosidase inhibitor cyclophellitol, from Phellinus sp. J Antibiot 43:49–53

Bachovchin DA, Brown SJ, Rosen H, Cravatt BF (2009) Identification of selective inhibitors of uncharacterized enzymes by high-throughput screening with fluorescent activity-based probes. Nat Biotechnol 27:387–394

Chauvigné-Hines LM, Anderson LN, Weaver HM, Brown JN, Koech PK, Nicora CD, Hofstad BA, Smith RD, Wilkins MJ, Callister SJ, Wright AT (2012) Suite of activity-based probes for cellulose-degrading enzymes. J Am Chem Soc 134:20521–20532

Cravatt BF, Wright AT, Kozarich JW (2008) Activity-based protein profiling: from enzyme chemistry to proteomic chemistry. Annu Rev Biochem 77:383–414

de Boer C, Armstrong Z, Lit VAJ, Barash U, Ruijgrok G, Boyango I, Weitzenberg MM, Schröder SP, Sarris AJC, Meeuwenoord NJ, Bule P, Kayal Y, Ilan N, Codée JDC, Vlodavsky I, Overkleeft HS, Davies GJ, Wu L (2022) Mechanism based heparanase inhibitors reduce cancer metastasis in vivo. Proc Natl Acad Sci USA 119:e2203167119

Dillon MBC, Bachovchin DA, Brown SJ, Finn MG, Rosen H, Cravatt BF, Mowen KA (2012) Novel inhibitors for PRMT1 discovered by high-throughput screening using activity-based fluorescence polarisation. ACS Chem Biol 7:1198–1204

Fang H, Peng B, Ong SY, Wu Q, Li L, Yao SQ (2021) Recent advances in activity-based probes (ABPs) and affinity-based probes (A*f*BPs) for profiling of enzymes. Chem Sci 12:8288–8310. See for a recent review on the development of the ABPP field

Fonovíc M, Bogyo M (2008) Activity-based probes as a tool for functional proteomic analysis of proteases. Exp Rev Proteomics 5:721–730

Gandy MN, Debowski AW, Stubbs KA (2011) A general method for affinity-based proteomic profiling of *exo*-α-glycosidases. Chem Commun 47:5037–5039

Greenbaum F, Medzihradszky KF, Burlingame A, Bogyo M (2000) Epoxide electrophiles as activity-dependent cysteine protease profiling and discovery tools. Chem Biol 7:569–581

Hansen FG, Bundgaard E, Madsen R (2005) A short synthesis of (+)-cyclophellitol. J Org Chem 70:10139–10142

Harrak Y, Barra CM, Delgado A, Castaño AR, Llebaria A (2011) Galacto-configured aminocyclitol phytoceramides are potent in vivo invariant natural killer T cell stimulators. J Am Chem Soc 133:12079–12084

Henrissat B, Davies G (1997) Structural and sequence-based classification of glycoside hydrolases. Curr Opin Struct Biol 7:637–644

Jain N, Attia MA, Offen WA, Davies GJ, Brumer H (2018) Synthesis and application of a highly branched, mechanism-based 2-deoxy-2-fluoro-oligosaccharide inhibitor of *endo*-xyloglucanases. Org Biomol Chem 16:8732–8741

Jariwala PB, Pellock SJ, Goldfarb D, Cloer EC, Artola M, Simpson JB, Bhatt AP, Walton WG, Roberts LR, Major MB, Davies GJ, Overkleeft HS, Redinbo MR (2020) Discovering the microbial enzymes driving drug toxicity with activity-based protein profiling. ACS Chem Biol 15:217–225

Jiang J, Kallemeijn WW, Wright DW, van den Nieuwendijk AMCH, Coco Rohde V, Folch EC, van den Elst H, Florea BI, Scheij S, Donker-Koopman WE, Verhoek M, Li N, Schürmann M, Mink D, Boot RG, Codée JDC, van der Marel GA, Davies GJ, Aerts JMFG, Overkleeft HS (2015) *In vitro* and *in vivo* comparative and competitive activity-based protein profiling of GH29 alpha-L-fucosidases. Chem Sci 6:2782–2789

Jiang J, Artola M, Beenakker TJM, Schröder SP, Petracca R, de Boer C, Aerts JMFG, van der Marel GA, Codée JDC, Overkleeft HS (2016a) The synthesis of cyclophellitol aziridine and its configurational and functional isomers. Eur J Org Chem 2016:3671–3678. See for a review on cyclophellitol and cyclophellitol aziridine syntheses

Jiang J, Kuo C-L, Wu L, Franke C, Kallemeijn WW, Florea BI, van Meel E, van der Marel GA, Codée JDC, Boot RG, Davies GJ, Overkleeft HS, Aerts JMFG (2016b) Detection of active mammalian GH31 alpha-glucosidases in health and disease using in-class, broad-spectrum activity-based probes. ACS Cent Sci 2:351–358

Kallemeijn WW, Li K-Y, Witte MD, Marques ARA, Aten J, Scheij S, Jiang J-B, Willems LI, Voorn-Brouwer TM, van Roomen CPAA, Ottenhoff R, Boot RG, van den Elst H, Walvoort MTC, Florea BI, Codée JDC, van der Marel GA, Aerts JMFG, Overkleeft HS (2012) Novel activity-based probes for broad-spectrum profiling of retaining beta-exoglucosidases *in situ* and *in vivo*. Angew Chem Int Ed 51:12529–12533

Koshland D (1953) Stereochemistry and the mechanism of enzymatic reactions. Biol Rev 28:416–436

Kuo CL, Su Q, van den Nieuwendijk AMCH, Beenakker TJM, Offen WA, Willems LI, Boot RG, Sarris AJ, Marques ARA, Codée JDC, van der Marel GA, Florea BI, Davies GJ, Overkleeft HS, Aerts JMFG (2023) The development of a broad-spectrum retaining beta-exogalactosidase activity-based probe. Org Biomol Chem 21:7813–7820

Lahav D, Liu B, van den Berg RJBHN, van den Nieuwendijk AMCH, Wennekes T, Ghisaidoobe AT, Breen I, Ferraz MJ, Kuo C-L, Wu L, Geurink PP, Ovaa H, van der Marel GA, van der Stelt M, Boot RG, Davies GJ, Aerts JMFG, Overkleeft HS (2017) A fluorescence polarization activity-based protein profiling assay in the discovery of potent, selective inhibitors for human non-lysosomal glucosylceramidase. J Am Chem Soc 139:14192–14197

Legler G (1966) Studies on the action mechanism of glycoside splitting enzymes, I. Presentation and properties of specific inhibitors, *Hoppe Seylers Z.* Physiol Chem 345:197–214

Legler G (1968) Labelling of the active centre of a beta-glucosidase. Biochim Biophys Acta 151: 728–729

Li K-Y, Jiang J, Witte MD, Kallemeijn WW, van den Elst H, Wong C-S, Chander SD, Hoogendoorn S, Beenakker TJM, Codée JDC, Aerts JMFG, van der Marel GA, Overkleeft HS (2014) Synthesis of cyclophellitol, cyclophellitol aziridine, and their tagged derivatives. Eur J Org Chem 2014:6030–6043

Li Z, Pickles IB, Sharma M, Melling B, Pallasdies L, Codée JDC, Williams SJ, Overkleeft HS, Davies GJ (2024) Detection of sulfoquinovosidase activity in cell lysates using activity-based probes. Angew Chem Int Ed 63:e202401358

Liu Y, Patricelli MP, Cravatt BF (1999) Activity-based protein profiling: the serine hydrolases. Proc Natl Acad Sci 96:14694–14699

Lombard V, Golaconda Ramulu H, Drula E, Coutinho PM, Henrissat B (2014) The carbohydrate active enzymes database (CAZy) in 2013. Nucleic Acids Res 42:D490–D495

Marco-Contelles J (2001) Cyclohexane epoxides – chemistry and biochemistry of (+)-cyclophellitol. Eur J Org Chem 2001:1607–1618. See for a review on cyclophellitol syntheses shortly following its discovery

McGregor N, Artola M, Nin-Hill A, Linzel D, Haon M, Reijngoud J, Ram AFJ, Rosso M-N, van der Marel GA, Codée JDC, van Wezel GP, Berrin J-G, Rovira C, Overkleeft HS, Davies GJ (2020) Rational design of mechanism-based inhibitors and activity-based probes for the identification of retaining α-L-arabinofuranosidases. J Am Chem Soc 142:4648–4662

McGregor NGS, Kuo C-L, Beenakker TJM, Wong C-S, Armstrong Z, Florea BI, Codée JDC, Overkleeft HS, Aerts JMFG, Davies GJ (2022) Synthesis of broad-specificity activity-based probes for exo-β-mannosidases. Org Biomol Chem 20:877–886

McGregor NGS, de Boer C, Foucart QPO, Beenakker T, Offen WA, Codée JDC, Willems LI, Overkleeft HS, Davies GJ (2023) A multiplexing activity-based protein profiling platform for dissection of a native bacterial xyloglucan-degrading system. ACS Cent Sci 9:2306–2314

Ofman TP, Küllmer F, van der Marel GA, Codée JDC, Overkleeft HS (2021) An orthogonally protected cyclitol for the construction of nigerose- and dextran-mimetic cyclophellitols. Org Lett 23:9516

Schröder SP, de Boer C, McGregor NGS, Rowland RJ, Moroz O, Blagova E, Reijngoud J, Arentshorst M, Osborn D, Morant MD, Abbate E, Stringer MA, Krogh KBRM, Raich L,

Rovira C, Berrin J-G, van Wezel GP, Ram AFJ, Florea BI, van der Marel GA, Codée JDC, Wilson KS, Wu L, Davies GJ, Overkleeft HS (2019) Dynamic and functional profiling of xylan-degrading enzymes in *Aspergillus* secretomes using activity-based probes. ACS Cent Sci 5: 1067–1078

Stubbs KA, Vocadlo DJ (2009) Affinity-based proteomics probes; tools for studying carbohydrate-processing enzymes. Aust J Chem 62:521–527. See also the reviews cited in references 8 and 9

Stubbs KA, Scaffidi A, Debowski AW, Mark BL, Stick RV, Vocadlo DJ (2008) Synthesis and use of mechanism-based protein-profiling probes for retaining β-D-glucosaminidases facilitate identification of pseudomonas aeruginosa NagZ. J Am Chem Soc 130:327–335

Tsai C-S, Yen H-Y, Lin M-I, Tsai T-I, Wang S-Y, Huang W-I, Hsu T-L, Cheng Y-SE, Fang J-M, Wong C-H (2013) Cell-permeable probe for identification and imaging of sialidases. Proc Natl Acad Sci USA 7:2466–2471

van der Gracht D, Rowland RJ, Roig-Zamboni V, Ferraz MJ, Louwerse M, Geurink PP, Aerts JMFG, Sulzenbacher G, Davies GJ, Overkleeft HS, Artola M (2023) Fluorescence polarisation activity-based protein profiling for the identification of deoxynojirimycin-type inhibitors selective for lysosomal retaining alpha- and beta-glucosidases. Chem Sci 14:9136–9144

van Scherpenzeel M, van den Berg RJBHN, Donker-Koopman WE, Liskamp RMJ, Aerts JMFG, Overkleeft HS, Pieters RJ (2010) Nanomolar affinity, iminosugar-based chemical probes for specific labeling of lysosomal glucocerebrosidase. Bioorg Med Chem 18:267–273

Vocadlo DJ, Bertozzi CR (2004) A strategy for functional proteomic analysis of glycosidase activity from cell lysates. Angew Chem Int Ed 43:5338–4342

Willems LI, van der Linden WA, Li N, Li K-Y, Liu N, Hoogendoorn S, van der Marel GA, Florea BI, Overkleeft HS (2011) Bioorthogonal chemistry: applications in activity-based protein profiling. Acc Chem Res 44:718–929

Willems LI, Jiang J, Li K-Y, Witte MD, Kallemeijn WW, Beenakker TJN, Schröder SP, Aerts JMFG, van der Marel GA, Codée JDC, Overkleeft HS (2014a) From covalent glycosidase inhibitors to activity-based glycosidase probes. Chem Eur J 20:10864–10872

Willems LI, Beenakker TJM, Murray B, Scheij S, Kallemeijn WW, Boot RG, Verhoek M, Donker-Koopman WE, Ferraz MJ, van Rijssel ER, Florea BI, Codée JDC, van der Marel GA, Aerts JMFG, Overkleeft HS (2014b) Potent and selective activity-based probes for GH27 human retaining alpha-galactosidases. J Am Chem Soc 136:11622–11625

Williams SJ, Hekmat O, Withers SG (2006) Synthesis and testing of mechanism-based protein profiling probes for retaining endo-glucosidases. Chembiochem 7:116–124

Withers SG, Umezawa K (1991) Cyclophellitol: a naturally occurring mechanism-based inactivator of beta-glucosidases. Biochem Biophys Res Commun 177:532–537

Withers SG, Street IP, Bird P, Dolphin DH (1987) 2-Deoxy-2-fluoroglucosides: a novel class of mechanism-based glucosidase inhibitors. J Am Chem Soc 109:7530–7531

Witte MD, Kallemeijn WW, Aten J, Li K-Y, Strijland A, Donker-Koopman WE, Blijlevens B, Kramer G, van den Nieuwendijk AMCH, Florea BI, Hooibrink B, Hollak CEM, Ottenhoff R, Boot RG, van der Marel GA, Overkleeft HS, Aerts JMFG (2010) Ultrasensitive *in situ* visualization of active glucocerebrosidase molecules. Nat Chem Biol 6:907–913

Wu L, Jiang J, Jin Y, Kallemeijn WW, Kuo C-L, Artola M, Dai W, van Elk C, van Eijk M, van der Marel GA, Codée JDC, Florea BI, Aerts JMFG, Overkleeft HS, Davies GJ (2017) Activity-based probes for functional interrogation of retaining beta-glucuronidases. Nat Chem Biol 13: 867–873

Wu L, Armstrong Z, Schröder SP, de Boer C, Artola M, Aerts JMFG, Overkleeft HS, Davies GJ (2019) An overview of activity-based probes for glycosidases. Curr Opin Chem Biol 53:25–36

Part II

Analytical Methods of Complex Carbohydrates

Advanced Mass Spectrometry Techniques for the Characterization of Carbohydrates

Niklas Geue, Caitlin Walton-Doyle, Eleonora Renzi, Mathew Bejoy, and Kevin Pagel

Contents

Abstract

Mass spectrometry-based techniques have emerged as a gamechanger for the analysis of carbohydrates. Here, we summarize state-of-the-art glycan analysis in vacuo, demonstrating how mass spectrometry, tandem mass spectrometry, ion mobility spectrometry, and gas-phase infrared spectroscopy, supported by chromatography methods and computational modelling, facilitate the structural characterization of isolated glycans and the analysis of the entire glycome. After an introduction to glycans and techniques, the sugar classes of *N*-glycans, *O*-glycans, human milk oligosaccharides, and glycosaminoglycans will be discussed in detail, before the chapter ends with an outlook on the future of mass spectrometry-based glycan analysis.

Authors Niklas Geue and Caitlin Walton-Doyle have equally contributed to this chapter.

N. Geue · C. Walton-Doyle · E. Renzi · M. Bejoy · K. Pagel (✉)
Institute of Chemistry and Biochemistry, Freie Universität Berlin, Berlin, Germany

Department of Molecular Physics, Fritz-Haber-Institut der Max-Planck-Gesellschaft, Berlin, Germany
e-mail: kevin.pagel@fu-berlin.de

F. Pfrengle et al. (eds.), *Complex Carbohydrates in Health and Disease*, Handbook of Experimental Pharmacology 288, https://doi.org/10.1007/164_2025_749

Keywords

Gas-phase infrared spectroscopy · Glycosaminoglycans · Human milk oligosaccharides · Ion mobility spectrometry · *N*-glycans · *O*-glycans · Tandem mass spectrometry

1 Introduction

Carbohydrates (or glycans) represent the most abundant biopolymers on earth. In addition to their roles as energy storage and structural scaffold, they are involved in a range of biological processes – either in their isolated form or when attached to other biomacromolecules such as lipids, peptides or proteins (Varki 2017). As for proteins, the biological function of glycans is directly linked to their structure. However, in contrast to proteins, their biosynthesis is not template driven and is instead based on a range of assembly steps that result in highly diverse structures that differ in composition, connectivity, configuration, and branching (Fig. 1) (Varki et al. 2017).

In comparison to other biomacromolecules, the influence of glycan structure on biological function is still poorly understood. This is mostly due to their high heterogeneity and technical limitations, leaving the elucidation of the sugar code as one of the major bottlenecks in glycosciences (Safferthal et al. 2024). NMR spectroscopy often identifies glycans and their isomers unambiguously but requires high sample concentrations and is not fully suitable for high-throughput analysis (Geue et al. 2022). In contrast, mass spectrometry (MS) allows the analysis of μM solutions and offers a robust and sensitive analysis, in particular when combined with liquid chromatography (LC-MS). The high degree of isomerization and long analysis times limit the use of LC-MS for glycan analysis, and orthogonal techniques

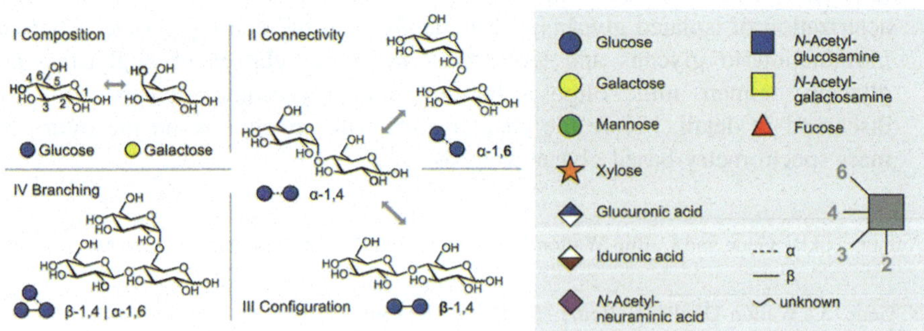

Fig. 1 Left: Different types of isomerism in glycans: (I) composition, (II) connectivity, (III) configuration, and (IV) branching. Right: Symbol nomenclature for glycans including their monosaccharides, connectivity, and configuration. (Adapted with permission from Grabarics et al. (2022), Copyright © 2021 The Authors. Published by American Chemical Society)

such as tandem mass spectrometry (MS^2), ion mobility spectrometry (IMS), and gas-phase infrared (IR) spectroscopy are often required (Grabarics et al. 2022; Safferthal et al. 2024). After an introduction of these analytical methods, we will discuss advances and case examples of the in vacuo analysis of *N*-glycans, *O*-glycans, human milk oligosaccharides (HMOs), and glycosaminoglycans (GAGs).

2 Techniques

Mass spectrometry separates isolated gas-phase ions based on their mass-to-charge ratio (*m/z*) (Gross 2017). Prior to MS analysis, sample complexity can be reduced *via* purification and enrichment, usually through chromatographic separation. Here, the analytes are contained in a mobile phase which carries them through a stationary phase. As the mobile phase passes over the stationary phase, the compounds interact with the stationary phase, eluting at different "retention" times (Miller 2005). Liquid chromatography (LC) is commonly used for glycan separation (Lei et al. 2010; Zauner et al. 2011; Vreeker and Wuhrer 2017), and the hydrophilic nature of glycans lends itself to approaches such as hydrophilic interaction liquid chromatography (HILIC) and porous graphitized carbon (PGC) chromatography. HILIC separation depends on size, charge, composition, linkage, and branching of the analytes, whereas PGC allows interaction and separation of glycans with the porous graphite surface through induced dipole dispersive forces (Chen et al. 2020). Sialic acids, sulfates, and other acidic moieties often give glycans negative charges, further lending them to separation by anion exchange (AEX) chromatography. Capillary electrophoresis (CE) uses an electric field to separate ions, and a recent review by Lu et al. describes its applicability for glycans (Lu et al. 2018). For reverse phase (RP) separation, derivatization is often implemented to achieve retention (Vreeker and Wuhrer 2017). Details of all these techniques are beyond the scope of this chapter and can be found in the respective literature.

Upon sample injection to the mass spectrometer, the sample is ionized and transferred to the gas phase. For biological applications, MS became widely applicable with the introduction of soft ionization techniques that induce little or no fragmentation, namely matrix-assisted laser desorption ionization (MALDI) (Karas and Hillenkamp 1988) and electrospray ionization (ESI) (Konermann et al. 2013; Fenn et al. 1989). While for MALDI the analytes are embedded in a solid matrix, and gas-phase ions are formed by exposure to a laser pulse, the ESI process starts with a dissolved sample. A voltage difference between the capillary containing the solution and the entrance of the mass spectrometer causes the creation of charged droplets. The detailed ionization mechanism depends on the analyte size and structure, and to date, this process remains debated (Konermann et al. 2013). For carbohydrates, the charged residue model (CRM) has been observed for native molecules, whereas the ion evaporation model (IEM) has been proposed for permethylated carbohydrates (Calixte et al. 2020). Nanoelectrospray ionization (nESI) is an adaptation of ESI, where glass capillaries with an opening in the submicron range are used (Jordan et al. 2022). Compared to other techniques, nESI only requires nanoliters of sample, though it is less commonly implemented due to clogging issues and irreproducible tip diameters.

ESI and MALDI are both routinely used for analysis of glycans, but in the past decade, ESI has become the dominant ionization method (Grabarics et al. 2022). The latest developments in glycan analysis with a specific focus on MALDI can be found in the literature (Harvey 2023). In ESI, glycans do not always yield abundant ion signals due to their hydrophilicity and the absence of basic sites that can be protonated. Thus, ionization efficiency can be increased by derivatization such as permethylation, peracetylation, or labeling at the reducing end (Grabarics et al. 2022; Zhou et al. 2017). Mass-to-charge ratios are measured in the mass analyzer, and the most common analyzers used in modern commercial mass spectrometers are quadrupoles, time-of-flight devices, and ion traps (such as the orbitrap), which differ in their mass accuracy, mass range, acquisition speed, and resolving power. More detailed information about mass analyzers can be found elsewhere (De Hoffmann and Stroobant 2007).

Mass analyzers can be mounted in tandem allowing multiple stages of *m/z* analyses. This technique is known as tandem mass spectrometry (MS/MS or MS^2) where the superscript 2 refers to the number of stages. Higher order mass analyses are referred to as MS^3, MS^4, or generally as MS^n (Gross 2017). MS/MS experiments allow selection of a parent/precursor ion, followed by its activation which often leads to ion dissociation. The resulting product ions are transmitted to a second mass analyzer for mass analysis. For glycans, MS^2 can allow the elucidation of the sugar sequence through top-down experiments. Analysis of glycan MS/MS spectra requires expert knowledge, and libraries are routinely used to annotate complex spectra. The UniCarb database represents one example, in which a library of diagnostic fragment ions was compiled from a LC-ESI-MS/MS analysis of over 200 *N*- and *O*-glycans (Campbell et al. 2014). IM-MS derived CCS values were also implemented in the database (Struwe et al. 2016a).

Several techniques are applied to activate glycan ions, which differ in the amount of energy deposited into the precursor ion, as well as the energy deposition time scale (Grabarics et al. 2022; Gross 2017). Collision-induced dissociation (CID) is the most widely used method (Grabarics et al. 2022). In CID, precursor ions are accelerated by an electric field and undergo inelastic collisions with an inert gas, leading to partial conversion of kinetic energy to internal energy. CID employs collision energies in the eV range, resulting in long activation times (Fig. 2). CID is a slow heating method, largely leading to cleavage of the weakest bonds. Higher-energy collisional dissociation (HCD) uses collision energy in the keV range, resulting in fewer collisions and faster ion activation. This is particularly relevant for the activation of higher *m/z*-ions, such as glycoconjugates.

Other techniques to induce dissociation include collision with photons, e.g., infrared multiphoton dissociation (IRMPD) and ultraviolet photon dissociation (UVPD), which activate ions with light from different regions of the electromagnetic spectrum (Bayat et al. 2020). IRMPD is a slow heating method based on the sequential absorption of IR photons, usually provided by a CO_2 laser and using a tunable light source instead of a fixed wavelength (Grabarics et al. 2022). IRMPD

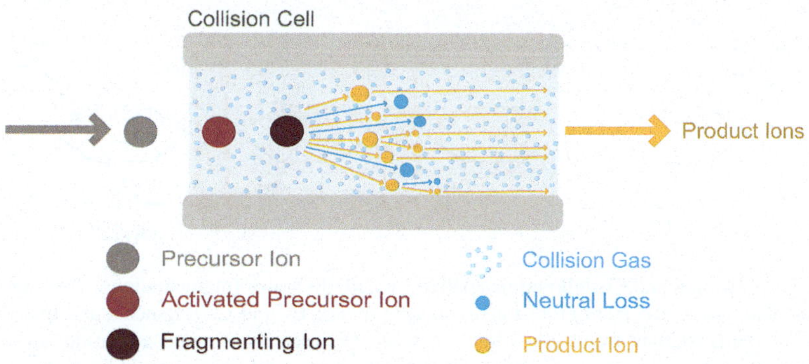

Fig. 2 Schematic principle of collision-induced dissociation (CID) showing fragmentation of a precursor ion to product ions

commonly creates similar fragments to CID and can also be employed to record gas-phase IR spectra, which will be discussed further below. UVPD employs photons in the UV range, which are energetic enough to induce single-photon dissociation, occurring *via* electronic excitation of the precursor ion.

Various electron-mediated activation methods have been developed, namely electron transfer dissociation (ETD, or ETnoD, when no dissociation occurs), electron capture dissociation (ECD), electron detachment dissociation (EDD), and electron excitation dissociation (EED) (Bayat et al. 2020). In ETD/ETnoD, electrons are transferred from a transfer reagent to the ion, whereas ECD describes the direct ion reduction with electrons. Both methods require multiply charged cations. In contrast, EED can be applied for singly charged cations, as there is no charge reduction, and with EDD, negative ion polarity can be used. In general, electron-mediated methods lead to faster activation and cleavage of the bond closest to the site of electronic disturbance, resulting in different fragments compared to slow activation methods such as CID and IRMPD.

Upon glycan activation, the gas-phase ions can undergo specific fragmentation and two different types of cleavages can be distinguished for carbohydrates: glycosidic cleavage and cross-ring cleavage. The resulting fragments are usually annotated with the Domon and Costello nomenclature depicted in Fig. 3. Information about the glycan sequence is primarily provided by the glycosidic cleavage, whereas cross-ring cleavages are important to deduce information about linkage and branching.

A frequent challenge during MS/MS experiments of glycans are rearrangement reactions, most prominently fucose migration (Lettow et al. 2023). During MS^2, a fucose residue is moved from the nonreducing end of a glycan to an adjacent or remote site of the same molecule. This and other rearrangement reactions can lead to unexpected *m/z* fragments, resulting in incorrect sequence assignments (Wuhrer et al. 2011).

In MS experiments of complex glycans, matrix effects can lead to ionization suppression/enhancement, giving misleading results. MS glycomics workflows can

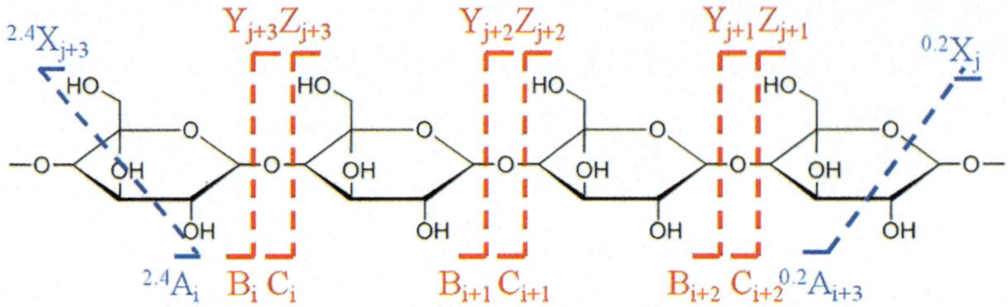

Fig. 3 The Domon and Costello nomenclature for carbohydrates fragmentations, showing A- and X- fragments produced from cross-ring cleavages (blue); B- and C- fragments and their Y- and Z- counterparts derive from glycosidic cleavage between two monosaccharide units (red). Superscripts indicate cleavages within carbohydrate rings, and subscripts indicate the position relative to the termini. (Reproduced with permission from Gray et al. 2016)

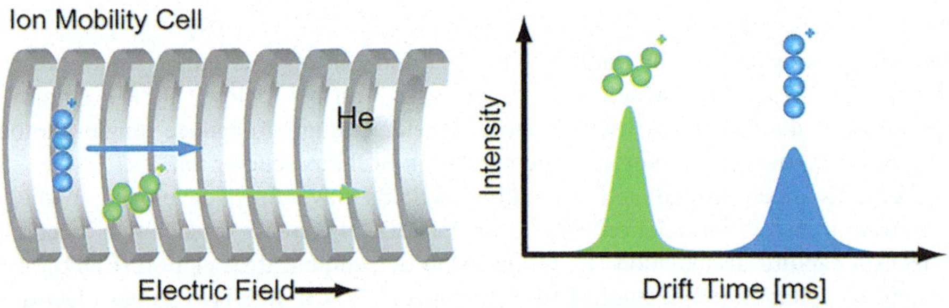

Fig. 4 Schematic of ion mobility (IM) separation. Ion separation occurs in the IM cell which is filled with an inert buffer gas (here helium) and separates analytes based on their size, charge and shape. The electric field applied pushes the ions through the drift cell, and the green more compact ions have shorter drift time compared to the blue larger ions. (Reproduced from Mucha et al. (2019) with permission from the Royal Society of Chemistry)

further lack the ability to differentiate stereoisomers, and thus, other orthogonal techniques are required, enabling in-depth structural investigation to obtain more comprehensive linkage and stereoisomeric information (Grabarics et al. 2022; Manz and Pagel 2018). Ion mobility spectrometry (IMS) can be coupled to MS (together: IM-MS) and is able to differentiate isomeric compounds on the millisecond time scale, becoming a key element in glycomics (Ruhaak et al. 2018; Christofi and Barran 2023; Geue et al. 2024). In the simplest form, drift tube IMS (DTIMS), ions are pushed through a gas-filled drift cell by application of an electric field (Manz and Pagel 2018), in which they undergo collisions with a buffer gas (Fig. 4). Within the drift tube, ions are separated based on their charge, shape and size. The time the ions take to traverse the drift cell (drift time) is measured, with smaller, and more compact ions reaching the detector first (Manz and Pagel 2018; Gabelica 2021).

The drift time can be converted to collision cross-section (CCS) values, which is the molecular descriptor correlated to the size and shape of an ion, enabling comparisons across different IMS instruments (Geue et al. 2024). To facilitate identifications, CCS values can be deposited in databases and calculated computationally for a comparison with experimental CCS (Picache et al. 2019). Traveling wave IMS (TWIMS) involves stacked ring ion guides, allowing high-resolution instrumental geometry modifications such as cyclic arrangements or structure for lossless ion manipulations (SLIM). Trapped IMS (TIMS) experiments are based on trapping the ions by a field gradient, opposed to the gas flow. In contrast to the previous techniques, ions with lower mobilities will elute first (Grabarics et al. 2022). Although DTIMS allows to determine CCS directly, TWIMS and TIMS require CCS calibration (Christofi and Barran 2023; Geue et al. 2024).

Even though experimental CCS values enable analyte identification, they are largely related to the shape and size of the ions without conveying full structural information. Theoretical approaches are indispensable and experimentally determined CCS values are compared to those calculated from computed structures. Empirical force field, semiempirical and first-principles methods are commonly used for glycan modelling, where the choice of method is usually a trade-off between accuracy and computational expense. Once a computational structure is predicted, CCS values can be simulated, most commonly through the trajectory method (Grabarics et al. 2022).

Spectroscopic techniques can be applied for the structural characterization of glycans. Electromagnetic radiation in the infrared (IR) range can excite molecules if the energy of the radiation is in resonance with vibrational transitions (Grabarics et al. 2022). Classical approaches are based on direct transmission of IR radiation through a solid, liquid or gaseous sample, but in the high vacuum of a mass spectrometer the ion density is limited (Fig. 5a). Action spectroscopy is required, which indirectly measures the absorption of photons by following a measurable "action", such as the dissociation of the intact ion or electron detachment (Fig. 5b) (Mucha et al. 2018). The most common IR action spectroscopy method is IRMPD, where irradiation with an IR laser induces fragmentation of weakly bound clusters or ions with low energetic barriers (Fig. 5c) (Manz and Pagel 2018). By measuring the fragmentation yield as a function of the wavelength, an IR spectrum is recorded (Grabarics et al. 2022). The obtained IR spectra can be compared to those simulated from candidate structures, often enabling unambiguous identification.

IRMPD is mostly limited to mono- and disaccharides due to red-shifting of absorption bands and spectral congestion, and thus improved IR spectroscopic gas-phase techniques at cryogenic temperatures, such as messenger tagging spectroscopy and action spectroscopy in helium nanodroplets, were developed (Grabarics et al. 2022; Khanal et al. 2017). Messenger tagging is a linear single-photon process, in which, upon resonant irradiation detachment of a weakly bound noninteracting messenger atom or tag occurs (e.g., He and N_2). Prerequisites for the attachment of the tag are temperatures close to the boiling point of the tag, which limits the number of glycan conformers present, thereby reducing spectral broadening. In IR action spectroscopy in helium nanodroplets, gas-phase ions are selected and accumulated inside an ion trap (Mucha et al. 2017). Traversing helium droplets pick up the trapped ions, which are cooled to the droplet temperature (0.37 K). Upon IR laser irradiation, the ions are vibrationally excited and cooled again to the ground

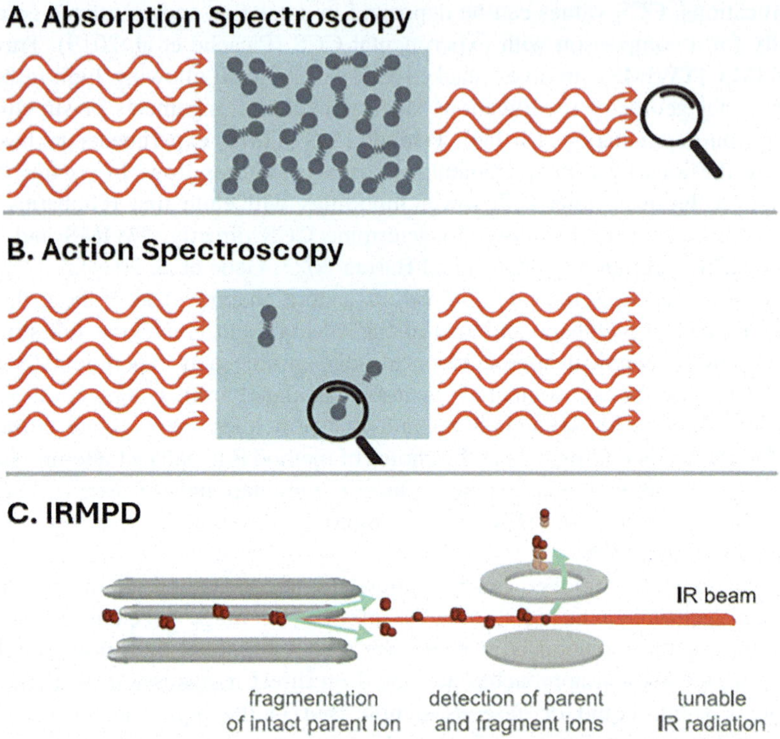

Fig. 5 (**A**) Absorption spectroscopy relates the intensity of the transmitted radiation to the intensity of the incident radiation. (**B**) Action spectroscopy of *m/z*-selected ions in the high vacuum of a mass spectrometer limits the ion density due to Coulomb repulsion. (**C**) Schematic of infrared multiple photon dissociation (IRMPD) spectroscopy, in which multiple photons excite the parent ion leading to fragmentation. (Adapted with permission from Grabarics et al. (2022). Copyright © 2021 The Authors. Published by American Chemical Society)

state by helium evaporation. Cold-ion spectroscopy spectra have unprecedented resolving power, providing highly diagnostic absorption bands.

3 *N*-Glycans

The two main glycan classes are *N*- and *O*-glycans, which differ in their connectivity to the protein or peptide backbone. *N*-glycans are commonly attached to an asparagine (Asn) residue by β-1 *N*-glycosidic linkage from an acetyl-D-glucosamine (GlcNAc). They are branched oligosaccharides found in most living organisms and play important roles in regulating cellular functions, immune response and protein folding (Reily et al. 2019). As a result of the high influence on physiological functions, *N*-glycans are of interest in biomarker and biotherapeutic fields (Lu et al. 2018). Compared to *O*-glycans, they usually contain more monosaccharide units and

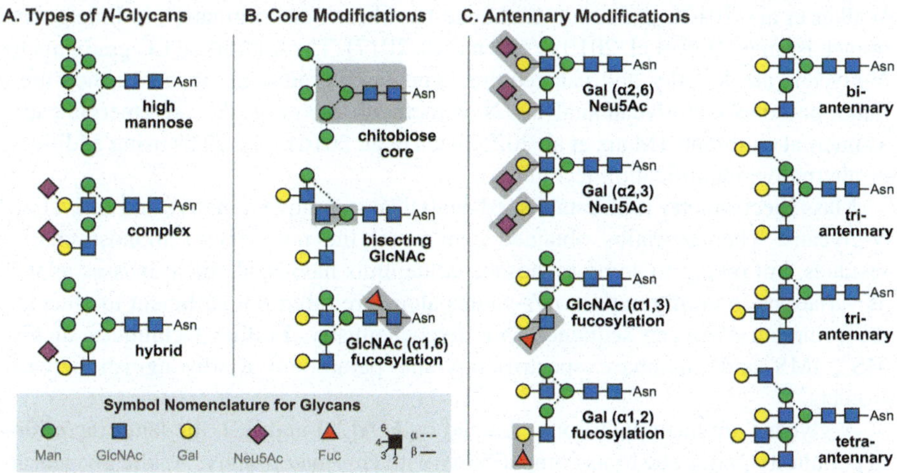

Fig. 6 Structure of the three *N*-glycan types (**A**), the possible modifications of the core (**B**) and occurring antennary modifications (**C**). The core structures are highlighted in (**B**). (Adapted with permission from Grabarics et al. (2022). Copyright © 2021 The Authors. Published by American Chemical Society)

have a conserved core structure. This core structure is made up of two GlcNAc units (one of which is linked to the protein backbone) and three D-mannose (Man) units. A central mannose residue is connected to GlcNAc and has two further mannose antennae by α-1,3- and α-1,6- linkages. This is displayed in Fig. 6b – the chitobiose core.

N-glycans are categorized into three types, displayed in Fig. 6a. First is the high mannose type, where the two mannose antennae are built upon with further mannose residues, second is the complex type in which the mannose antennae are built on by a GlcNAc unit and subsequently extended by other monosaccharides and third is the hybrid type, where the C6 mannose antenna is a high mannose moiety and the C3 is complex. The complexity increases with the presence of fucose, galactose, branching sialic acid capping and other modifications such as extension from the core (Fig. 6b) and increased and modified antennae (Fig. 6c) (Lu et al. 2018). *N*-glycans are often in complex matrices, and thus prior to analysis, chemical or enzymatic extraction and purification are necessary. Commonly this is achieved through enzymatic cleavage from proteins or peptides by glycosidases such as PNGase F.

Chromatographically, HILIC is commonly utilized in large-scale glycomics studies (Ruhaak et al. 2011; Knežević et al. 2009; Royle et al. 2008), and PGC has previously demonstrated separation of isomers (Pabst et al. 2007). *N*-glycans can conveniently be labelled with a fluorophore by reductive amination, which can improve ionization, as well as flurophores providing a charge, allowing compatibility for separation by CE. RP separation of derivatized glycans has been shown to outperform HILIC both in number of glycans and peak capacity in a study by Hunter

Walker et al. (2012), and anion exchange has shown discrimination in linkage and branch isomers (Lei et al. 2010; Veillon et al. 2017). The traditional LC gradients for *N*-glycans are lengthy and can require large sample amounts meaning they are a major bottleneck in glycan analysis. New methods to combat this are emerging such as internal calibrants (Manz et al. 2022; Aich et al. 2016), as well as using additional or alternative identification techniques.

Mass spectrometry is an established analytical technique of native and derivatized *N*-glycans. The *m/z*-ratios obtained can reveal the numbers of monosaccharide residues, but owing to some monosaccharide units having identical masses as well as variations in configuration, *m/z*-values alone are often not sufficient to elucidate the complete *N*-glycan sequence and stereochemistry. Further techniques such as MS^n, IMS and gas-phase spectroscopy are paramount to distinguish between isomers.

N-glycans are most commonly ionized by MALDI and ESI, the latter increasing in popularity partly due to its compatibility with chromatography. Acidic glycans are more easily analyzed in negative ionization mode (Harvey 2020); however, derivatizing tools such as permethylation not only increase volatility and improve separation but also make glycans easier to analyze in positive ionization mode by improving the positive ion affinity (Zhou et al. 2016). Though, derivatization can be labor intensive, and the high pH of reagents may lead to oxidative degradation or destruction of analytes (Kang et al. 2005).

Connectivity and branching of glycans can be investigated by fragmentation through MS^2. CID has given information about the connectivity of units with glycosidic bond cleavage as well as some cross-ring (A- and X-type, Fig. 3) fragments being observed (Reinhold et al. 1995), the same can be seen by IRMPD (Lancaster et al. 2006). The fragmentation observed is dependent on the adduct, ionization polarity and type of glycan, and in complex mixtures the informative A- and X-type cross-ring fragments are observed at much lower abundances (Grabarics et al. 2022; Harvey 2000). Those detected, even with HCD, are often ${}^{1,5}X_n$ cleavages which cannot be used to elucidate detailed structures (Han and Costello 2013).

Many faster activating, electron-based techniques require multiple charges, often realized with charged metal adduct cations. ECD, EED, and EDD have all been applied to study *N*-glycans (Zhou and Håkansson 2011; Yu et al. 2012; Adamson and Håkansson 2007a). By ECD, *N*-glycans have proven harder to fragment compared to a linear oligosaccharide, with metal adducts giving more fragments (Adamson and Håkansson 2007a). By 'hot' ECD (employed using higher energy electrons) C-, Z-fragments were dominant, with A- and X- cross-ring cleavages also observed for sodiated adducts (Zhao et al. 2008). In a comparison with ECD, the higher energy EED gave more cross-ring fragments for lithium adducts of glycans, respectively, as well as diagnostic information on glycosidic link positions, as shown in Fig. 7 (Yu et al. 2012). Here, Fig. 7a demonstrates cleavage maps by ECD at 1.5 eV and 7b displays EED at 14 eV.

EDD has produced both glycosidic and cross-ring fragmentation, with the label altering fragmentation type (Zhou and Håkansson 2011). UVPD gives rise to cross-ring fragments and can yield richer fragmentation patterns than CID (Ko and

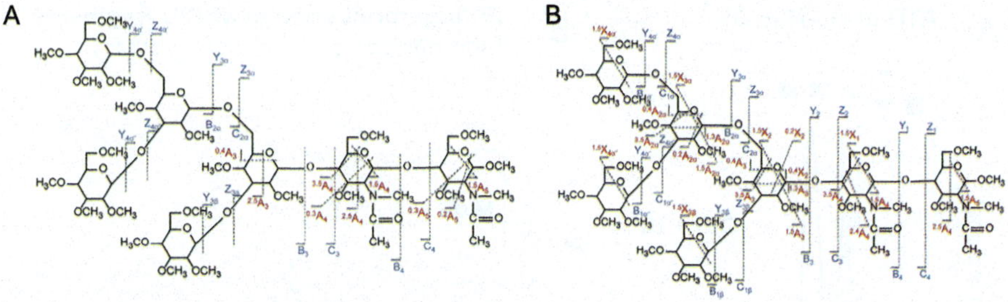

Fig. 7 Cleavage maps of $Man_5GlcNAc_2$ $[M + 2\,Li]^{2+}$ when fragmented with (**a**) 1.5 eV ECD and (**b**) 14 eV EED. It can be seen that the higher energy electrons produce more fragments including cross-ring. (Figure adapted with permission from Yu et al. (2012). Copyright © 2012 American Chemical Society)

Brodbelt 2011). In a study by Devakumar et al., a high mannose glycan gives rise to abundant A- and X- fragments with a 157 nm laser, with B-, C-, Y-, and Z- also observed (Devakumar et al. 2007). However, the properties of some isomeric glycans, especially in complex mixtures, leads to limited information from fragmentation patterns alone.

IM-MS of *N*-Glycans can discriminate between isomers when combined with fragmentation, as is shown both by TWIMS (Harvey et al. 2012, 2013) and DTIMS (Zhu et al. 2012). By DTIMS, a comprehensive study on high mannose glycans reported CCS values for different adducts, polarities, and drift gases, where further investigation by TWIMS revealed gas phase conformers of $[M\text{-}H]^-$ (Struwe et al. 2015). When high mannose glycans were analyzed as doubly charged metal cations, ATDs suggested different conformers arising from the metal location (Zhu et al. 2015).

Harvey et al. found that isomeric separation by IM was slightly better in positive ionization compared to negative, but that the more informative fragmentation in negative mode could also deduce isomers. Further, the appearance of broad ATD peaks indicating isomeric glycans, which could not be separated, was overcome by fragmentation in the transfer cell, where extracted fragment ATDs displayed some separation (Harvey et al. 2016). In another study, prior to separation, $[M + H_2PO_4]^-$ adducts were fragmented and unique CCS of fragments could discriminate between (precursor) isomers (Harvey et al. 2018). Some studies have considered both intact and fragmented glycans which were separated by IMS allowing differential detection of isomers (Pallister et al. 2020).

These approaches have been used in applications such as serum analysis, which inferred that DT separated isomers could show discrimination between control and disease, showing importance of IM in the glycomics field (Gaye et al. 2012; Isailovic et al. 2008). The increasing resolution of instrumentation has exhibited better ATD separation of glycan isomers (Pagel and Harvey 2013), and with the advances of SLIM and cyclic ion mobility, the resolving power of IMS is constantly increasing (Geue et al. 2024). Oganesyan et al. implemented multipass cyclic IM which

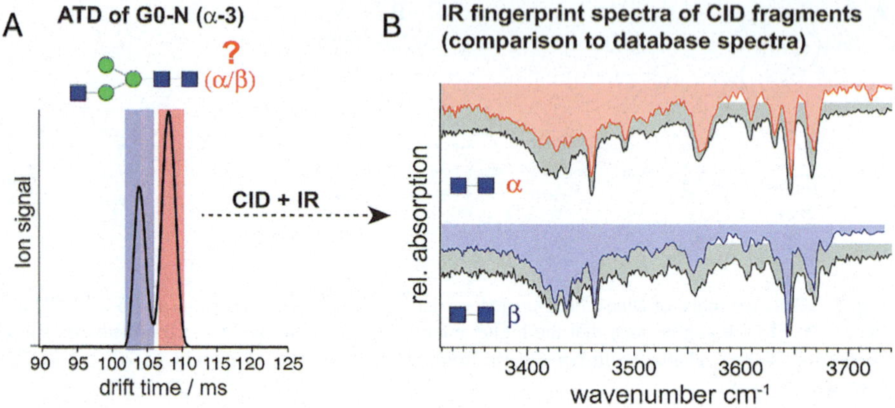

Fig. 8 (**A**) The arrival times of mobility separated α- and β- anomers. (**B**) CID fragments corresponding to the peaks in (**A**) gave distinct IR spectra, which when matched to databases (grey) could distinguish between the reducing ends and characterized the mobility peaks. (Reproduced with permission from Ben Faleh et al. (2022). Copyright © 2022 The Authors. Published by American Chemical Society)

demonstrated peak splitting after multiple cycles possibly attributed to different gas-phase conformations (Oganesyan et al. 2022).

Gas-phase IR spectroscopy is another identification method gaining interest in *N*-glycan characterization. In combination with IM-MS, cryogenic IR spectroscopy has been used to compare reducing end anomers of the core, where comparisons to databases could discriminate between α- and β- anomers of a diacetyl chitobiose core as displayed in Fig. 8. Here, ATDs displayed two mobility peaks, and using CID, the diacetyl core fragment for each peak had IR spectra recorded and database matched (Ben Faleh et al. 2022).

Cryogenic messenger tagging IR of isolated *N*-glycans has successfully matched spectra to a database of standards for identification, showing great potential (Dyukova et al. 2020). The advent of SLIM-IMS led to subsequent studies with ultrahigh IMS resolution. Here, cryogenic vibrational spectroscopy distinguished between positional isomers (Dyukova et al. 2021), and IMS^n separated and identified fragments by IMS and IR (Bansal et al. 2022). From fragment identification, positional and configurational isomers could be assigned. Further, the faster speeds of SLIM compared to LC can replace the bottleneck in LC-MS, on the scale of milliseconds instead of minutes (Safferthal 2024).

4 *O*-Glycans

The second major group of carbohydrates that are attached to proteins and peptides are *O*-glycans. In contrast to *N*-glycans, they are linked to oxygen atoms, usually of the amino acids serine (Ser) or threonine (Thr). Most *O*-glycans are found in mucins and are hence often referred to as mucin-type *O*-glycans, where they can make up to 50–80% of the glycoprotein mass (Varki et al. 2017). Their hydrophilic nature, in combination with their often negative charge, results in a high degree of solvation and salt adduction, leading to an increased viscosity that is typical for mucins (Grabarics et al. 2022). Changes in *O*-glycosylation can be associated with diseases, the most prominent example being cystic fibrosis (CF). Here, people with CF show significantly altered sulfation and sialylation in their sputum glycans, giving rise to target sugars for diagnostics and treatments (Venkatakrishnan et al. 2013).

O-glycans follow four major core structures, starting from the monosaccharide GalNAc, which is linked to Ser or Thr (Fig. 9) (Grabarics et al. 2022). Attachment of β-1,3-Gal results in core 1, and further attachment of a β-1,6-linked GlcNAc leads to core 2. These two units are by far the most typical core structures in *O*-glycosylation and are found throughout the human body. Core 3 is built by attachment of a β-1,3-GlcNAc to GalNAc, and an additional β-1,6-linked GlcNAc unit leads to core 4. Four other, less common core structures have been found in *O*-glycans. Extension of the core structures occurs for example via β-1,3- and β-1,6-linked *N*-acetyllactosamine (LacNAc = Gal-GlcNAc) extensions, sulfation, fucosylation or sialylation. The resultant *O*-glycans often contain antigenic structures at their non-reducing ends, for example the Lewis blood group determinants (Fig. 9c).

O-glycans are generally more difficult to analyze than *N*-glycans, due to the absence of defined sequons that allow *O*-glycosylation, and the lack of a specific cleavage enzyme such as PNGase F for *N*-glycans (Grabarics et al. 2022). Details of the complex procedure of *O*-glycan release are beyond the scope of this chapter and can be found in the literature (Wilkinson and Saldova 2020). Upon release, *O*-glycans are commonly purified and separated with chromatography methods such as HILIC (Buszewski and Noga 2012) and PGC (Thomsson et al. 1999).

As for most compound classes, CID and HCD are the most common MS^2 methods for *O*-glycan characterization. Due to the high degree of sulfation and sialylation, negative polarity is preferred, which additionally has the advantage of more diagnostic cross-ring fragments (A- and X-ions) (Grabarics et al. 2022). An early example is the work by Zhang et al., who used CID to identify *O*-glycosylation features in egg jelly (Zhang et al. 2004). Later approaches were based on the sequential release of *O*-glycans from the glycoproteins before CID (Jensen et al. 2012). More recently, Carlson and coworkers applied CID to investigate the structural diversity of human gastric mucin glycans (Jin et al. 2017). The authors found up to 14 *O*-glycans being unique to the specific patient, and a consistent increase in sialylation and sulfation was observed for those with cancer. HCD is particularly useful when CID is not sufficient for fragmentation, especially for higher mass ions, with HCD fragments found to discriminate between sulfate modification regiochemistry (Issa et al. 2018). Another interesting approach is the use of MS^n, which for both underivatized and permethylated *O*-glycans has helped to elucidate more complex glycan structures (Grabarics et al. 2022).

O-glycans have rarely been studied by IMS and gas-phase IR spectroscopy, due to the complexity of the structures and difficulties in release and sample preparation. In an extensive study by Baker and colleagues, no clear trends were found for

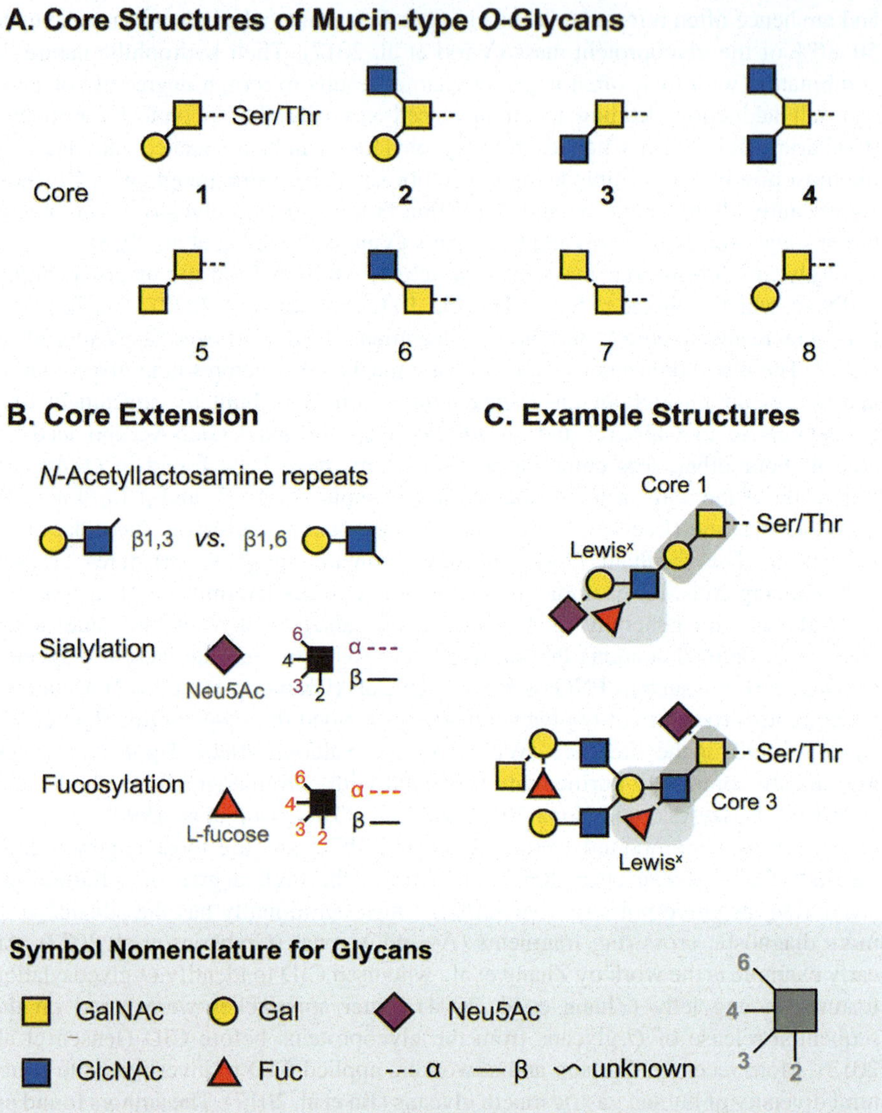

Fig. 9 (**A**) Abundant (1–4) and less abundant (5–8) *O*-glycan core structures. (**B**) Typical extensions of glycan cores are the β-1,3- or β-1,6- attachment of *N*-acetyllactosamine, *L*-fucose and sialic acid units. (**C**) Examples for *O*-glycans based on core 1 (top) and core 3 (bottom) including their linkage site to the glycoprotein. At the non-reducing end, *O*-glycans often contain antigenic structures such as the Lewis blood group determinants, highlighted by gray boxes. (Reproduced with permission from Grabarics et al. (2022). Copyright © 2021 The Authors. Published by American Chemical Society)

separation conditions regarding ion polarity and adduct formation, suggesting the need for optimization in each sample mixture (Zheng et al. 2017). Jin et al. applied a combination of LC-MS and IMS for the separation of *O*-glycans from porcine gastric and human salivary mucins, demonstrating that IMS can resolve isomeric differences for which LC is not sufficient (Jin et al. 2019). More recently, Vos et al. characterized sialic acid *O*-acetylation patterns and glycosidic linkage type using DTIMS (Vos et al. 2023). Last year, Bechtella et al. introduced a TIMS-MS^2 workflow for mucin-type *O*-glycans, showing that TIMS can distinguish isomers that LC and TWIMS are unable to separate (Fig. 10) (Bechtella et al. 2024).

Gas-phase IR spectroscopy on *O*-glycans has largely been limited to proof-of-principle investigations (Mucha et al. 2017; Masellis et al. 2017); however, studies that aim to decipher fucose migration, highly relevant for Lewis and blood group epitopes in *O*-glycans, have been performed as discussed in the techniques section (Grabarics et al. 2022).

5 Human Milk Oligosaccharides

Human milk oligosaccharides (HMOs) are unconjugated carbohydrates that represent the third-largest fraction in human milk (5–25 g L^{-1}), following lactose and lipids (Grabarics et al. 2022). While lactose largely serves as an energy source for infants, the functions of HMOs are more diverse and complex. With prebiotic, anti-infectious, and anti-inflammatory effects, they support the development of a healthy microbiome for newborns, reducing pathogenic infections (Bode 2012). HMOs have further been found to influence the formation of the immune system (Donovan and Comstock 2017).

Over 200 HMO structures have been elucidated to date, with many of them being isomers (Soyyılmaz et al. 2021). Despite their complexity, they follow a strict assembly that is primarily based on five monosaccharides – D-glucose (Glu), D-galactose (Gal), *N*-acetyl-D-glucosamine (GlcNAc), L-fucose (Fuc), and N-acetyl-D-neuraminic acid (Neu5Ac) – as displayed in Fig. 11 (Safferthal et al. 2024). The HMO structures begin with the lactose core unit (Gal β-1,4 Glc) at the reducing end and can be elongated by N-acetyllactosamine (LacNAc) building blocks with β-1,3- or β-1,6-linkages and terminated by either lacto-N-biose (type 1) or LacNAc (type 2) units. Further modifications with α-1,2-, α-1,3-, and/or α-1,4-linked Fuc residues occur in the majority of HMOs, whereas α-2,3- or α-2,6-linked acidic Neu5Ac blocks only appear in 10–30% (Bode 2012). As a consequence of these decorations, HMOs often display a diverse range of Lewis and blood group epitopes (Grabarics et al. 2022).

The separation and characterization of HMOs has traditionally been performed with NMR spectroscopy and chromatography methods; however, these are increasingly considered impractical due to the required high sample volumes and lack of reference standards. LC-MS(/MS) has taken over as a highly diagnostic and sensitive technique, although the full separation of HMO isomers with LC has proven

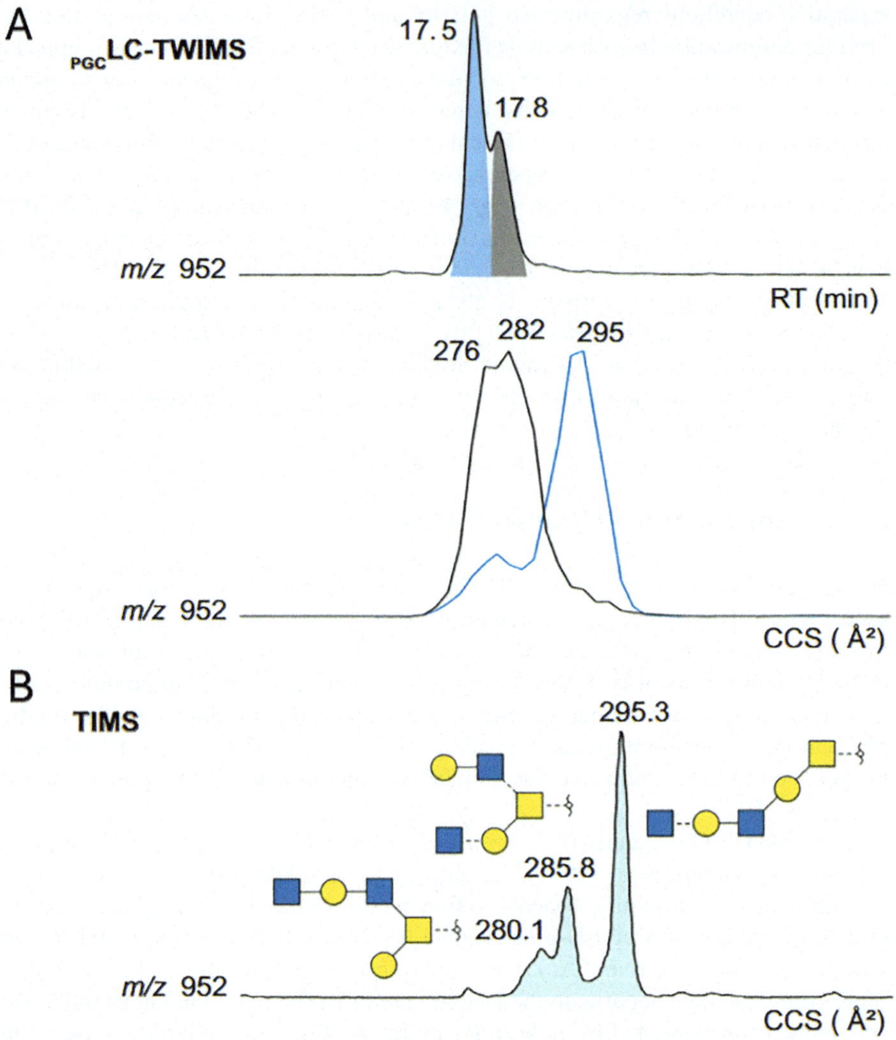

Fig. 10 Separation of isomeric *O*-glycans at $m/z = 952$ with (**A**) LC-MS (top) and TWIMS (bottom) as well as (**B**) TIMS. (Reproduced with permission from Bechtella et al. (2024). Copyright © 2024 The Authors)

challenging (Safferthal et al. 2024), and hence, the recent application of modern gas-phase separation and identification techniques.

Analyzing HMOs with MS can readily elucidate the monosaccharide composition of a given sample as all the building blocks have mass differences. The anomeric linkage is fixed for all monosaccharide blocks, which leaves connectivity and branching as challenges for the MS-based analysis of glycans (Grabarics et al. 2022). As discussed in the techniques section, MS^2, IMS and gas-phase IR

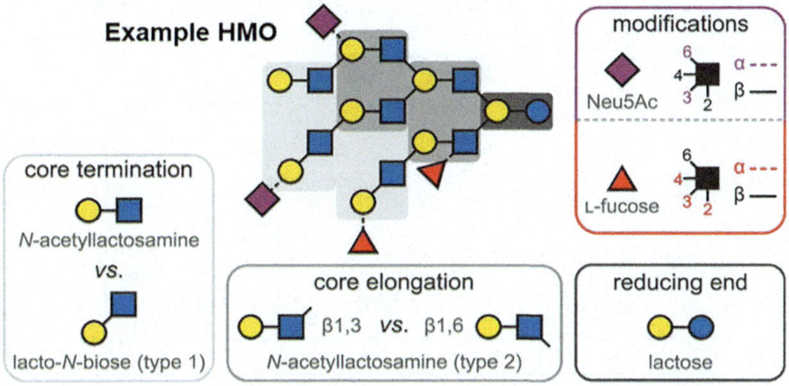

Fig. 11 Schematic structure of HMOs, beginning at the reducing end lactose. Core elongation, termination, and possible modifications are illustrated, as well as an exemplary HMO structure. (Adapted with permission from Grabarics et al. (2022). Copyright © 2021 The Authors. Published by American Chemical Society)

spectroscopy are suitable tools to distinguish such isomers, and we will discuss studies that focus on the HMO analysis with these techniques.

Distinguishing the connectivity and branching of HMOs with MS^2 relies largely on the formation of cross-ring fragments (A- and X-ions, Fig. 3). CID and IRMPD largely lead to formation of glycosyl ions (B-, C-, Y- and Z-ions, Fig. 3), and although this can be circumvented through permethylation and metal adduction, the information from CID and IRMPD remains limited for HMO sequencing (Grabarics et al. 2022). In 2007, Adamson and Håkansson applied ECD for doubly positively charged as well as EDD for doubly negatively charged, sialylated HMOs (Adamson and Håkansson 2007a, b). They found larger shares of A- and X-ions than with CID and IRMPD, which facilitated the elucidation of linkage and branching patterns. These pioneering studies demonstrated how electron-based dissociation reactions can be complementary to information obtained from slow-heating MS^2 methods. Four years later, Han and Costello used ETD for the first time to distinguish HMO isomers (Han and Costello 2011). While the adduction of the glycans with singly charged alkali metals proved to be unsuitable due to the small number of resulting fragments, ETD of adducts with Mg^{2+} and Ca^{2+} resulted in highly informative MS^2 spectra that could be used to assign the respective isomer. In some cases, secondary CID fragmentation of singly charged ETD fragments was required to elucidate connectivity unambiguously, once again illustrating the complementarity of slow- and fast-heating MS^2 methods for glycan analysis. Most recently, Tang et al. added EED to the repertoire of MS^2 methods for HMO characterization in combination with fixed charge derivatization (Tang et al. 2018).

An alternative fast-heating fragmentation method is UVPD, which suffers from the lack of strong UV chromophores in native HMOs, hence requiring the use of high photon energies. After irradiating acidic HMOs at $\lambda = 193$ nm, Ko and

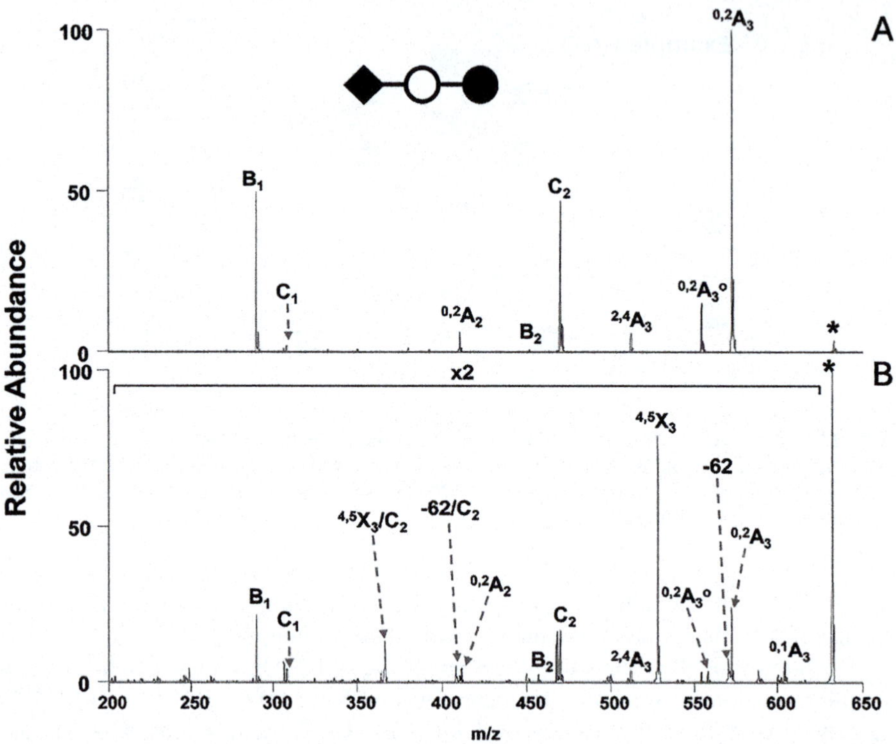

Fig. 12 MS^2 spectra of a deprotonated exemplary HMO ($m/z = 632$) with (**A**) CID and (**B**) UVPD. A superscript o represents the loss of H_2O. The precursor ion is labelled with an asterisk. (Adapted with permission from Ko and Brodbelt (2011). Copyright © 2011 American Chemical Society)

Brodbelt found not only better sequence coverage but also more cross-ring fragments compared to CID, which enabled the identification of connectivity isomers (Fig. 12) (Ko and Brodbelt 2011). Alternative approaches using UVPD for HMOs even at low photon energies involve the covalent attachment of aromatic chromophores prior to fragmentation (Wilson and Brodbelt 2008), or labelling with the aromatic radical precursor 2-iodoanilline prior to irradiation (radical-directed dissociation) (Zhang and Julian 2014).

The introduction of commercial IM-MS platforms has facilitated the separation and identification of isomeric HMOs; however, the differences in ion mobilities can be low. This can be addressed with improving either selectivity, through chemical modifications that increase the relative CCS differences between isomers, or the usage of more recently developed instruments with higher resolving power. Selectivity can be increased through the use of metal adducts that induce different conformations (Geue et al. 2023). For example, Huang and Dodds probed HMOs with different alkali metals ranging from Li^+ to Cs^+, yielding the best CCS separation with $[M + Na]^+$ adducts (Huang and Dodds 2013). For the same HMOs, the authors extended their findings to alkaline earth metals in combination with ETnoD,

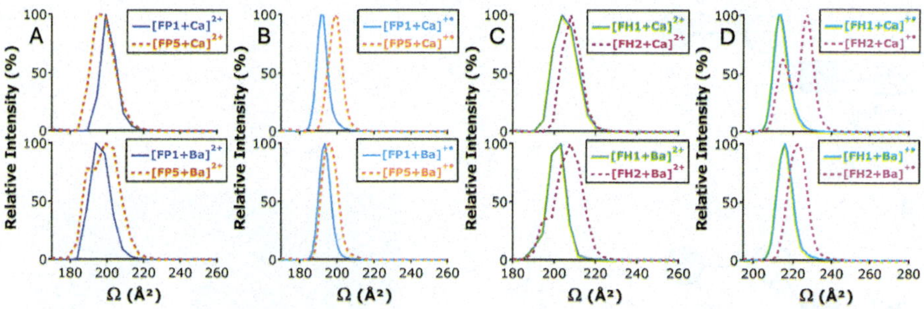

Fig. 13 $^{TW}CCS_{N2}$ distributions of the isomers lacto-*N*-fucopentaose I (FP1) and V (FP5), (**A**) and (**B**) and lacto-N-difucohexaose I (FH1) and II (FH2), (**C**) and (**D**). (**A**) FP1 and FP5 as their Ca^{2+} (top) and Ba^{2+} (bottom) adducts. (**B**) FP1 and FP5 as the ET products of Ca^{2+} (top) and Ba^{2+} (bottom) adducts. (**C**) FH1 and FH2 as their Ca^{2+} (top) and Ba^{2+} (bottom) adducts. (**D**) FH1 and FH2 as the ET products of their Ca^{2+} (top) and Ba^{2+} (bottom) adducts. (Adapted with permission from Huang and Dodds (2015). Copyright © 2015 American Chemical Society)

showing that non-dissociative electron transfer improves the separation for Ca^{2+} adducts, but not for Ba^{2+} (Fig. 13) (Huang and Dodds 2015). The influence of ion adduct choice was later confirmed by other studies, as well as the effect of ion polarity which suggested both had a consistent influence on CCS values of HMOs (Zheng et al. 2017; Struwe et al. 2016b). An elegant way of linkage-specific characterization of HMOs was recently suggested by Habibi and Nagy, who found that selectivity was improved upon encapsulation of HMOs in cucurbiturils and cyclodextrin macrocycles (Habibi and Nagy 2023).

Recent advances in high-resolution IM-MS instrumentation have further improved the separation of HMO isomers. For example, Pu et al. separated HMO tetrasaccharides using TIMS on a Fourier-transform ion cyclotron resonance MS and identified the linkage position through subsequent fragmentation with EED, providing diagnostic cross-ring cleavages not observed in other MS^2 methods (Pu et al. 2016). More recently, Alves and co-workers used TIMS in combination with a time-of-flight mass analyzer, separating HMO standard isomers with CCS differences as small as 1.5% and identifying HMOs directly from breast milk (Delvaux et al. 2021; Rathahao-Paris et al. 2022). With a travelling-wave SLIM device, Smith and co-workers not only base-line resolved HMO isomers but also unraveled new features that have not previously been observed on platforms with lower IM resolving powers (Deng et al. 2017; Nagy et al. 2018). While the nature of these distributions has not been fully characterized, their existence highlights the importance of high-resolution IM instrumentation (Grabarics et al. 2022). More recently, Nagy and co-workers utilized a cyclic IMS device for the separation and characterization of HMOs, showing full separation of seven sets of sialylated, fucosylated galactosyllactose and lactosamine HMO building block isomers over a pathlength of >40 m (Peterson and Nagy 2021). Last year, the same group introduced an internal reference database for high-resolution $^{TW}CCS_{N2}$ values of

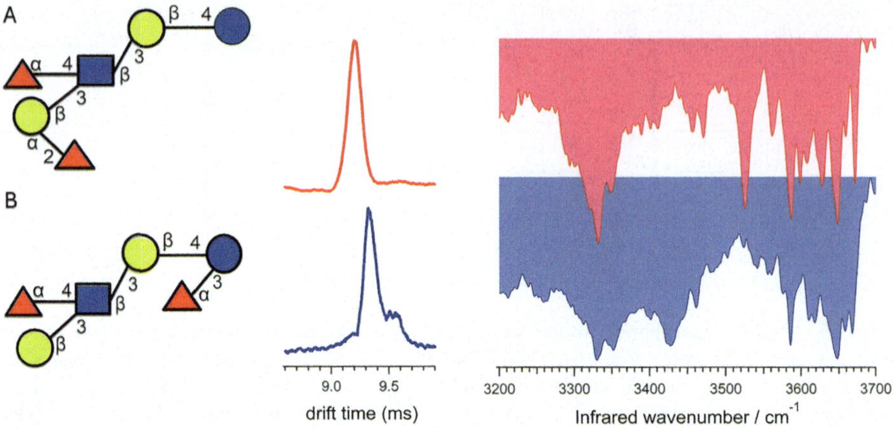

Fig. 14 Gas-phase IR fingerprints and arrival times of isomeric HMO hexasaccharides lacto-N-difucohexaose I (**A**) and lacto-N-difucohexaose II (**B**). (Reproduced from Khanal et al. (2018) with permission of the Royal Society of Chemistry)

(permethylated) HMOs, aiding the structural characterization of unknown HMO mixtures (Habibi et al. 2024).

The characterization of HMOs using gas-phase IR spectroscopy has only recently emerged, and due to the availability of instrumentation, it is so far only accessible to expert groups. The first high-resolution IR fingerprints were acquired in helium nanodroplets, unambiguously distinguishing the linkage of the galactose unit in lacto-N-tetraose (LNT) and lacto-N-neotetraose (LNnT) (Mucha et al. 2017). Shortly after, Depland et al. identified sialic acid linkage isomers using IRMPD (Depraz Depland et al. 2018). Particularly interesting is the combination of cryogenic gas-phase IR spectroscopy with IMS as shown by Rizzo and co-workers, who characterized HMOs ranging from trisaccharides to hexasaccharides with messenger-tagging IR spectroscopy (Fig. 14) (Khanal et al. 2018). Even more promising is messenger-tagging IR spectroscopy of mobility-separated glycan isomers (IMS-IR) for isomeric mixtures (Warnke et al. 2021), or even of mobility-separated glycan fragments (IMS-CID-IMS-IR) (Abikhodr et al. 2023). The latter greatly benefits the characterization of larger and more complex HMOs, for which the precursor IR fingerprints are not sufficiently resolved (Safferthal et al. 2024).

6 Glycosaminoglycans

Glycosaminoglycans (GAGs) are linear, highly acidic, and sulfated polysaccharides composed of repeated disaccharide units and expressed by all multicellular organisms (Zappe et al. 2022). Their biological functions are determined by the structural composition such as the number of disaccharide units, glycosidic linkage, and sulfate modifications (Mattson et al. 2017). GAGs are classified into four families: heparin/heparan sulfate (HP/HS), chondroitin sulfate/dermatan sulfate (CS/DS), keratan sulfate (KS), and hyaluronic acid (HA, Fig. 15).

Heparin/heparan sulfate (HP/HS) is the most acidic and structurally heterogeneous class. The core chain of GlcA-β-4-GlcNAc-α-4 is extensively modified during biosynthesis. The hexuronic acid residue can be modified by epimerization of GlcA to IdoA (Fig. 15) and 2-*O*-sulfation. The glucosamine residue can be modified by *N*-sulfation (GlcNS), *N*-acetylation (GlcNAc) and *O*-sulfation both at the C6, and rarely at the C3 position (Thacker et al. 2014; Jones et al. 2011). Heparin (HP) is a shorter (50–200 disaccharides) and more complex structure with a higher order of sulfate modifications (on average 2.7 per disaccharide) (Taylor et al. 2019). Chondroitin sulfate (CS) and dermatan sulfate (DS) have similar structures, differing in hexuronic acid residue. Chondroitin sulfates (CS) have a core structure of GlcA and GalNAc, exclusively alternated to form GlcA-β-3-GalNAc-β-4 chains. Demantan sulfate (DS) alters from CS only by also containing IdoA, caused by C5 epimerization of GlcA. CS and DS are extensively sulfated on both components of the core structure (Sugahara et al. 2003). The third group, keratan sulfate (KS) is a unique GAG as it does not possess a hexuronic acid monomer. The disaccharide unit is instead composed of Gal and GlcNAc linked as Gal-β-4–GlcNAc-β-3 and sulfation can be observed at the 6-*O*-position of both monomers (Caterson and Melrose 2018). Finally, hyaluronic acid (HA) has the simplest structure of the GAG classes, consisting of repeating disaccharide units of GlcA and GlcNAc in a GlcA-β-3-GlcNAc-β-4 chain which is not modified by sulfation or epimerization (Grabarics et al. 2022).

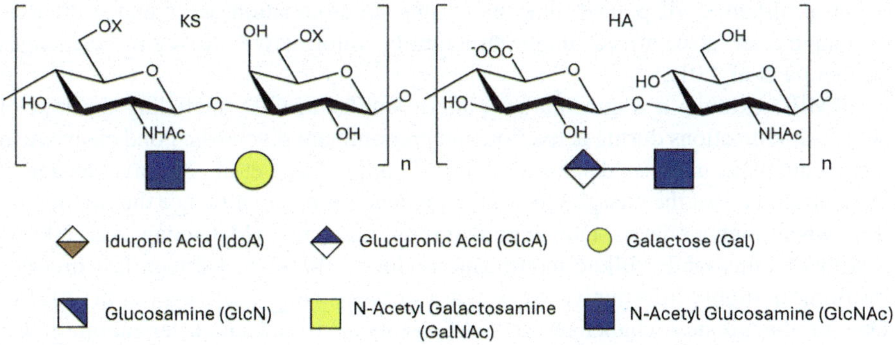

Fig. 15 Disaccharide units of the four glycosaminoglycan families. For the HS/HP class, only one of the core structures is displayed. (Adapted with permission from Solakyildirim 2019)

GAGs consist of hundreds of disaccharide units requiring depolymerization into oligosaccharides to simplify characterization. This is achieved either enzymatically (Ernst et al. 1995) or chemically (Vismara et al. 2017) with β-elimination commonly being employed, as the resulting unsaturated oligosaccharides are UV detectable (Grabarics et al. 2022). Depolymerized oligomers can be separated using a range of liquid chromatography techniques. SEC is a common and robust method; however, it has been shown to be unable to separate some isomers – specifically sulfation sites (Pepi et al. 2021a). The anionic nature lends GAGs to AEX and HILIC separation, both of which have been used to analyze low molecular weight heparins (Miller et al. 2016a; Sun et al. 2017); however, high salt concentrations and long methods can limit their applications. Capillary zone electrophoresis (CZE), a type of CE, classified more than 80 compositions of heparin oligosaccharides (dp6–dp12) in a study by Sanderson et al. (Sanderson et al. 2018) This method is highly sensitive to salt concentration which can increase peak broadening.

The structural analysis of GAGs is challenging due to their non-template-driven enzymatic biosynthesis and modifications. MS can provide the composition; however, analysis is complicated by the loss of labile sulfate modifications. This is facilitated by protonation in the gas phase, though it can be combatted by using negative ionization mode ESI (Pepi et al. 2021a). MS/MS is often employed for sequencing GAG polymers, and can aid in determining diastereoisomers (Pepi et al. 2021a; Zaia 2008). CID leads to the fragmentation of the most labile bonds, which often results in sulfate loss (Zaia et al. 2001). This can be minimized using derivatization such as permethylation (Huang et al. 2016), H^+/metal exchange (Shi and Zaia 2009) or deprotonation of sulfate groups. Cleavage of labile glycosidic bonds is also common, while cross-ring fragments are minimal or absent (Nguan et al. 2022). Liang et al. introduced a propionylation method to replace sulfate groups, which allowed them to chromatographically separate isomers of derivatized HS disaccharides and achieve full sequencing using CID (Liang et al. 2018). Further, Zaia et al. demonstrated the influence of acidic groups in CID fragmentation. Where sulfated oligosaccharides without hexuronic acids exhibit C-type ions, sulfated and uronic acids together produce B- and Y-ions (Fig. 3) (Zaia et al. 2007). Analysis of a heparin oligomer displayed that increasing deprotonation or metal cation/H^+ exchange can also afford more structurally informative cross-ring cleavages (Kailemia et al. 2012).

IRMPD exhibits a fragmentation pattern similar to CID, including the loss of sulfate modifications during activation and predominant glycosidic bond cleavage in the product spectrum, as displayed in Fig. 16b, c (Wolff et al. 2008a). McClellan et al. observed that the charge state of the precursor ion can influence the fragmentation, where higher charge states fragment more efficiently (McClellan et al. 2002).

EDD retains labile sulfate modifications due to the radical-driven fast process, resulting in highly informative cross-ring fragments (Fig. 16a) (Pepi et al. 2021a). One of the greatest challenges in GAG analysis is the characterization of C5 epimerization, which cannot be distinguished using CID or IRMPD. Wolff et al. successfully elucidated isomerism of uronic acids in HS tetrasaccharides using EDD. The diagnostic ${}^{0,2}A_3$, B_3 and B_3-CO_2 fragments were observed exclusively for GlcA,

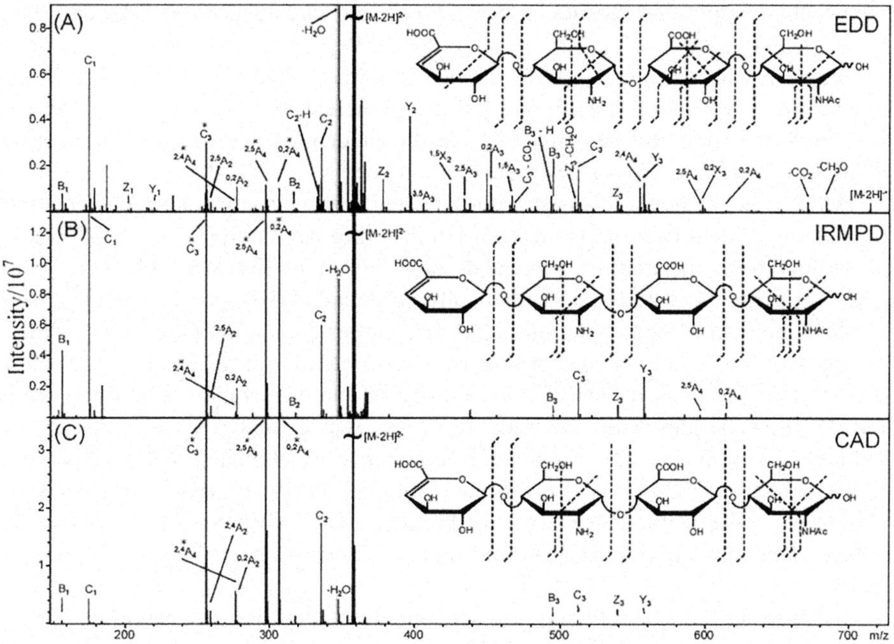

Fig. 16 MS/MS spectrum of a tetrasaccharide of HS. (**A**) EDD activation observed both glycosidic and cross-ring cleavages. (**B**) IRMPD and (**C**) CID/CAD spectrum displays predominantly glycosidic bond cleavage (B-, C-, Y-, Z-) and small amounts of cross-ring A- cleavages. (Reproduced with permission from Wolff et al. (2007a). Copyright © 2007 American Society for Mass Spectrometry)

enabling identification of the acid moiety (Wolff et al. 2007b). Leach et al. also observed that EDD is most efficient for sulfated GAGs when the precursor ion's charge state exceeds the number of sulfate groups by at least one (Leach et al. 2012). However, due to the negative charge of the ions, excitation by irradiation takes longer, which makes it difficult to combine with other methods of separation.

Negative electron transfer dissociation (NETD) is the negative ion complementary of electron transfer dissociation (ETD). It has a duty cycle of 10–100 ms and hence can be hyphenated with online separation methods. Stickney et al. implemented CE paired with NETD-MS to distinguish diastereomers of HS tetrasaccharides (Stickney et al. 2019) and Leach et al. sequenced sulfated oligosaccharides (HS and CS) with NETD and observed both glycosidic and cross-ring fragment ions without losses of the sulfate (Leach et al. 2017). It was also shown by Wu et al. that NETD could distinguish between positional isomers of 3-*O* and 6-*O* of HS without derivatization (Wu et al. 2018).

UVPD has been applied to derive informative cross-ring fragmentation of heparin disaccharides, demonstrated using a 220 nm laser and an ion trap mass spectrometer (Racaud et al. 2009). The Brodbelt group introduced a 193 nm laser for the fragmentation of HS tetrasaccharides resulting in the cleavage of all glycosidic bonds, along with a significant number of cross-ring fragments observed without the loss of sulfate modifications (Klein et al. 2019). Pepi et al. conducted an extensive study on CS with a 193 nm and 213 nm laser, varying number of pulses, wavelength and precursor ionization state. It was found that increasing the number of pulses can give more informative fragments and that precursor ions with higher charge states exhibit more diverse cross-ring fragmentation (Pepi et al. 2021b).

IMS has been shown to assist with the separation of isomers and gives another dimension of identification. Isomers of HS differing only at the C5 center of uronic acids have been successfully separated using AEX and TWIMS (Schenauer et al. 2009). An extensive study on synthetically produced HP/HS octasaccharides using TWIMS and MS/MS observed that oligosaccharides containing GlcA adopt a more compact structure compared to those with IdoA (Miller et al. 2015; Miller et al. 2016b). Pagel et al. used TIMS for separation and characterization of partly isomeric HP/HS disaccharides from complex mixtures, where analyses only took a few minutes, displaying major advantages compared to chomatographic separation (Polewski et al. 2024). The Lin group combined TIMS to NETD MS/MS for a qualitative and quantitative analysis of highly sulfated HP/HS without loss of the sulfate. This method yielded distinguishable CCS for isomers (Fig. 17) (Wei et al. 2019).

Miller et al. developed a shotgun sequencing method using a modified Synapt instrument with a linear drift tube, enabling fragmentation prior to IM. This approach allowed for the direct calculation of CCS values for both precursor and fragment ions of HS, and the fragment data was reconstructed to characterize longer chains (Miller et al. 2020). Kailemia et al. separated isomers and isobars of HS and CS by combining FAIMS (field asymmetric ion mobility spectrometry), with EDD fragmentation, displaying the complementary nature of the analysis types (Kailemia et al. 2014).

Gas-phase IR spectroscopy adds another dimension to GAG characterization (Greis et al. 2022). IRMPD has been used to characterize glucuronic and iduronic acid (Cagmat et al. 2010), as well as positional isomers, such as glucosamine 6-sulfate and glucosamine 3-sulfate (Schindler et al. 2017). Khanal et al. has also determined isomers of sodium adducts of CS and HS disaccharides using messenger-tagging spectroscopy with H_2 (Khanal et al. 2017). Helium droplet cryo-IR spectroscopy has been compared to classic IRMPD and displayed much higher spectral resolution, even at lower wavenumbers for synthetic HP pentasaccharides (Lettow et al. 2020a). Even small changes such as single stereocenters of HS tetrasaccharides could be resolved (Fig. 18) (Lettow et al. 2020b).

The mass spectrometric analysis of GAGs is particularly difficult due to the presence of labile bonds and complex isomeric structures. To overcome these challenges, additional separation techniques such as chromatography and ion mobility, along with advanced characterization methods like gas-phase infrared (IR) spectroscopy, are required for a more comprehensive analysis. Riedel et al. recently developed a random forest model to predict GAG classification and modification. Data from cryogenic gas-phase helium droplet IR spectroscopy of

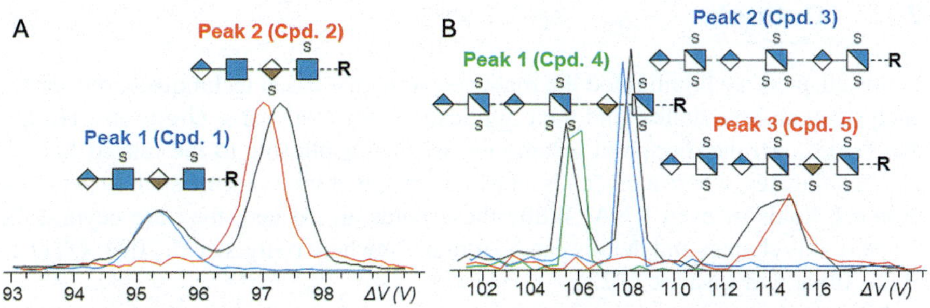

$[M-3H]^{3-}$	CCS (N_2, Å²)
Compound 1	334.6 ± 0.4
Compound 2	342.7 ± 0.5

$[M-3H]^{3-}$	CCS (N_2, Å²)
Compound 3	362.3 ± 0.2
Compound 4	354.3 ± 1.0
Compound 5	379.9 ± 0.6

Fig. 17 Separation of tetramer and hexamer isomers by TIMS. (**A**) Extracted IM spectra of isomers of two tetramers ($[M-3H]^{3-}$) (blue and red trace) and their mixture (black trace) as well as the respective CCS values. (**B**) Extracted IM spectra of three hexasaccharides isomers ($[M-3H]^{3-}$) (blue, green, and red trace) and their mixture (black trace) displaying different CCS values. (Figure reproduced with permission from Wei et al. (2019). Copyright © 2019 American Chemical Society)

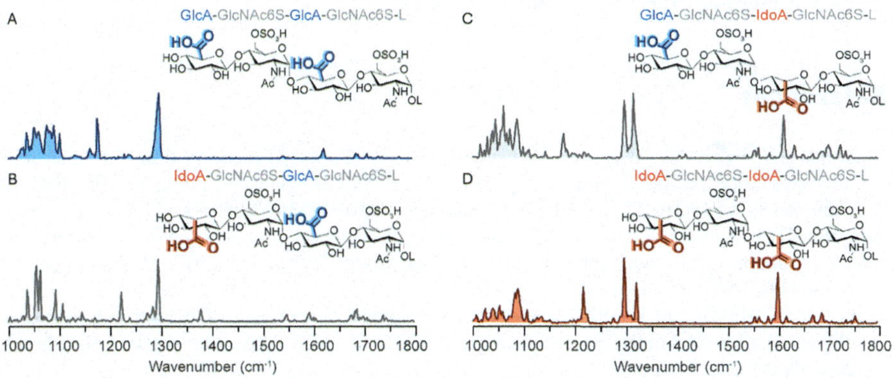

Fig. 18 Cyro-IR spectra of HS tetrasacharides. In (**A**), both uronic acids are GlcA (blue), (**B** and **C**) have one GlcA and one IdoA (red) residue and (**D**) has both uronic acids as IdoA. (Reproduced with permission from Lettow et al. (2020b). Copyright © 2020 American Chemical Society)

m/z-selected heparan sulfate oligosaccharides was used, demonstrating how machine learning techniques can support the structural characterization of GAGs (Riedel et al. 2023).

7 Outlook

In this chapter, we highlighted the mass spectrometry-based techniques used for the analysis of carbohydrates, and their applications for *N*-glycans, *O*-glycans, HMOs, and GAGs. All the discussed techniques are also applicable to the related field of glycoconjugates, where a glycan is attached to another biomolecule such as peptides, proteins, lipids or even RNA. While these fields are related, they are beyond the scope of this chapter and details can be found elsewhere (Girgis et al. 2024; Oh et al. 2025; Bojar and Lisacek 2022).

MS-based glycan characterization has developed rapidly over the last decades, through advances in activation techniques, ion mobility and gas-phase IR spectroscopy. We believe that the combination of developing instrumentation, for example through methods such as ion-soft landing and single molecule microscopy (Wu et al. 2020), and the advancement of artificial intelligence tools (Urban et al. 2024) will drive the development of the field even further and eventually enable the complete elucidation of the sugar code.

Acknowledgements N.G., C.W.D. and K.P. are grateful for funding through the ERC Consolidator Grant "Glycospec" (ERC-2019-CoG-863934-GlycoSpec). E.R., M.B. and K.P. acknowledge support from the Marie Skłodowska-Curie actions doctoral network "MobiliTraIN".

Conflict of Interests There are no conflicts to declare.

References

Abikhodr AH, Faleh AB, Warnke S, Yatsyna V, Rizzo TR (2023) Identification of human milk oligosaccharide positional isomers by combining IMS-CID-IMS and cryogenic IR spectroscopy. Analyst 148(10):2277–2282. https://doi.org/10.1039/D3AN00407D

Adamson JT, Håkansson K (2007a) Electron capture dissociation of oligosaccharides ionized with alkali, alkaline earth, and transition metals. Anal Chem 79(7):2901–2910. https://doi.org/10.1021/ac0621423

Adamson JT, Håkansson K (2007b) Electron detachment dissociation of neutral and sialylated oligosaccharides. J Am Soc Mass Spectrom 18(12):2162–2172. https://doi.org/10.1016/j.jasms.2007.09.007

Aich U, Lakbub J, Liu A (2016) State-of-the-art technologies for rapid and high-throughput sample preparation and analysis of N-glycans from antibodies. Electrophoresis 37(11):1468–1488. https://doi.org/10.1002/elps.201500551

Bansal P, Faleh AB, Warnke S, Rizzo TR (2022) Identification of N-glycan positional isomers by combining IMS and vibrational fingerprinting of structurally determinant CID fragments. Analyst 147(4):704–711. https://doi.org/10.1039/D1AN01861B

Bayat P, Lesage D, Cole RB (2020) Tutorial: ion activation in tandem mass spectrometry using ultra-high resolution instrumentation. Mass Spectrom Rev 39(5–6):680–702. https://doi.org/10.1002/mas.21623

Bechtella L, Chunsheng J, Fentker K, Ertürk GR, Safferthal M, Polewski Ł, Götze M, Graeber SY, Vos GM, Struwe WB, Mall MA, Mertins P, Karlsson NG, Pagel K (2024) Ion mobility-tandem mass spectrometry of mucin-type O-glycans. Nat Commun 15(1):2611. https://doi.org/10.1038/s41467-024-46825-4

Ben Faleh A, Warnke S, Bansal P, Pellegrinelli RP, Dyukova I, Rizzo TR (2022) Identification of mobility-resolved N-glycan isomers. Anal Chem 94(28):10101–10108. https://doi.org/10.1021/acs.analchem.2c01181

Bode L (2012) Human milk oligosaccharides: every baby needs a sugar mama. Glycobiology 22(9): 1147–1162. https://doi.org/10.1093/glycob/cws074

Bojar D, Lisacek F (2022) Glycoinformatics in the artificial intelligence era. Chem Rev 122(20): 15971–15988. https://doi.org/10.1021/acs.chemrev.2c00110

Buszewski B, Noga S (2012) Hydrophilic interaction liquid chromatography (HILIC)—a powerful separation technique. Anal Bioanal Chem 402(1):231–247. https://doi.org/10.1007/s00216-011-5308-5

Cagmat EB, Szczepanski J, Pearson WL, Powell DH, Eyler JR, Polfer NC (2010) Vibrational signatures of metal-chelated monosaccharide epimers: gas-phase infrared spectroscopy of Rb+-tagged glucuronic and iduronic acid. Phys Chem Chem Phys 12(14):3474–3479. https://doi.org/10.1039/B924027F

Calixte EI, Liyanage OT, Kim HJ, Ziperman ED, Pearson AJ, Gallagher ES (2020) Release of carbohydrate–metal adducts from electrospray droplets: insight into glycan ionization by electrospray. J Phys Chem B 124(3):479–486. https://doi.org/10.1021/acs.jpcb.9b10369

Campbell MP, Nguyen-Khuong T, Hayes CA, Flowers SA, Alagesan K, Kolarich D, Packer NH, Karlsson NG (2014) Validation of the curation pipeline of UniCarb-DB: building a global glycan reference MS/MS repository. Biochim Biophys Acta BBA Proteins and Proteomics 1844(1):108–116. https://doi.org/10.1016/j.bbapap.2013.04.018

Caterson B, Melrose J (2018) Keratan sulfate, a complex glycosaminoglycan with unique functional capability. Glycobiology 28(4):182–206. https://doi.org/10.1093/glycob/cwy003

Chen C-H, Lin Y-P, Ren C-T, Shivatare SS, Lin N-H, Wu C-Y, Chen C-H, Lin J-L (2020) Enhancement of fucosylated *N*-glycan isomer separation with an ultrahigh column temperature in porous graphitic carbon liquid chromatography-mass spectrometry. J Chromatogr A 1632: 461610. https://doi.org/10.1016/j.chroma.2020.461610

Christofi E, Barran P (2023) Ion mobility mass spectrometry (IM-MS) for structural biology: insights gained by measuring mass, charge, and collision cross section. Chem Rev 123(6): 2902–2949. https://doi.org/10.1021/acs.chemrev.2c00600

De Hoffmann E, Stroobant V (2007) Mass spectrometry: principles and applications. Wiley, Incorporated

Delvaux A, Rathahao-Paris E, Guillon B, Cholet S, Adel-Patient K, Fenaille F, Junot C, Alves S (2021) Trapped ion mobility spectrometry time-of-flight mass spectrometry for high throughput and high resolution characterization of human milk oligosaccharide isomers. Anal Chim Acta 1180:338878. https://doi.org/10.1016/j.aca.2021.338878

Deng L, Webb IK, Garimella SVB, Hamid AM, Zheng X, Norheim RV, Prost SA, Anderson GA, Sandoval JA, Baker ES, Ibrahim YM, Smith RD (2017) Serpentine ultralong path with extended routing (SUPER) high resolution traveling wave ion mobility-MS using structures for lossless ion manipulations. Anal Chem 89(8):4628–4634. https://doi.org/10.1021/acs.analchem.7b00185

Depraz Depland A, Renois-Predelus G, Schindler B, Compagnon I (2018) Identification of sialic acid linkage isomers in glycans using coupled InfraRed multiple photon dissociation (IRMPD) spectroscopy and mass spectrometry. Int J Mass Spectrom 434:65–69. https://doi.org/10.1016/j.ijms.2018.09.005

Devakumar A, Mechref Y, Kang P, Novotny MV, Reilly JP (2007) Laser-induced photofragmentation of neutral and acidic glycans inside an ion-trap mass spectrometer. Rapid Commun Mass Spectrom 21(8):1452–1460. https://doi.org/10.1002/rcm.2981

Donovan SM, Comstock SS (2017) Human milk oligosaccharides influence neonatal mucosal and systemic immunity. Ann Nutr Metab 69(Suppl. 2):41–51. https://doi.org/10.1159/000452818

Dyukova I, Carrascosa E, Pellegrinelli RP, Rizzo TR (2020) Combining cryogenic infrared spectroscopy with selective enzymatic cleavage for determining glycan primary structure. Anal Chem 92(2):1658–1662. https://doi.org/10.1021/acs.analchem.9b04776

Dyukova I, Faleh AB, Warnke S, Yalovenko N, Yatsyna V, Bansal P, Rizzo TR (2021) A new approach for identifying positional isomers of glycans cleaved from monoclonal antibodies. Analyst 146(15):4789–4795. https://doi.org/10.1039/D1AN00780G

Ernst S, Langer R, Cooney CL, Sasisekharan R (1995) Enzymatic degradation of glycosaminogiycans. Crit Rev Biochem Mol Biol 30(5):387–444. https://doi.org/10.3109/10409239509083490

Fenn JB, Mann M, Meng CK, Wong SF, Whitehouse CM (1989) Electrospray ionization for mass spectrometry of large biomolecules. Science 246(4926):64–71. https://doi.org/10.1126/science.2675315

Gabelica V (2021) Ion mobility – mass spectrometry: fundamentals and applications. The Royal Society of Chemistry. https://doi.org/10.1039/9781839162886

Gaye MM, Valentine SJ, Hu Y, Mirjankar N, Hammoud ZT, Mechref Y, Lavine BK, Clemmer DE (2012) Ion mobility-mass spectrometry analysis of serum N-linked glycans from esophageal adenocarcinoma phenotypes. J Proteome Res 11(12):6102–6110. https://doi.org/10.1021/pr300756e

Geue N, Winpenny REP, Barran PE (2022) Structural characterization methods for supramolecular chemistry that go beyond crystallography. Chem Soc Rev 51(1):8–27. https://doi.org/10.1039/d0cs01550d

Geue N, Bennett TS, Ramakers LAI, Timco GA, McInnes EJL, Burton NA, Armentrout PB, Winpenny REP, Barran PE (2023) Adduct ions as diagnostic probes of metallosupramolecular complexes using ion mobility mass spectrometry. Inorg Chem 62(6):2672–2679. https://doi.org/10.1021/acs.inorgchem.2c03698

Geue N, Winpenny REP, Barran PE (2024) Ion mobility mass spectrometry for large synthetic molecules: expanding the analytical toolbox. J Am Chem Soc 146(13):8800–8819. https://doi.org/10.1021/jacs.4c00354

Girgis M, Petruncio G, Russo P, Peyton S, Paige M, Campos D, Sanda M (2024) Analysis of N- and O-linked site-specific glycosylation by ion mobility mass spectrometry: state of the art and future directions. Proteomics 24(12–13):2300281. https://doi.org/10.1002/pmic.202300281

Grabarics M, Lettow M, Kirschbaum C, Greis K, Manz C, Pagel K (2022) Mass spectrometry-based techniques to elucidate the sugar code. Chem Rev 122(8):7840–7908. https://doi.org/10.1021/acs.chemrev.1c00380

Gray CJ, Thomas B, Upton R, Migas LG, Eyers CE, Barran PE, Flitsch SL (2016) Applications of ion mobility mass spectrometry for high throughput, high resolution glycan analysis. Biochim Biophys Acta BBA Gen Subj 1860(8):1688–1709. https://doi.org/10.1016/j.bbagen.2016.02.003

Greis K, Kirschbaum C, von Helden G, Pagel K (2022) Gas-phase infrared spectroscopy of glycans and glycoconjugates. Curr Opin Struct Biol 72:194–202. https://doi.org/10.1016/j.sbi.2021.11.006

Gross JH (2017) Mass spectrometry. Springer International Publishing, Cham. https://doi.org/10.1007/978-3-319-54398-7

Habibi SC, Nagy G (2023) Assessing the use of host-guest chemistry in conjunction with cyclic ion mobility separations for the linkage-specific characterization of human milk oligosaccharides. Int J Mass Spectrom 483:116977. https://doi.org/10.1016/j.ijms.2022.116977

Habibi SC, Bradford VR, Baird SC, Lucas SW, Chouinard CD, Nagy G (2024) Development of a cyclic ion mobility spectrometry-mass spectrometry-based collision cross-section database of permethylated human milk oligosaccharides. J Mass Spectrom 59(8):e5076. https://doi.org/10.1002/jms.5076

Han L, Costello CE (2011) Electron transfer dissociation of milk oligosaccharides. J Am Soc Mass Spectrom 22(6):997–1013. https://doi.org/10.1007/s13361-011-0117-9

Han L, Costello CE (2013) Mass spectrometry of glycans. Biochem Biokhimiia 78(7):710–720. https://doi.org/10.1134/S0006297913070031

Harvey DJ (2000) Collision-induced fragmentation of underivatized N-linked carbohydrates ionized by electrospray. J Mass Spectrom 35(10):1178–1190. https://doi.org/10.1002/1096-9888(200010)35:10<1178::AID-JMS46>3.0.CO;2-F

Harvey DJ (2020) Negative ion mass spectrometry for the analysis of N-linked glycans. Mass Spectrom Rev 39(5–6):586–679. https://doi.org/10.1002/mas.21622

Harvey DJ (2023) Analysis of carbohydrates and glycoconjugates by matrix-assisted laser desorption/ionization mass spectrometry: an update for 2019–2020. Mass Spectrom Rev 42(5): 1984–2206. https://doi.org/10.1002/mas.21806

Harvey DJ, Scarff CA, Crispin M, Scanlan CN, Bonomelli C, Scrivens JH (2012) MALDI-MS/MS with traveling wave ion mobility for the structural analysis of N-linked glycans. J Am Soc Mass Spectrom 23(11):1955–1966. https://doi.org/10.1007/s13361-012-0425-8

Harvey DJ, Scarff CA, Edgeworth M, Crispin M, Scanlan CN, Sobott F, Allman S, Baruah K, Pritchard L, Scrivens JH (2013) Travelling wave ion mobility and negative ion fragmentation for the structural determination of N-linked glycans. Electrophoresis 34(16):2368–2378. https://doi.org/10.1002/elps.201200669

Harvey DJ, Scarff CA, Edgeworth M, Pagel K, Thalassinos K, Struwe WB, Crispin M, Scrivens JH (2016) Travelling-wave ion mobility mass spectrometry and negative ion fragmentation of hybrid and complex N-glycans. J Mass Spectrom 51(11):1064–1079. https://doi.org/10.1002/jms.3828

Harvey DJ, Seabright GE, Vasiljevic S, Crispin M, Struwe WB (2018) Isomer information from ion mobility separation of high-mannose glycan fragments. J Am Soc Mass Spectrom 29(5): 972–988. https://doi.org/10.1007/s13361-018-1890-5

Huang Y, Dodds ED (2013) Ion mobility studies of carbohydrates as group I adducts: isomer specific collisional cross section dependence on metal ion radius. Anal Chem 85(20): 9728–9735. https://doi.org/10.1021/ac402133f

Huang Y, Dodds ED (2015) Discrimination of isomeric carbohydrates as the electron transfer products of group II cation adducts by ion mobility spectrometry and tandem mass spectrometry. Anal Chem 87(11):5664–5668. https://doi.org/10.1021/acs.analchem.5b00759

Huang R, Zong C, Venot A, Chiu Y, Zhou D, Boons G-J, Sharp JS (2016) De Novo sequencing of complex mixtures of heparan sulfate oligosaccharides. Anal Chem 88(10):5299–5307. https://doi.org/10.1021/acs.analchem.6b00519

Isailovic D, Kurulugama RT, Plasencia MD, Stokes ST, Kyselova Z, Goldman R, Mechref Y, Novotny MV, Clemmer DE (2008) Profiling of human serum glycans associated with liver cancer and cirrhosis by IMS−MS. J Proteome Res 7(3):1109–1117. https://doi.org/10.1021/pr700702r

Issa SMA, Vitiazeva V, Hayes CA, Karlsson NG (2018) Higher energy collisional dissociation mass spectrometry of sulfated O-linked oligosaccharides. J Proteome Res 17(9):3259–3267. https://doi.org/10.1021/acs.jproteome.8b00376

Jensen PH, Karlsson NG, Kolarich D, Packer NH (2012) Structural analysis of N- and O-glycans released from glycoproteins. Nat Protoc 7(7):1299–1310. https://doi.org/10.1038/nprot.2012.063

Jin C, Kenny DT, Skoog EC, Padra M, Adamczyk B, Vitizeva V, Thorell A, Venkatakrishnan V, Lindén SK, Karlsson NG (2017) Structural diversity of human gastric mucin glycans*. Mol Cell Proteomics 16(5):743–758. https://doi.org/10.1074/mcp.M117.067983

Jin C, Harvey DJ, Struwe WB, Karlsson NG (2019) Separation of isomeric O-glycans by ion mobility and liquid chromatography–mass spectrometry. Anal Chem 91(16):10604–10613. https://doi.org/10.1021/acs.analchem.9b01772

Jones CJ, Beni S, Limtiaco JFK, Langeslay DJ, Larive CK (2011) Heparin characterization: challenges and solutions. Annu Rev Anal Chem 4:439–465. https://doi.org/10.1146/annurev-anchem-061010-113911

Jordan JS, Xia Z, Williams ER (2022) Tips on making tiny tips: secrets to submicron nanoelectrospray emitters. J Am Soc Mass Spectrom 33(3):607–611. https://doi.org/10.1021/jasms.1c00372

Kailemia MJ, Li L, Ly M, Linhardt RJ, Amster IJ (2012) Complete mass spectral characterization of a synthetic ultralow-molecular-weight heparin using collision-induced dissociation. Anal Chem 84(13):5475–5478. https://doi.org/10.1021/ac3015824

Kailemia MJ, Park M, Kaplan DA, Venot A, Boons G-J, Li L, Linhardt RJ, Amster IJ (2014) High-field asymmetric-waveform ion mobility spectrometry and electron detachment dissociation of isobaric mixtures of glycosaminoglycans. J Am Soc Mass Spectrom 25(2):258–268. https://doi.org/10.1007/s13361-013-0771-1

Kang P, Mechref Y, Klouckova I, Novotny MV (2005) Solid-phase permethylation of glycans for mass spectrometric analysis. Rapid Commun Mass Spectrom 19(23):3421–3428. https://doi.org/10.1002/rcm.2210

Karas M, Hillenkamp F (1988) Laser desorption ionization of proteins with molecular masses exceeding 10,000 daltons. Anal Chem 60(20):2299–2301. https://doi.org/10.1021/ac00171a028

Khanal N, Masellis C, Kamrath MZ, Clemmer DE, Rizzo TR (2017) Glycosaminoglycan analysis by cryogenic messenger-tagging IR spectroscopy combined with IMS-MS. Anal Chem 89(14): 7601–7606. https://doi.org/10.1021/acs.analchem.7b01467

Khanal N, Masellis C, Kamrath MZ, Clemmer DE, Rizzo TR (2018) Cryogenic IR spectroscopy combined with ion mobility spectrometry for the analysis of human milk oligosaccharides. Analyst 143(8):1846–1852. https://doi.org/10.1039/C8AN00230D

Klein DR, Leach FE, Amster IJ, Brodbelt JS (2019) Structural characterization of glycosaminoglycan carbohydrates using ultraviolet photodissociation. Anal Chem 91(9):6019–6026. https://doi.org/10.1021/acs.analchem.9b00521

Knežević A, Polašek O, Gornik O, Rudan I, Campbell H, Hayward C, Wright A, Kolčić I, O'Donoghue N, Bones J, Rudd PM, Lauc G (2009) Variability, heritability and environmental determinants of human plasma N-glycome. J Proteome Res 8(2):694–701. https://doi.org/10.1021/pr800737u

Ko BJ, Brodbelt JS (2011) 193 nm ultraviolet photodissociation of deprotonated sialylated oligosaccharides. Anal Chem 83(21):8192–8200. https://doi.org/10.1021/ac201751u

Konermann L, Ahadi E, Rodriguez AD, Vahidi S (2013) Unraveling the mechanism of electrospray ionization. Anal Chem 85(1):2–9. https://doi.org/10.1021/ac302789c

Lancaster KS, An HJ, Li B, Lebrilla CB (2006) Interrogation of N-linked oligosaccharides using infrared multiphoton dissociation in FT-ICR mass spectrometry. Anal Chem 78(14): 4990–4997. https://doi.org/10.1021/ac0600656

Leach FE, Arungundram S, Al-Mafraji K, Venot A, Boons G-J, Amster IJ (2012) Electron detachment dissociation of synthetic heparan sulfate glycosaminoglycan tetrasaccharides varying in degree of sulfation and hexuronic acid stereochemistry. Int J Mass Spectrom 330–332: 152–159. https://doi.org/10.1016/j.ijms.2012.07.002

Leach FEI, Riley NM, Westphall MS, Coon JJ, Amster IJ (2017) Negative electron transfer dissociation sequencing of increasingly sulfated glycosaminoglycan oligosaccharides on an orbitrap mass spectrometer. J Am Soc Mass Spectrom 28(9):1844–1854. https://doi.org/10.1007/s13361-017-1709-9

Lei M, Novotny MV, Mechref Y (2010) Sequential enrichment of sulfated glycans by strong anion-exchange chromatography prior to mass spectrometric measurements. J Am Soc Mass Spectrom 21(3):348–357. https://doi.org/10.1016/j.jasms.2009.09.017

Lettow M, Grabarics M, Mucha E, Thomas DA, Polewski Ł, Freyse J, Rademann J, Meijer G, von Helden G, Pagel K (2020a) IR action spectroscopy of glycosaminoglycan oligosaccharides. Anal Bioanal Chem 412(3):533–537. https://doi.org/10.1007/s00216-019-02327-7

Lettow M, Grabarics M, Greis K, Mucha E, Thomas DA, Chopra P, Boons G-J, Karlsson R, Turnbull JE, Meijer G, Miller RL, von Helden G, Pagel K (2020b) Cryogenic infrared spectroscopy reveals structural modularity in the vibrational fingerprints of heparan sulfate Diastereomers. Anal Chem 92(15):10228–10232. https://doi.org/10.1021/acs.analchem.0c02048

Lettow M, Greis K, Mucha E, Lambeth TR, Yaman M, Kontodimas V, Manz C, Hoffmann W, Meijer G, Julian RR, Von Helden G, Marianski M, Pagel K (2023) Decoding the fucose

migration product during mass-spectrometric analysis of blood group epitopes. Angew Chem Int Ed 62(24):e202302883. https://doi.org/10.1002/anie.202302883
Liang Q, Chopra P, Boons G-J, Sharp JS (2018) Improved De Novo sequencing of heparin/heparan sulfate oligosaccharides by propionylation of sites of Sulfation. Carbohydr Res 465:16–21. https://doi.org/10.1016/j.carres.2018.06.002
Lu G, Crihfield CL, Gattu S, Veltri LM, Holland LA (2018) Capillary electrophoresis separations of glycans. Chem Rev 118(17):7867–7885. https://doi.org/10.1021/acs.chemrev.7b00669
Manz C, Pagel K (2018) Glycan analysis by ion mobility-mass spectrometry and gas-phase spectroscopy. Curr Opin Chem Biol 42:16–24. https://doi.org/10.1016/j.cbpa.2017.10.021
Manz C, Götze M, Frank C, Zappe A, Pagel K (2022) Dextran as internal calibrant for N-glycan analysis by liquid chromatography coupled to ion mobility-mass spectrometry. Anal Bioanal Chem 414(17):5023–5031. https://doi.org/10.1007/s00216-022-04133-0
Masellis C, Khanal N, Kamrath MZ, Clemmer DE, Rizzo TR (2017) Cryogenic vibrational spectroscopy provides unique fingerprints for glycan identification. J Am Soc Mass Spectrom 28(10):2217–2222. https://doi.org/10.1007/s13361-017-1728-6
Mattson JM, Turcotte R, Zhang Y (2017) Glycosaminoglycans contribute to extracellular matrix fiber recruitment and arterial wall mechanics. Biomech Model Mechanobiol 16(1):213–225. https://doi.org/10.1007/s10237-016-0811-4
McClellan JE, Costello CE, O'Connor PB, Zaia J (2002) Influence of charge state on product ion mass spectra and the determination of 4S/6S sulfation sequence of chondroitin sulfate oligosaccharides. Anal Chem 74(15):3760–3771. https://doi.org/10.1021/ac025506+
Miller JM (2005) Chromatography: concepts and contrasts. Wiley
Miller RL, Wei W, Schwörer R, Zubkova OV, Tyler PC, Turnbull JE, Leary JA (2015) Composition, sequencing and ion mobility mass spectrometry of heparan sulfate-like octasaccharide isomers differing in glucuronic and Iduronic acid content. Eur J Mass Spectrom 21(3):245–254. https://doi.org/10.1255/ejms.1337
Miller RL, Guimond SE, Shivkumar M, Blocksidge J, Austin JA, Leary JA, Turnbull JE (2016a) Heparin isomeric oligosaccharide separation using volatile salt strong anion exchange chromatography. Anal Chem 88(23):11542–11550. https://doi.org/10.1021/acs.analchem.6b02801
Miller RL, Dykstra AB, Wei W, Holsclaw C, Turnbull JE, Leary JA (2016b) Enrichment of two isomeric heparin oligosaccharides exhibiting different affinities toward monocyte chemoattractant Protein-1. Anal Chem 88(23):11551–11558. https://doi.org/10.1021/acs.analchem.6b02803
Miller RL, Guimond SE, Schwörer R, Zubkova OV, Tyler PC, Xu Y, Liu J, Chopra P, Boons G-J, Grabarics M, Manz C, Hofmann J, Karlsson NG, Turnbull JE, Struwe WB, Pagel K (2020) Shotgun ion mobility mass spectrometry sequencing of heparan sulfate saccharides. Nat Commun 11(1):1481. https://doi.org/10.1038/s41467-020-15284-y
Mucha E, González Flórez AI, Marianski M, Thomas DA, Hoffmann W, Struwe WB, Hahm HS, Gewinner S, Schöllkopf W, Seeberger PH, von Helden G, Pagel K (2017) Glycan fingerprinting via cold-ion infrared spectroscopy. Angew Chem Int Ed 56(37):11248–11251. https://doi.org/10.1002/anie.201702896
Mucha E, Marianski M, Xu F-F, Thomas DA, Meijer G, von Helden G, Seeberger PH, Pagel K (2018) Unravelling the structure of glycosyl cations via cold-ion infrared spectroscopy. Nat Commun 9(1):4174. https://doi.org/10.1038/s41467-018-06764-3
Mucha E, Stuckmann A, Marianski M, Struwe WB, Meijer G, Pagel K (2019) In-depth structural analysis of glycans in the gas phase. Chem Sci 10(5):1272–1284. https://doi.org/10.1039/C8SC05426F
Nagy G, Attah IK, Garimella SVB, Tang K, Ibrahim YM, Baker ES, Smith RD (2018) Unraveling the isomeric heterogeneity of glycans: ion mobility separations in structures for lossless ion manipulations. Chem Commun 54(83):11701–11704. https://doi.org/10.1039/C8CC06966B
Nguan H-S, Tsai S-T, Ni C-K (2022) Collision-induced dissociation of cellobiose and maltose. J Phys Chem A 126(9):1486–1495. https://doi.org/10.1021/acs.jpca.1c10046

Oganesyan I, Hajduk J, Harrison JA, Marchand A, Czar MF, Zenobi R (2022) Exploring gas-phase MS methodologies for structural elucidation of branched N-glycan isomers. Anal Chem 94(29): 10531–10539. https://doi.org/10.1021/acs.analchem.2c02019

Oh MJ, Seo Y, Seo N, An HJ (2025) MS-based glycome characterization of biotherapeutics with N- and O-glycosylation. Mass Spectrom Rev. https://doi.org/10.1002/mas.21925

Pabst M, Bondili JS, Stadlmann J, Mach L, Altmann F (2007) Mass + retention time = structure: a strategy for the analysis of *N* -glycans by carbon LC-ESI-MS and its application to fibrin *N* -glycans. Anal Chem 79(13):5051–5057. https://doi.org/10.1021/ac070363i

Pagel K, Harvey DJ (2013) Ion mobility–mass spectrometry of complex carbohydrates: collision cross sections of sodiated N-linked glycans. Anal Chem 85(10):5138–5145. https://doi.org/10.1021/ac400403d

Pallister EG, Choo MSF, Walsh I, Tai JN, Tay SJ, Yang YS, Ng SK, Rudd PM, Flitsch SL, Nguyen-Khuong T (2020) Utility of ion-mobility spectrometry for deducing branching of multiply charged glycans and glycopeptides in a high-throughput positive ion LC-FLR-IMS-MS workflow. Anal Chem 92(23):15323–15335. https://doi.org/10.1021/acs.analchem.0c01954

Pepi LE, Sanderson P, Stickney M, Amster IJ (2021a) Developments in mass spectrometry for glycosaminoglycan analysis: a review. Mol Cell Proteomics MCP 20:100025. https://doi.org/10.1074/mcp.R120.002267

Pepi LE, Leach FEI, Klein DR, Brodbelt JS, Amster IJ (2021b) Investigation of the experimental parameters of ultraviolet photodissociation for the structural characterization of chondroitin sulfate glycosaminoglycan isomers. J Am Soc Mass Spectrom 32(7):1759–1770. https://doi.org/10.1021/jasms.1c00119

Peterson TL, Nagy G (2021) Toward sequencing the human milk glycome: high-resolution cyclic ion mobility separations of core human milk oligosaccharide building blocks. Anal Chem 93(27):9397–9407. https://doi.org/10.1021/acs.analchem.1c00942

Picache JA, Rose BS, Balinski A, Leaptrot KL, Sherrod SD, May JC, McLean JA (2019) Collision cross section compendium to annotate and predict multi-omic compound identities. Chem Sci 10(4):983–993. https://doi.org/10.1039/C8SC04396E

Polewski L, Moon E, Zappe A, Götze M, Szekeres GP, Roth C, Pagel K (2024) Ion mobility mass spectrometry-based disaccharide analysis of glycosaminoglycans. Chem Eur J 30(35): e202400783. https://doi.org/10.1002/chem.202400783

Pu Y, Ridgeway ME, Glaskin RS, Park MA, Costello CE, Lin C (2016) Separation and identification of isomeric glycans by selected accumulation-trapped ion mobility spectrometry-electron activated dissociation tandem mass spectrometry. Anal Chem 88(7):3440–3443. https://doi.org/10.1021/acs.analchem.6b00041

Racaud A, Antoine R, Joly L, Mesplet N, Dugourd P, Lemoine J (2009) Wavelength-tunable ultraviolet photodissociation (UVPD) of heparin-derived disaccharides in a linear ion trap. J Am Soc Mass Spectrom 20(9):1645–1651. https://doi.org/10.1016/j.jasms.2009.04.022

Rathahao-Paris E, Delvaux A, Li M, Guillon B, Venot E, Fenaille F, Adel-Patient K, Alves S (2022) Rapid structural characterization of human milk oligosaccharides and distinction of their isomers using trapped ion mobility spectrometry time-of-flight mass spectrometry. J Mass Spectrom 57(10):e4885. https://doi.org/10.1002/jms.4885

Reily C, Stewart TJ, Renfrow MB, Novak J (2019) Glycosylation in health and disease. Nat Rev Nephrol 15(6):346–366. https://doi.org/10.1038/s41581-019-0129-4

Reinhold VN, Reinhold BB, Costello CE (1995) Carbohydrate molecular weight profiling, sequence, linkage, and branching data: ES-MS and CID. Anal Chem 67(11):1772–1784. https://doi.org/10.1021/ac00107a005

Riedel J, Lettow M, Grabarics M, Götze M, Miller RL, Boons G-J, Meijer G, von Helden G, Szekeres GP, Pagel K (2023) Predicting structural motifs of glycosaminoglycans using cryogenic infrared spectroscopy and random Forest. J Am Chem Soc 145(14):7859–7868. https://doi.org/10.1021/jacs.2c12762

Royle L, Campbell MP, Radcliffe CM, White DM, Harvey DJ, Abrahams JL, Kim Y-G, Henry GW, Shadick NA, Weinblatt ME, Lee DM, Rudd PM, Dwek RA (2008) HPLC-based analysis of serum N-glycans on a 96-well plate platform with dedicated database software. Anal Biochem 376(1):1–12. https://doi.org/10.1016/j.ab.2007.12.012

Ruhaak LR, Uh H-W, Beekman M, Hokke CH, Westendorp RGJ, Houwing-Duistermaat J, Wuhrer M, Deelder AM, Slagboom PE (2011) Plasma protein N-glycan profiles are associated with calendar age, familial longevity and health. J Proteome Res 10(4):1667–1674. https://doi.org/10.1021/pr1009959

Ruhaak LR, Xu G, Li Q, Goonatilleke E, Lebrilla CB (2018) Mass spectrometry approaches to glycomic and glycoproteomic analyses. Chem Rev 118(17):7886–7930. https://doi.org/10.1021/acs.chemrev.7b00732

Safferthal M, Polewski Ł, Chang CW, Greis K, Pagel K (2024) Cold-ion spectroscopy of carbohydrates. https://doi.org/10.1039/9781839166433-00076

Sanderson P, Stickney M, Leach FE, Xia Q, Yu Y, Zhang F, Linhardt RJ, Amster IJ (2018) Heparin/heparan sulfate analysis by covalently modified reverse polarity capillary zone electrophoresis-mass spectrometry. J Chromatogr A 1545:75–83. https://doi.org/10.1016/j.chroma.2018.02.052

Schenauer MR, Meissen JK, Seo Y, Ames JB, Leary JA (2009) Heparan sulfate separation, sequencing, and isomeric differentiation: ion mobility spectrometry reveals specific Iduronic and glucuronic acid-containing hexasaccharides. Anal Chem 81(24):10179. https://doi.org/10.1021/ac902186h

Schindler B, Barnes L, Gray CJ, Chambert S, Flitsch SL, Oomens J, Daniel R, Allouche AR, Compagnon I (2017) IRMPD spectroscopy sheds new (infrared) light on the sulfate pattern of carbohydrates. J Phys Chem A 121(10):2114–2120. https://doi.org/10.1021/acs.jpca.6b11642

Shi X, Zaia J (2009) Organ-specific heparan sulfate structural phenotypes. J Biol Chem 284(18):11806–11814. https://doi.org/10.1074/jbc.M809637200

Solakyildirim K (2019) Recent advances in glycosaminoglycan analysis by various mass spectrometry techniques. Anal Bioanal Chem 411(17):3731–3741. https://doi.org/10.1007/s00216-019-01722-4

Soyyılmaz B, Mikš MH, Röhrig CH, Matwiejuk M, Meszaros-Matwiejuk A, Vigsnæs LK (2021) The mean of milk: a review of human milk oligosaccharide concentrations throughout lactation. Nutrients 13(8):2737. https://doi.org/10.3390/nu13082737

Stickney M, Sanderson P, Leach FE, Zhang F, Linhardt RJ, Amster IJ (2019) Online capillary zone electrophoresis negative electron transfer dissociation tandem mass spectrometry of glycosaminoglycan mixtures. Int J Mass Spectrom 445:116209. https://doi.org/10.1016/j.ijms.2019.116209

Struwe WB, Benesch JL, Harvey DJ, Pagel K (2015) Collision cross sections of high-mannose N -Glycans in commonly observed adduct states – identification of gas-phase conformers unique to [M − H] − ions. Analyst 140(20):6799–6803. https://doi.org/10.1039/C5AN01092F

Struwe WB, Pagel K, Benesch JLP, Harvey DJ, Campbell MP (2016a) GlycoMob: an ion mobility-mass spectrometry collision cross section database for glycomics. Glycoconj J 33(3):399–404. https://doi.org/10.1007/s10719-015-9613-7

Struwe WB, Baldauf C, Hofmann J, Rudd PM, Pagel K (2016b) Ion mobility separation of deprotonated oligosaccharide isomers – evidence for gas-phase charge migration. Chem Commun 52(83):12353–12356. https://doi.org/10.1039/C6CC06247D

Sugahara K, Mikami T, Uyama T, Mizuguchi S, Nomura K, Kitagawa H (2003) Recent advances in the structural biology of chondroitin sulfate and dermatan sulfate. Curr Opin Struct Biol 13(5):612–620. https://doi.org/10.1016/j.sbi.2003.09.011

Sun X, Guo Z, Yu M, Lin C, Sheng A, Wang Z, Linhardt RJ, Chi L (2017) Hydrophilic interaction chromatography-multiple reaction monitoring mass spectrometry method for basic building block analysis of low molecular weight heparins prepared through nitrous acid depolymerization. J Chromatogr A 1479:121–128. https://doi.org/10.1016/j.chroma.2016.11.061

Tang Y, Pu Y, Gao J, Hong P, Costello CE, Lin C (2018) De Novo glycan sequencing by electronic excitation dissociation and fixed-charge derivatization. Anal Chem 90(6):3793–3801. https://doi.org/10.1021/acs.analchem.7b04077

Taylor SL, Hogwood J, Guo W, Yates EA, Turnbull JE (2019) By-products of heparin production provide a diverse source of heparin-like and heparan sulfate glycosaminoglycans. Sci Rep 9(1): 2679. https://doi.org/10.1038/s41598-019-39093-6

Thacker BE, Xu D, Lawrence R, Esko JD (2014) Heparan sulfate 3-*O*-sulfation: a rare modification in search of a function. Matrix Biol 35:60–72. https://doi.org/10.1016/j.matbio.2013.12.001

Thomsson KA, Karlsson NG, Hansson GC (1999) Liquid chromatography–electrospray mass spectrometry as a tool for the analysis of sulfated oligosaccharides from mucin glycoproteins. J Chromatogr A 854(1):131–139. https://doi.org/10.1016/S0021-9673(99)00625-1

Urban J, Jin C, Thomsson KA, Karlsson NG, Ives CM, Fadda E, Bojar D (2024) Predicting glycan structure from tandem mass spectrometry via deep learning. Nat Methods 21(7):1206–1215. https://doi.org/10.1038/s41592-024-02314-6

Varki A (2017) Biological roles of glycans. Glycobiology 27(1):3–49. https://doi.org/10.1093/glycob/cww086

Varki A, Cummings RD, Esko JD, Stanley P, Hart GW, Aebi M, Darvill AG, Kinoshita T, Packer NH, Prestegard JH, Schnaar RL, Seeberger PH (2017) Essentials of glycobiology. Cold Spring Harbor Laboratory Press

Veillon L, Huang Y, Peng W, Dong X, Cho BG, Mechref Y (2017) Characterization of isomeric glycan structures by LC-MS/MS. Electrophoresis 38(17):2100–2114. https://doi.org/10.1002/elps.201700042

Venkatakrishnan V, Packer NH, Thaysen-Andersen M (2013) Host mucin glycosylation plays a role in bacterial adhesion in lungs of individuals with cystic fibrosis. Expert Rev Respir Med 7(5):553–576. https://doi.org/10.1586/17476348.2013.837752

Vismara E, Pierini M, Mascellani G, Liverani L, Lima M, Guerrini M, Torri G (2017) Low-molecular-weight heparin from Cu2+ and Fe2+ fenton type depolymerisation processes. Thromb Haemost 103:613–622. https://doi.org/10.1160/TH09-02-0084

Vos GM, Hooijschuur KC, Li Z, Fjeldsted J, Klein C, de Vries RP, Toraño JS, Boons G-J (2023) Sialic acid O-acetylation patterns and glycosidic linkage type determination by ion mobility-mass spectrometry. Nat Commun 14(1):6795. https://doi.org/10.1038/s41467-023-42575-x

Vreeker GCM, Wuhrer M (2017) Reversed-phase separation methods for glycan analysis. Anal Bioanal Chem 409(2):359–378. https://doi.org/10.1007/s00216-016-0073-0

Walker SH, Carlisle BC, Muddiman DC (2012) Systematic comparison of reverse phase and hydrophilic interaction liquid chromatography platforms for the analysis of N-linked glycans. Anal Chem 84(19):8198–8206. https://doi.org/10.1021/ac3012494

Warnke S, Ben Faleh A, Rizzo TR (2021) Toward high-throughput cryogenic IR fingerprinting of mobility-separated glycan isomers. ACS Meas Sci Au 1(3):157–164. https://doi.org/10.1021/acsmeasuresciau.1c00018

Wei J, Wu J, Tang Y, Ridgeway ME, Park MA, Costello CE, Zaia J, Lin C (2019) Characterization and quantification of highly sulfated glycosaminoglycan isomers by gated-trapped ion mobility spectrometry negative electron transfer dissociation MS/MS. Anal Chem 91(4):2994–3001. https://doi.org/10.1021/acs.analchem.8b05283

Wilkinson H, Saldova R (2020) Current methods for the characterization of O-glycans. J Proteome Res 19(10):3890–3905. https://doi.org/10.1021/acs.jproteome.0c00435

Wilson JJ, Brodbelt JS (2008) Ultraviolet photodissociation at 355 nm of fluorescently labeled oligosaccharides. Anal Chem 80(13):5186–5196. https://doi.org/10.1021/ac800315k

Wolff JJ, Amster IJ, Chi L, Linhardt RJ (2007a) Electron detachment dissociation of glycosaminoglycan tetrasaccharides. J Am Soc Mass Spectrom 18(2):234–244. https://doi.org/10.1016/j.jasms.2006.09.020

Wolff JJ, Chi L, Linhardt RJ, Amster IJ (2007b) Distinguishing glucuronic from iduronic acid in glycosaminoglycan tetrasaccharides by using electron detachment dissociation. Anal Chem 79(5):2015–2022. https://doi.org/10.1021/ac061636x

Wolff JJ, Laremore TN, Busch AM, Linhardt RJ, Amster IJ (2008a) Influence of charge state and sodium cationization on the electron detachment dissociation and infrared multiphoton dissociation of glycosaminoglycan oligosaccharides. J Am Soc Mass Spectrom 19(6):790–798. https://doi.org/10.1016/j.jasms.2008.03.010

Wu J, Wei J, Hogan JD, Chopra P, Joshi A, Lu W, Klein J, Boons G-J, Lin C, Zaia J (2018) Negative electron transfer dissociation sequencing of 3-O-sulfation-containing heparan sulfate oligosaccharides. J Am Soc Mass Spectrom 29(6):1262–1272. https://doi.org/10.1007/s13361-018-1907-0

Wu X, Delbianco M, Anggara K, Michnowicz T, Pardo-Vargas A, Bharate P, Sen S, Pristl M, Rauschenbach S, Schlickum U, Abb S, Seeberger PH, Kern K (2020) Imaging single glycans. Nature 582(7812):375–378. https://doi.org/10.1038/s41586-020-2362-1

Wuhrer M, Deelder AM, van der Burgt YEM (2011) Mass spectrometric glycan rearrangements. Mass Spectrom Rev 30(4):664–680. https://doi.org/10.1002/mas.20337

Yu X, Huang Y, Lin C, Costello CE (2012) Energy-dependent electron activated dissociation of metal-adducted permethylated oligosaccharides. Anal Chem 84(17):7487–7494. https://doi.org/10.1021/ac301589z

Zaia J (2008) Mass spectrometry and the emerging field of glycomics. Chem Biol 15(9):881–892. https://doi.org/10.1016/j.chembiol.2008.07.016

Zaia J, McClellan JE, Costello CE (2001) Tandem mass spectrometric determination of the 4S/6S sulfation sequence in chondroitin sulfate oligosaccharides. Anal Chem 73(24):6030–6039. https://doi.org/10.1021/ac015577t

Zaia J, Miller MJC, Seymour JL, Costello CE (2007) The role of mobile protons in negative ion CID of oligosaccharides. J Am Soc Mass Spectrom 18(5):952–960. https://doi.org/10.1016/j.jasms.2007.01.016

Zappe A, Miller RL, Struwe WB, Pagel K (2022) State-of-the-art glycosaminoglycan characterization. Mass Spectrom Rev 41(6):1040–1071. https://doi.org/10.1002/mas.21737

Zauner G, Deelder AM, Wuhrer M (2011) Recent advances in hydrophilic interaction liquid chromatography (HILIC) for structural Glycomics. Electrophoresis 32(24):3456–3466. https://doi.org/10.1002/elps.201100247

Zhang X, Julian RR (2014) Radical mediated dissection of oligosaccharides. Int J Mass Spectrom 372:22–28. https://doi.org/10.1016/j.ijms.2014.07.045

Zhang J, Lindsay LL, Hedrick JL, Lebrilla CB (2004) Strategy for profiling and structure elucidation of mucin-type oligosaccharides by mass spectrometry. Anal Chem 76(20):5990–6001. https://doi.org/10.1021/ac049666s

Zhao C, Xie B, Chan S-Y, Costello CE, O'Connor PB (2008) Collisionally activated dissociation and electron capture dissociation provide complementary structural information for branched permethylated oligosaccharides. J Am Soc Mass Spectrom 19(1):138–150. https://doi.org/10.1016/j.jasms.2007.10.022

Zheng X, Zhang X, Schocker NS, Renslow RS, Orton DJ, Khamsi J, Ashmus RA, Almeida IC, Tang K, Costello CE, Smith RD, Michael K, Baker ES (2017) Enhancing glycan isomer separations with metal ions and positive and negative polarity ion mobility spectrometry-mass spectrometry analyses. Anal Bioanal Chem 409(2):467–476. https://doi.org/10.1007/s00216-016-9866-4

Zhou W, Håkansson K (2011) Electron detachment dissociation of fluorescently labeled sialylated oligosaccharides. Electrophoresis 32(24):3526–3535. https://doi.org/10.1002/elps.201100327

Zhou S, Hu Y, Mechref Y (2016) High-temperature LC-MS/MS of permethylated glycans derived from glycoproteins. Electrophoresis 37(11):1506–1513. https://doi.org/10.1002/elps.201500568

Zhou S, Veillon L, Dong X, Huang Y, Mechref Y (2017) Direct comparison of derivatization strategies for LC-MS/MS analysis of *N*-glycans. Analyst 142(23):4446–4455. https://doi.org/10.1039/C7AN01262D

Zhu F, Lee S, Valentine SJ, Reilly JP, Clemmer DE (2012) Mannose7 glycan isomer characterization by IMS-MS/MS analysis. J Am Soc Mass Spectrom 23(12):2158–2166. https://doi.org/10.1007/s13361-012-0491-y

Zhu F, Glover MS, Shi H, Trinidad JC, Clemmer DE (2015) Populations of metal-glycan structures influence MS fragmentation patterns. J Am Soc Mass Spectrom 26(1):25–35. https://doi.org/10.1007/s13361-014-1000-2

The Road Toward Nanopore Sequencing of Glycosaminoglycans

Wesley Pietsch, Tom Schumann, Marc Safferthal, Niklas Geue, Kevin Pagel, and Michael Götze

Contents

Abstract

Nanopores have emerged as a powerful, label-free technique for single molecule analysis, offering high sensitivity and rapid analysis capabilities. Originally developed for DNA sequencing, nanopores have shown promise not only for the characterization of other biomolecules, such as RNA, proteins, and glycans but also of small inorganic compounds, such as nanoparticles. Glycosaminoglycans (GAGs) are a linear, highly charged subclass of glycans, which play essential roles in cell signaling, tissue development, and inflammation processes. The immense structural complexity of GAGs involving unique

W. Pietsch · T. Schumann
Institute of Chemistry and Biochemistry, Freie Universität Berlin, Berlin, Germany

M. Safferthal · N. Geue · K. Pagel · M. Götze (✉)
Institute of Chemistry and Biochemistry, Freie Universität Berlin, Berlin, Germany

Department of Molecular Physics, Fritz-Haber-Institut der Max-Planck-Gesellschaft, Berlin, Germany
e-mail: m.goetze@fu-berlin.de

F. Pfrengle et al. (eds.), *Complex Carbohydrates in Health and Disease*, Handbook of Experimental Pharmacology 288, https://doi.org/10.1007/164_2025_750

sulfation patterns renders their analysis challenging. This chapter provides a comprehensive overview on the application of biological and solid-state nanopores for the analysis of GAGs.

Keywords

Biological nanopores · Carbohydrates · Molecule translocation · Sequencing · Solid-state nanopores

1 Introduction

In 1989, Deamer proposed the idea to sequence a single strand of deoxyribonucleic acid (DNA) in one read (Deamer et al. 2016; Branton et al. 2008). Based on the principle of the Coulter Counter (Don 2003), the DNA must translocate through a nanoscopic hole inside a membrane to read tiny fluctuations in a current signal. Initial experiments with Kasianowicz confirmed the capability of the transmembrane protein pore α-Hemolysin (α-HL) to translocate a single strand of RNA (Kasianowicz et al. 1996). The concept was further improved by several groups (Akeson et al. 1999; Meller et al. 2000; Ashkenasy et al. 2005; Stoddart et al. 2009; Lieberman et al. 2010), until Manrao and coworkers obtained well-resolved current levels which could be related to a known DNA sequence in 2012 (Manrao et al. 2012). They utilized a *Mycobacterium smegmatis* porin A (MspA) nanopore coupled to a phi29 DNA polymerase to slowly push the strand trough the pore.

The working principle of nanopore measurements is schematically illustrated in Figure 1 (Deamer et al. 2016; Shi et al. 2017; Kono and Arakawa 2019). An isolating membrane separates two chambers filled with an aqueous salt solution. The membrane contains a single, nanometer-sized hole that is the sole passage between the two chambers. When a voltage is applied, ions can move through the hole to the corresponding electrode, which generates a current. If a molecule is drawn into the

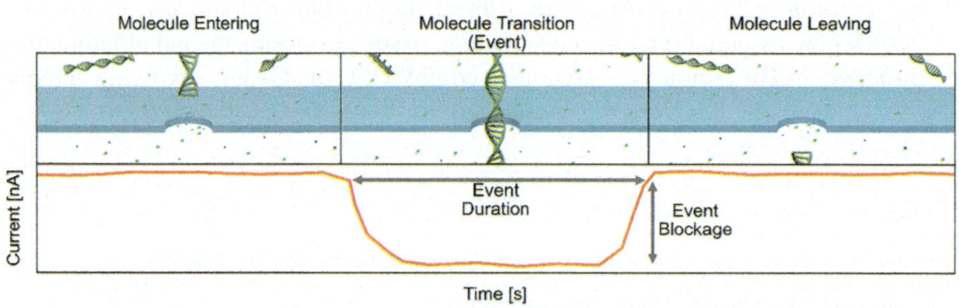

Fig. 1 Schematic illustration of the working principle of nanopores. The top part shows the three steps of a molecule translocation, and the bottom part shows the corresponding current graph. Before entering and after leaving the pore, the current is on its baseline level. Simplified, the event duration depends on the length of the molecule, while the blockage depth correlates with the molecule thickness

pore by electrophoretic or electroosmotic forces, it occupies space inside the pore and hinders the ion flow, which results in a current drop. After the translocation event, the current returns to its baseline value. This depth (event blockage) and length (event duration) of the current drop are specific for the dimensions of the molecule.

2 Types of Nanopores

Nanopores can be divided into two subclasses: biological and solid-state nanopores (ssNPs, Fig. 2). The term biological nanopore refers to transmembrane proteins that are typically embedded in lipid bilayers (Gurnev and Nestorovich 2014). The transmembrane proteins play crucial roles in maintaining the homoeostasis of the cell transporting ions and small molecules (Mayer et al. 2022). Protein pores are composed of several subunits that form a lumen with a typical pore diameter of 1 to 10 nm, allowing small molecules to pass through the pore. The defined 3D structure of the lumen results in high data reproducibility and low background noise (Mayer et al. 2022; Akhtarian et al. 2021).

For biological nanopores, various transmembrane proteins have been deployed. The first and most well known is the pore-forming toxin (PFT) α-HL from *staphylococcus aureus*. Notably, spontaneous closing of the pore during data collection can lead to misinterpretation. Another example of PFTs is aerolysin (AeL), a heptameric protein that has an inner pore diameter of 1.8 nm and shows lower translocation times due to the charge distribution inside the lumen. It exhibits a particular high sensitivity for analyzing DNA and peptides. A third example, the *Mycobacterium smegmatis* porin A (MspA), is known for its high denaturation resistance and stability against several conditions, for example, pH values ranging from 2 to 14 and various salt concentrations. It excels through a small sensing region with only 1.2 nm, making it an excellent candidate for single-stranded DNA sequencing. Pores

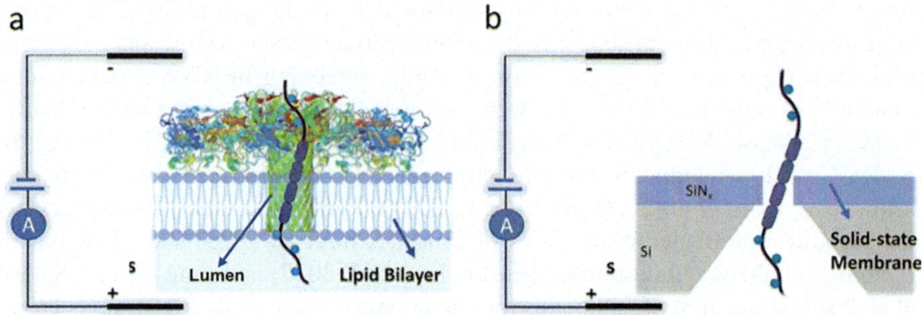

Fig. 2 Examples of a biological nanopore (**a**) and a solid-state nanopore (**b**). (**a**) A *Mycobacterium smegmatis* porin A (MspA) pore embedded into a lipid bilayer, translocating a linear molecule through its lumen. (**b**) A free-standing silicon nitride membrane supported by a silicon waver with a single fabricated pore. (Adapted with permission from Ref. Akhtarian et al. (2021). Rights managed by Taylor & Francis)

like cytolysin A possess a larger pore diameter of approximately 5.5 nm, which makes them suitable for protein analysis (Li et al. 2020).

In contrast to their biological counterparts, ssNPs have higher mechanical and chemical stability. The dimensions of the pore can be controlled with nanometer precision, which allows matching the diameter of the nanopore close to the size of the analyte. Such custom-made nanopores therefore experience high signal, but also low signal-to-noise ratios due to the high capacitance of the membrane material. The first and most used ssNPs are silicone-based devices, especially silicon nitride (SiN_x), due to their low capacitance and well-established procedures from the semiconductor industry. Other materials are Al_2O_3, ZnO, HfO_2, or TiO_2 as well as 2D materials like graphene or MoS_2, which are further discussed in the corresponding literature (Taniguchi 2022; He et al. 2021; Lin et al. 2022; Xue et al. 2020). Given the potential of ssNPs for precise control over the pore dimensions, different fabrication methods have been developed (Lin et al. 2022; Fried et al. 2021). The first ssNP with molecular dimensions used for the analysis of biomolecules was fabricated by ion beam sculpting in a SiN_x membrane in 2001 (Li et al. 2001). To date, it is more common to drill nanopores directly into a solid-state membrane using a transmission electron microscope (TEM). An alternative approach is the so-called controlled dielectric breakdown (CDB) (Kwok et al. 2014; Waugh et al. 2020). It is an in situ method, in which an isolating membrane is exposed to an electric field comparable to, but less than, the dielectric strength of the material. This results in a current flow, through a trap assisted tunneling mechanism. Eventually, a connected path of defects spans the material and breakdown occurs, leading to a sudden increase in current that physically damages the material.

3 Applications of Nanopores

While this chapter focusses on the application of nanopores to GAGs, we briefly set the scene by summarizing their use for nucleic acids, proteins, and glycans. After the first successful DNA translocation by Manroe and co-workers in 2012, nanopores were developed further to sequence the entire genome of prokaryotic and eukaryotic organisms (Loman et al. 2015). More recently, screening of RNA modifications, mutations in certain gene segments were developed, providing a significant enhancement for disease detection in clinical diagnostics (Wang et al. 2021). During the SARS-CoV-2 pandemic, the entire viral genome was decoded using nanopore sequencing (Kim et al. 2020; Nouri et al. 2021). Further, nanopores were used for the identification of numerous types of cancer (Chi et al. 2025) such as leukemia (Orsini et al. 2018), brain tumors (Euskirchen et al. 2017) and lung cancer (Suzuki et al. 2017). Detection of pathogens for pneumonia (Gorrie et al. 2018) and bacterial meningitis (Moon et al. 2019) was also realized by nanopore technology due to its rapid real-time sequencing ability. Various hereditary diseases, such as Parkinson's (Leija-Salazar et al. 2019) and Alzheimer's disease, were studied through genome analyses (De Roeck et al. 2019). While all these studies deploy biological nanopores, efforts have been made to combine and develop methods and techniques to achieve

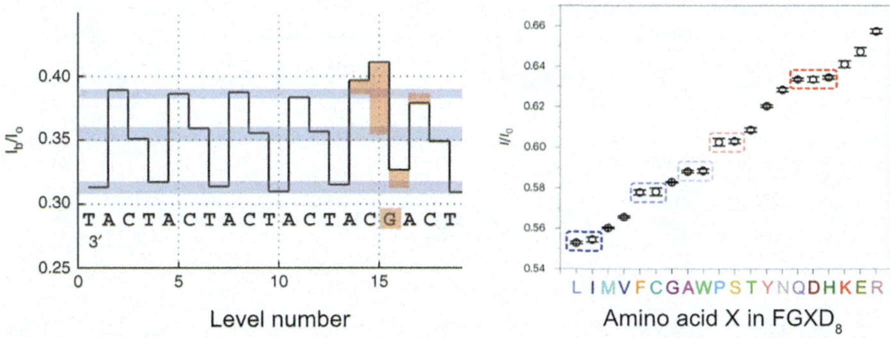

Fig. 3 Mean blockage ratios (blockage current divided by baseline current) of a repeating DNA sequence, with distinct signal levels corresponding to different nucleotide bases. A single substitution of thymine with guanine, highlighted in orange, results in a measurable shift in current. Right: Mean current levels of individual amino acids linked to an anionic tail. The signal differences are less pronounced, and not all amino acids can be reliably distinguished based on current blockage alone. (Adapted with permission from Refs. Manrao et al. (2012), Zhang et al. (2024a). Copyright © 2012 & 2023, The Author(s), under exclusive licence to Springer Nature America, Inc.)

sequencing in ssNPs (He et al. 2024; Zhou et al. 2025). To date, however, ssNPs lag their biological counterparts in terms of resolution (Akahori et al. 2017; He et al. 2011; Heerema et al. 2018).

The application of nanopore sequencing for protein analysis is limited by various factors. Instead of four base pairs, proteins consist of twenty proteinogenic amino acids, which are challenging to distinguish, not even considering posttranslational modifications (Fig. 3) (Lu et al. 2025; Dorey and Howorka 2023; Ritmejeris et al. 2024; Zhang et al. 2024a). In addition, polypeptides and proteins have an unsymmetrical charge distribution, which hinders transport toward the sensing region using an electric potential (Lu et al. 2025). In 2020, Ouldali and co-workers deployed an aerolysin nanopore to detect subtle molecular differences between different amino acids (Ouldali et al. 2020). Martin-Baniandres et al. utilized electro-osmotic flow to unfold and translocate proteins and detect post-translational modifications without the need for pore modifications or so-called motor proteins (Martin-Baniandres et al. 2023). A major step toward full peptide sequencing was taken by Zhang and co-workers (Zhang et al. 2024a). Using enzymatic cleavage and host-guest interactions, they were able to sequence short peptide chains obtained from digestion and drew conclusions about the order of digestion (Heerema et al. 2018). Developments in protein sequencing with ssNPs are currently limited by resolution, making it a mere detector of proteins in their native forms (Yusko et al. 2011; Yusko et al. 2017; Li et al. 2019). Kennedy and co-workers attempted to resolve the peptide sequence using subnanometer-diameter pores in 2016; however, they were unable to discriminate between all amino acids (Kennedy et al. 2016). To overcome these issues, optical techniques were coupled with nanopores to achieve single-molecule protein sequencing (Zhao et al. 2022).

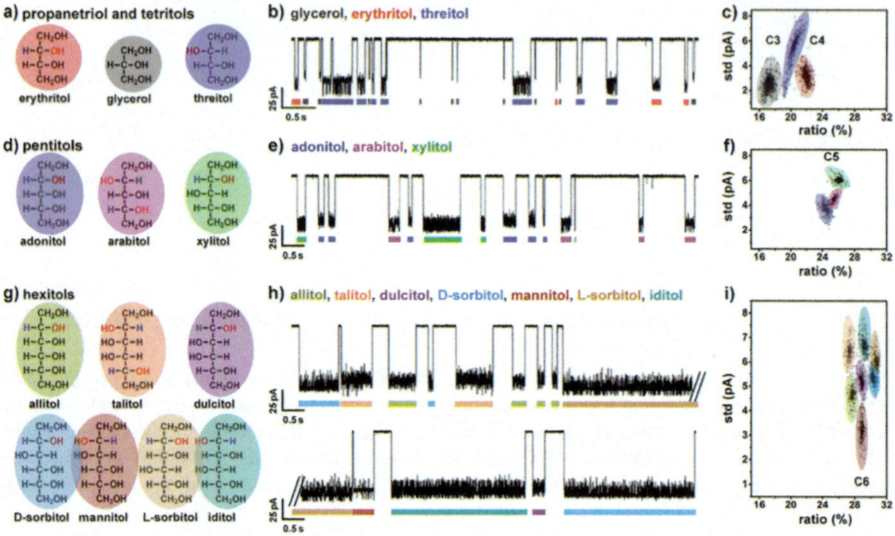

Fig. 4 Representation of tri- and tetroses (**a**), pentoses (**d**), and hexoses (**g**) alongside their corresponding current traces (**b**, **e**, **h**). (**c**, **f**, **i**) Events are distributed in a scatter plot by their blockage ratio versus the standard deviation and color-coded to match the respective aldose (number of events <500). The data indicates that aldoses with the same number of carbon atoms exhibit similar blockage ratios, requiring an additional dimension, here the standard deviation, to achieve differentiation. (Adapted with permission from Ref. Liu et al. (2022). Copyright © 2022, American Chemical Society)

Glycan sequences are more complex than those of proteins, not only differing in their monosaccharide composition, but also their connectivity, configuration and branching. These structural differences play an important role in their biological function, making the application of nanopore sequencing for the characterization of glycans both desirable and challenging (Varki et al. 2022). As early as 2011, first experiments showed differences for oligosaccharides that are 1, 4- or 1, 6-linked (Bacri et al. 2011). Using an engineered α-HL pore, D-glucose and D-fructose were distinguished in 2018 (Ramsay and Bayley 2018). Major steps toward glycan sequencing were achieved by the group of Huang, who differentiated nine types of reducing-end monosaccharides and all 15 monosaccharide alditols (Fig. 4) (Liu et al. 2022; Zhang et al. 2022). Further, they deciphered their connectivity by discriminating between the glycosidic linkages of disaccharide isomers (Zhang et al. 2024b). Additionally, the group of Gao was able to identify the distinct electrical signatures produced by acetylamino and carboxyl groups (Xia et al. 2023) and utilize the specificity of glycosidases to identify the glycan sequence by monitoring the hydrolysis products (Yao et al. 2025). Li and Lu from Qing's group employed a chemical derivatization strategy to facilitate glycan translocation through an aerolysin nanopore, enabling the differentiation of glycan isomers, including regio- and stereoisomers as well as branched glycans (Li et al. 2023; Lu et al. 2024). This represents the first successful nanopore-based discrimination of structurally complex glycans, paving the way for precise glycan structural analysis, as detailed elsewhere (Zhao et al. 2024; Yao et al. 2024; Yin et al. 2024).

While nanopore sequencing of DNA and proteins has revolutionized biomedical research, their application for glycans remains comparatively underexplored (Lu et al. 2024; Zhao et al. 2024; Yao et al. 2024). Here, particularly the class of glycosaminoglycans is suitable due to their linear nature, and we summarize the state-of-the-art of nanopore usage for GAGs below.

4 Application of Nanopores for Glycosaminoglycans

Glycosaminoglycans (GAGs) are a class of linear and highly negatively charged carbohydrates that are built of repeating disaccharide units. These disaccharides are composed of an amino sugar [D-glucosamine (GlcN) or D-galactosamine (GalN)] and uronic acid [D-glucuronic acid (GlcA) or L-iduronic acid (IdoA)] or galactose (Gal) (Zhang et al. 2010). Based on the type and combinations of these, GAGs can be further categorized into four sub-groups: hyaluronic acid (HA), keratan sulfate (KS), chondroitin sulfate or dermatan sulfate (CS/DS), as well as heparin (HP) or heparan sulfate (Fig. 5).

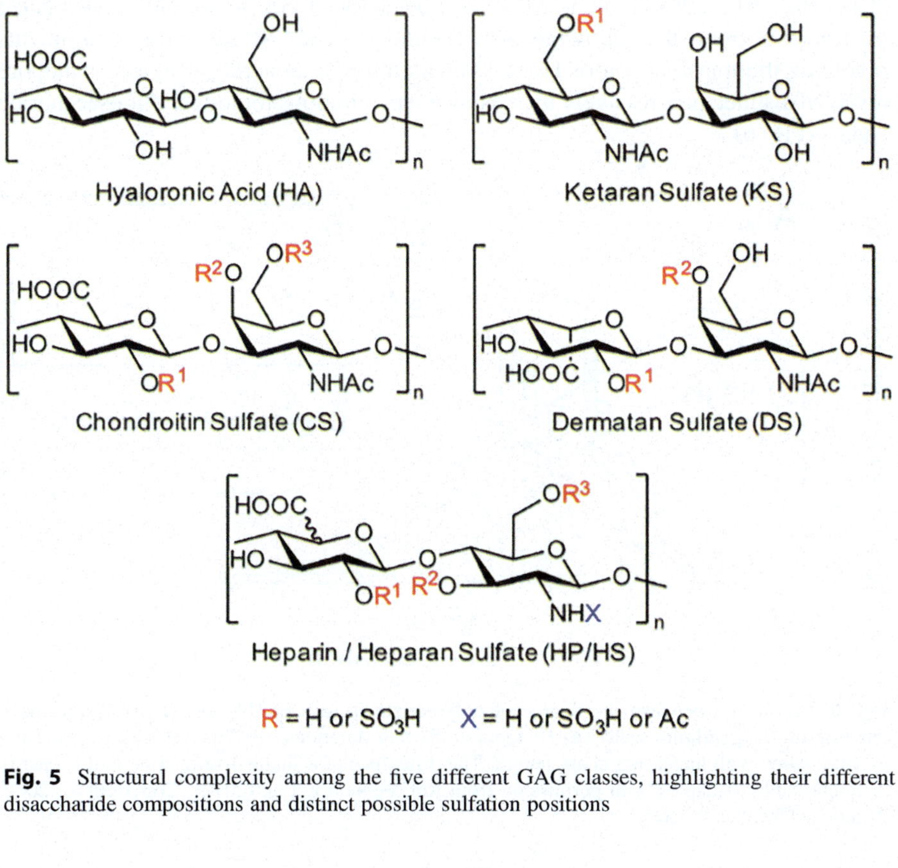

Fig. 5 Structural complexity among the five different GAG classes, highlighting their different disaccharide compositions and distinct possible sulfation positions

The high structural diversity of GAGs enables interactions with biomolecules to modulate various processes, such as cell adhesion, proliferation and migration, extracellular matrix assembly, tissue repair, coagulation, and immune responses (Aquino et al. 2010). These functions are largely carried out through their interaction with proteins, whereby the binding of proteins and GAGs is specific and requires a defined GAG sequence (Yang and Chi 2017). Since the biosynthesis of GAGs is non-template-driven and highly dependent on the environment and enzymes, the complete analysis of the structure and thus the understanding of the mechanisms of GAG functions remains challenging (Perez et al. 2023). Nuclear magnetic resonance (NMR) and mass spectrometry (MS) methods are the state-of-the-art techniques for GAG analysis; however, they both come with their own challenges. For example, NMR measurements require high sample amounts, whereas MS-based methods are not able to distinguish all possible isomers of GAG chains longer than few disaccharide units. Nanopore technology has shown first success in the analysis of GAGs and has potential to decipher GAG sequences fully.

4.1 Biological Nanopores

Following the successful demonstration of neutral oligosaccharide translocation in 2011 (Bacri et al. 2011), the translocation behavior of GAGs was first investigated by Daniel's group the following year (Fennouri et al. 2012). Initial experiments involving the translocation of HA oligomers through an aerolysin (AeL) nanopore revealed distinct, progressively increasing event durations for longer oligosaccharide chains (Fig. 6).

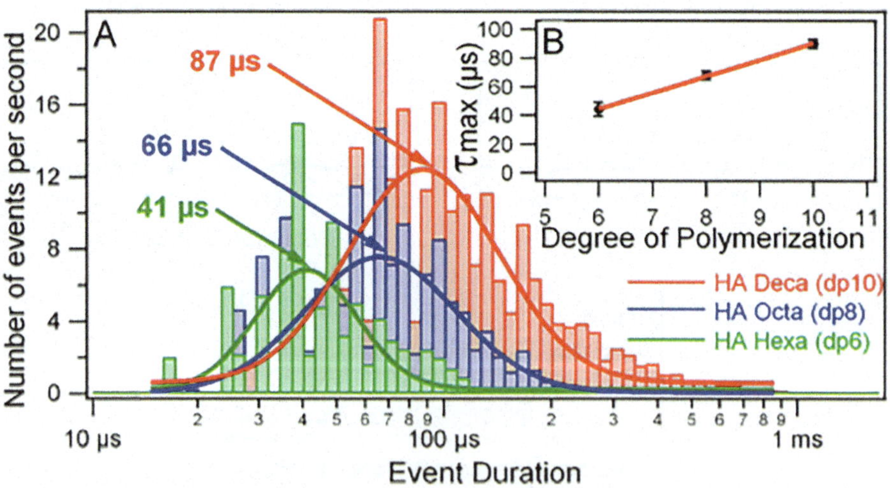

Fig. 6 Correlation between the degree of polymerization (dp) of HA and the measured event duration on a logarithmic scale. (**a**) Histogram of event durations. (**b**) Trend showing increasing event duration with increasing chain length. The difference of a single disaccharide unit is clearly distinguishable. (Adapted with permission from Ref. Fennouri et al. (2012). Copyright © 2012, American Chemical Society)

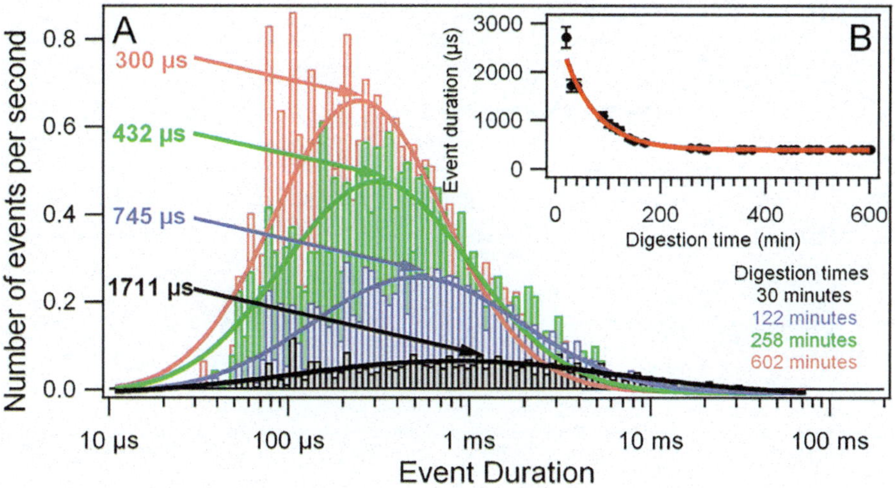

Fig. 7 Real-time nanopore recording of in situ HA digestion. (**a**) Histogram of event durations at different time points during digestion. (**b**) Trend of average event duration over time, showing a decrease in event length as digestion progresses, eventually reaching a plateau. (Reproduced with permission from Ref. Fennouri et al. (2012). Copyright © 2012, American Chemical Society)

Encouraged by these findings, Fennouri and co-workers proposed utilizing the translocation properties of GAGs for real-time monitoring of enzymatic HA cleavage. The digestion was carried out directly within one compartment of the nanopore, and the results indicated a progressive decrease in event duration over time (Fig. 7).

In 2018, Fennouri extended their work on HA sensing by comparing translocation behavior of HA through an AeL and an α-HL nanopore (Fennouri et al. 2018). When a decasaccharide was introduced into one compartment of the α-HL pore, no translocation events were observed. To further investigate this difference, an enzymatic in situ digestion of HA was performed using an α-HL pore, following the approach previously employed with the AeL pore (Fennouri et al. 2012). Similar results were obtained, though the event durations were an order of magnitude shorter, indicating a faster translocation of HA molecules through the α-HL pore. This suggests that smaller chain lengths translocate through the pore without being detected, highlighting the importance of pore choice.

Up to this point, only HA had been studied. Arisa Bayat, also part of Daniel's group, expanded the investigation to other GAG classes, focusing on differences in key structural attributes (Bayat et al. 2022). Using a wild-type AeL nanopore, the effects of chain length, degree of sulfation, glycosidic linkage, epimerization, and composition were analyzed. As observed in previous studies (Fennouri et al. 2012; Fennouri et al. 2018), the event duration increases almost linearly when HP and CS sulfate were translocated through the pore. The influence of sulfation degree was examined by comparing heparin with its non-sulfated counterpart, heparosan (HN). Differences in event populations were evident in the scatter plots. Heparin exhibited

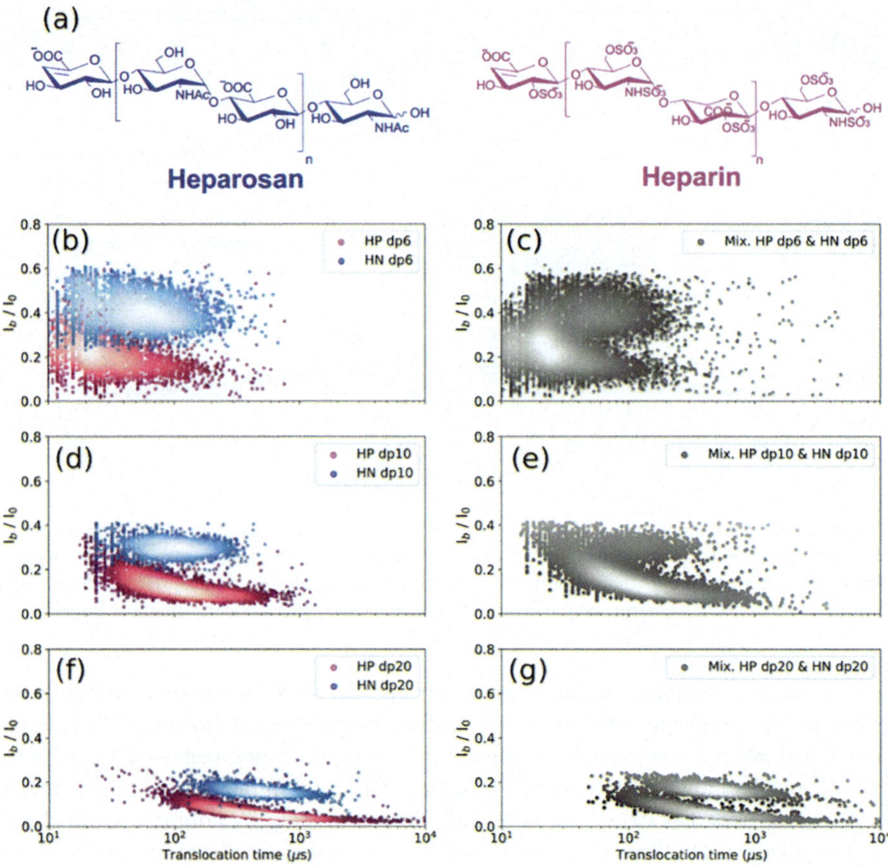

Fig. 8 Influence of sulfation degree on event length and blockage. (**a**) Chemical structures of HN and HP. To assess the impact of sulfation, the non-sulfated HN is compared to the fully sulfated HP. (**b**, **d**, **f**) Across varying chain lengths, HP exhibits a higher blockage than HN, while the event length is reduced. (**c**, **e**, **g**) This difference is sufficiently pronounced to be detected when measuring a mixture of HN and HP. (Reproduced with permission from Ref. Bayat et al. (2022). Copyright © 2022, The Author(s))

a higher event blockage, likely due to special demand from sulfate groups, along with a slight decrease in event duration, possibly attributed to its higher ionic charge (Fig. 8).

Differences in *glycosidic linkage* can be seen when HN is compared to HA. Although the event blockage is similar, the event duration shows a clear difference. The ability to differentiate between *epimers* was investigated by measuring CS and DS, which differ at the C5 carbon of the hexuronic acid. Slightly higher blockage and duration was seen for DS, becoming more prone with higher chain length. More significant differences were observed when changing the *building blocks* of the GAG chain. Here, the comparison between HA and CS yielded a

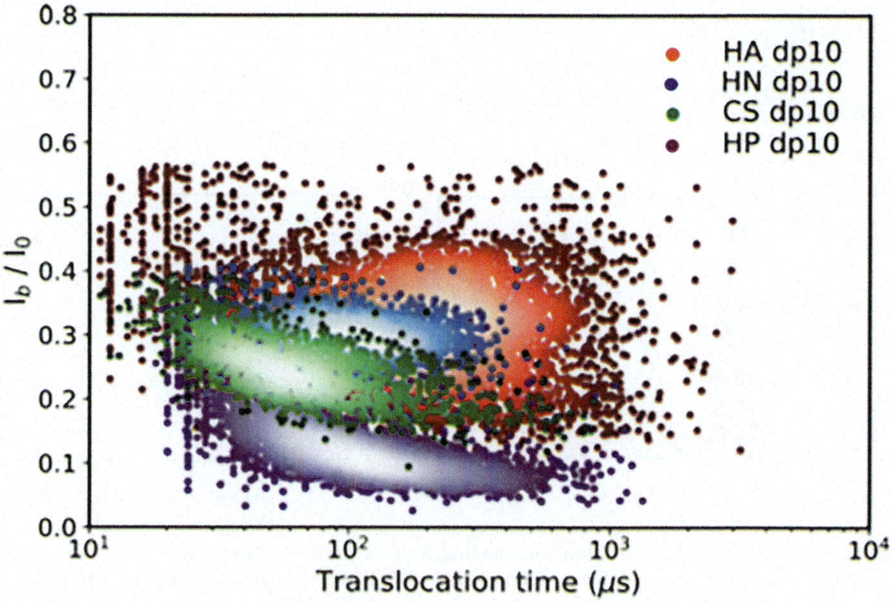

Fig. 9 Overlayed scatter plots of blockage ratio vs. translocation time (event duration) for hyaluronic acid, heparosan, chondroitin sulfate, and heparin dp10. (Adapted with permission from Ref. Bayat et al. (2022). Copyright © 2022, The Author(s))

higher blockage for CS and longer event duration for HA. This could be explained by the larger spatial volume occupied by the sulfate groups of the CS and the lower ionic charge of the HA. When *GAG classes* are compared with each other, distinct differences in event distribution are observed. This might be due to molecular size and charge. Heparin exhibits the highest blockage, followed by chondroitin sulfate, likely due to the spatial requirements of its sulfate groups, whereas HA shows the longest event duration, which may be attributed to its lower charge (Fig. 9).

4.2 Solid-State Nanopores

The first application of solid-state nanopores for GAG analysis was demonstrated in 2018 (Rivas et al. 2018). In this study, Felipe Rivas and colleagues investigated the detection and molecular weight (MW) discrimination of HA. After initial validation using polydisperse HA, size fractions ranging from 54 kDa to 2.4 MDa were analyzed. A linear correlation was observed between the mean event charge deficit (ECD), representing the integrated area of each translocation event (event blockage over event time), and the MW of HA (Fig. 10). To assess the method's applicability, synovial fluid samples from two horses (H1 and H2) were analyzed, as HA degradation in synovial fluid is associated with joint diseases such as osteoarthritis (OA). In H1, a reduction in MW of HA was observed five days after the surgical induction

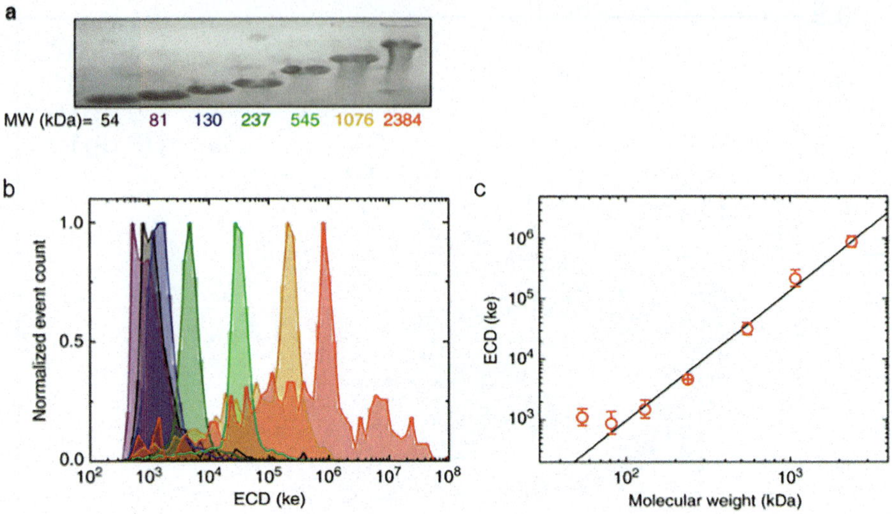

Fig. 10 Analysis of HA size fractions ranging from 54 to 2384 kDa. (**a**) Gel analysis of HA fractions. (**b**) Event charge deficit (ECD) plotted against the normalized event count from ssNP measurements. (**c**) Correlation between the ECD and the MW of HA. (Reproduced with permission from Ref. Rivas et al. (2018). Copyright © 2018, The Author(s))

of OA, consistent between gel electrophoresis and ssNP ECD measurements. In contrast, an increase in HA MW was detected in H2 after 12 days, possibly indicating upregulated HA synthesis during the acute post-traumatic phase.

Later in 2018, Karawdeniya and colleagues employed ssNPs to detect oversulfated chondroitin sulfate (OSCS) (Karawdeniya et al. 2018), motivated by the HP contamination incident in 2008 (Guerrini et al. 2008; Kishimoto et al. 2008). The primary difference between OSCS and HP was found in the event distribution, with OSCS events exhibiting a tail toward higher blockage values (lower f_b), while HP events displayed a tail toward longer event durations (Fig. 11). Based on these characteristics, an automated thresholding procedure was developed, enabling the detection of OSCS in HP samples, demonstrating the potential of ssNPs as a screening tool for clinical HP quality control.

Im and co-workers introduced a machine learning approach, using a support vector machine (SVM) for the identification and analysis of CS and HP samples (Im et al. 2019). Initial measurements of HP and CS oligosaccharides dp 20 revealed that primary event characteristics, namely their blockage depth and event duration, were insufficient for accurate discrimination. To overcome this limitation, multiple independent parameters were analyzed using SVM, achieving an average classification accuracy of 93.7%. The approach was further tested on HP-CS mixtures, where CS content as low as ~0.8% (w/w) was reliably detected. Polydisperse GAGs, such as unfractionated heparin (UFH) and enoxaparin, were classified with an accuracy of ca. 83%. Additionally, a calibration curve was established for quantitative analysis,

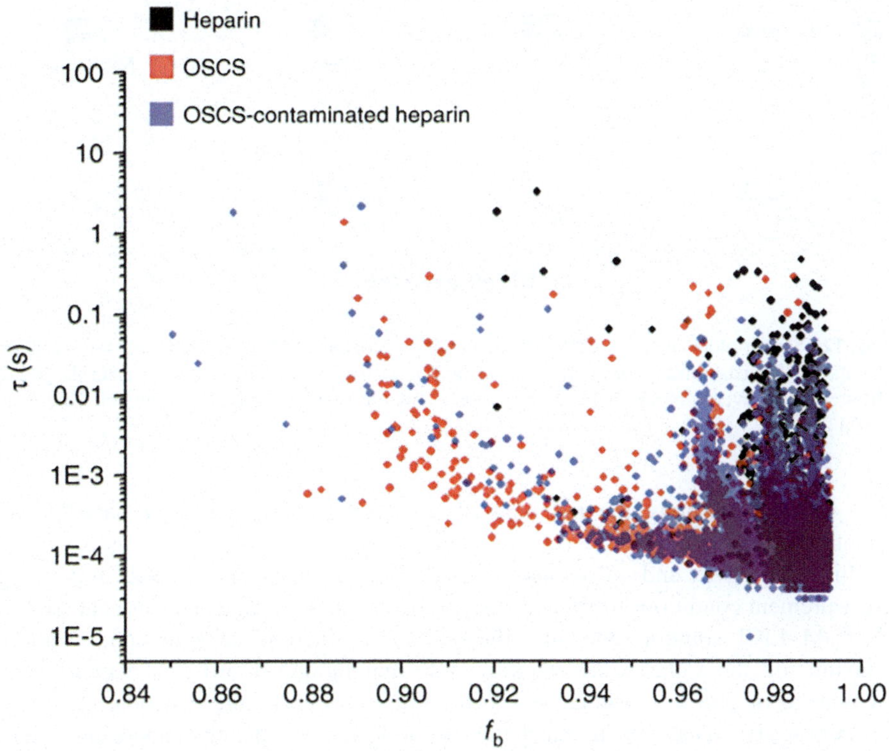

Fig. 11 Event distribution (top) of ssNP measurements of OSCS (red), HP (black), and HP with OSCS-contamination (purple). f_b represents the ratio of blockage current to baseline current. (Adapted with permission from Ref. Karawdeniya et al. (2018). Copyright © 2018, The Author(s))

with measurement deviations remaining below 10%, demonstrating the potential of ssNP-based GAG quantification using reference standards.

In 2021, Ke Xia et al. addressed a critical limitation in GAG analysis: the lack of pure, structurally defined standards. To overcome this, uniform heparan sulfate (HS) chains with distinct sulfation patterns and a defined length of ≈ 40 disaccharide units were synthesized (Fig. 12) (Xia et al. 2021). Scatter plots of translocation events were analyzed using Google's deep neural network (Godec et al. 2019), where event features were converted into vectors and classified using principal component analysis. With this method, all four synthetic GAG standards were successfully differentiated. Current traces from 20-min recordings were segmented into shorter reads, each containing an equal number of events. These data sets were subjected to a 100-repeats stratified random sampling approach, with 75% used for training and 25% for external validation. SVM was identified as the most accurate classification method, achieving an assignment accuracy of 91.2%. Control experiments with randomly labeled datasets yielded significantly lower accuracies

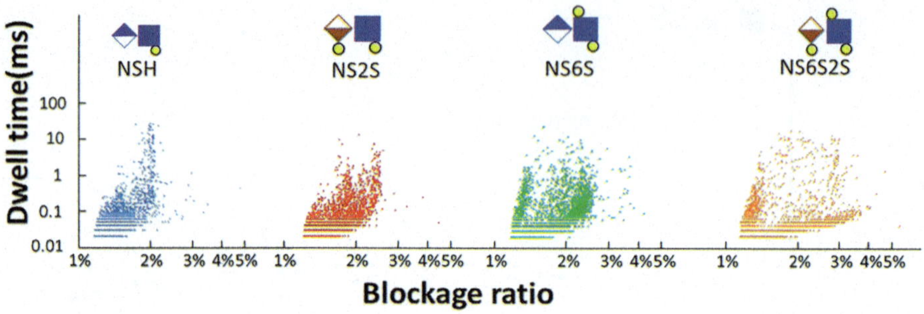

Fig. 12 Scatter plots visualizing the event duration vs blockage ratio of <2000 ssNP translocation events. Each measurement was performed using one of the four synthetic GAG standards. Slight differences in event distribution are observable. (Adapted with permission from Ref. Xia et al. (2021))

(25–45%), confirming the potential of machine learning for ssNP-based GAG identification.

In 2022, Rivas and co-workers refined their HA analysis method, optimizing measurement conditions to enhance resolution and signal quality (Rivas et al. 2022). A 6 M LiCl solution was identified as an optimal electrolyte concentration, maximizing the signal-to-noise ratio while maintaining a sufficient capture rate. A voltage of 200 mV was chosen to minimize device failure and pore clogging, and a pore size of ca. 7 nm to match the size of the analyte, thereby enhancing signal-to-noise ratio while reducing the chance of pore clogging (Fig. 13). The effect of salt asymmetry was further investigated, showing that a higher concentration gradient increased the event rate but at the expense of reduced resolution.

Erxleben and co-workers applied the HA MW discrimination method (Rivas et al. 2018) and developed a selective method for heavy chain-HA (HC-HA) complex extraction for MW nanopore sensing (Fig. 14) (Erxleben et al. 2024). Magnetic beads coated with antibodies against inter-alpha-inhibitor (IαI) were used to bind HC-HA complexes from biological samples. After washing, a broad protease mix was applied to digest IαI heavy chains, releasing HA, which allowed the HA content to be extracted using a simple organic solvent workup. When this method was applied to more complex biofluids like human synovial fluid spiked with the quasi-monodisperse HC-HA, the expected signal was absent, likely due to matrix interference from endogenous CS. To address this, an additional extraction step was introduced. Following pronase digestion, HA was captured using a second set of paramagnetic beads coated with VG1, a HA-specific binding protein (Clark et al. 2011). The final purified HA fraction was eluted using high-molarity LiCl and analyzed via ssNP measurements (Fig. 14). The improved method was applied to patient samples with joint disease and a reduction in HA MW was observed, agreeing with previous findings (Watkins et al. 2021).

In addition to GAG sequencing, ssNPs are used to better understand protein-GAG interactions. Thyashan and his colleagues investigated the interaction of HP and

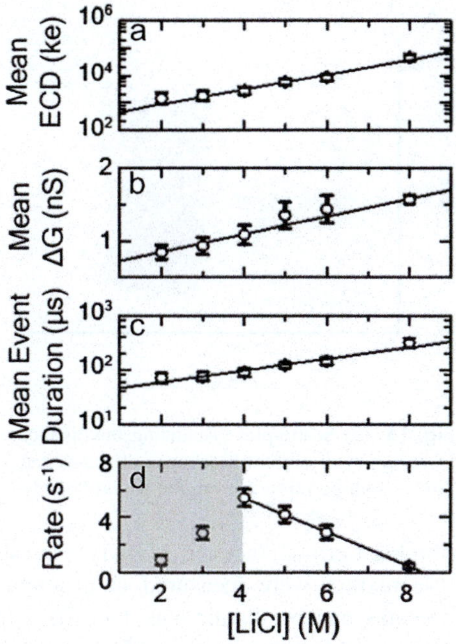

Fig. 13 Influence of LiCl concentration on HA translocation (237 kDa) at 200 mV. Nanopores were 7–9 nm in diameter. (**a**) Mean ECD increases exponentially. (**b**) Linear increase in mean blockage. (**c**) Exponential increase in event duration. The higher blockage is attributed to the increased ion count at higher LiCl concentrations, while the prolonged event duration may result from reduced net charge and lower electrophoretic force. (**d**) Event rate decreases for LiCl concentrations below 4 M due to reduced signal quality. (Adapted with permission from Ref. Rivas et al. (2022). Copyright © 2022, The Author(s))

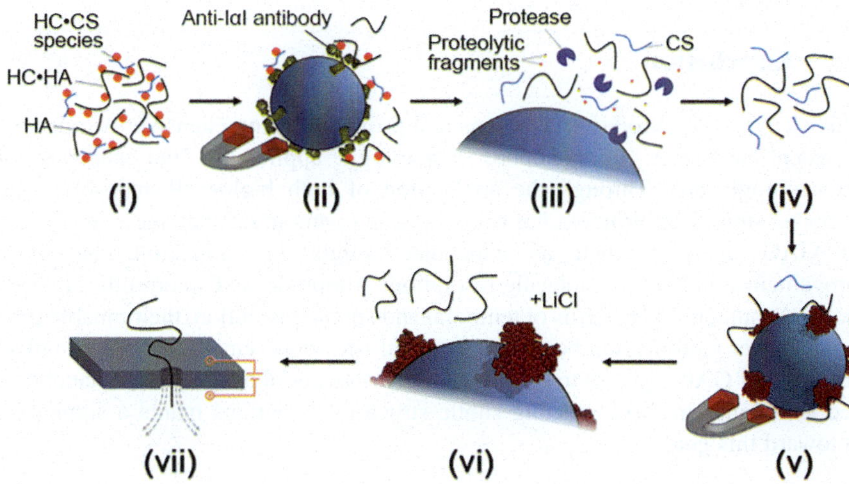

Fig. 14 Schematic illustration of the selective isolation protocol extract HA-HC complexes out of complex biological samples. The mixture with the covalently bound HA-HC complexes (**i**) is incubated with paramagnetic beads, which are coated with an antibody raised against IαI. Unbound material is washed away (**ii**). The antibody and HCs are digested (**iii**) and the free HA (**iv**) is bound to another set of paramagnetic beads coated with VG1 (that specifically binds to HA, (**v**). Using high-molarity LiCl, HA is eluted from the beads (**vi**) and pure HA for ssNP measurements is obtained (**vii**). (Adapted with permission from Ref. Erxleben et al. (2024). Copyright © 2024, American Chemical Society)

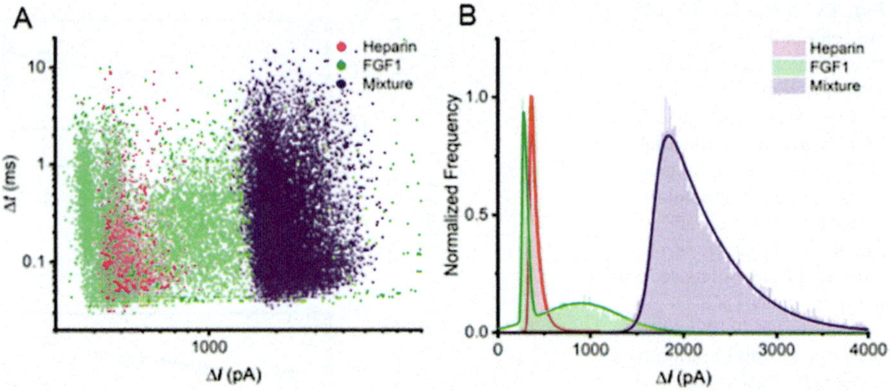

Fig. 15 (**a**) Scatterplot visualizing the distinct populations of HP, FGF-1 and their complex. (**b**) Distribution of current blockages of translocated events. (Reproduced from Ref. Thyashan et al. (2024) with permission from the Royal Society of Chemistry)

fibroblast growth factor 1 (FGF-1) (Thyashan et al. 2024). Heparin, FGF-1, and their 1:1 mixture were measured at different voltages. Clear blockage and duration domains can be distinguished in the scatter plot, concluding the formation of a protein-GAG complex (Fig. 15).

5 Conclusion

In this chapter, we introduce nanopores and provide a comprehensive overview of the recent advances in nanopore technology applied to the analysis of glycosaminoglycans. Through the application of both biological and solid-state nanopores, significant progress has been made in overcoming challenges associated with GAG analysis, such as molecular weight discrimination, sensitivity improvements, and single molecule resolution. Although ssNPs currently lag their biological counterparts in terms of temporal and spatial resolution, their tunable pore size and easier modification bear high potential for improvements. While complete sequencing of GAGs has yet to be achieved, the successful application of nanopore-based techniques in distinguishing subtle structural variations marks a significant step toward this goal.

Acknowledgments W. P. and M. S. thank the Deutsche Forschungsgemeinschaft (DFG, German Research Foundation) -Project IDs 372486779—SFB1340, 431232613—SFB1449, and 434130070—GRK2662. N. G. acknowledges support from the ERC Consolidator Grant "Glycospec" ERC-2019-CoG-863934-GlycoSpec.

Conflict of Interests There are no conflicts to declare.

References

Akahori R, Yanagi I, Goto Y, Harada K, Yokoi T, Takeda K (2017) Discrimination of three types of homopolymers in single-stranded DNA with solid-state nanopores through external control of the DNA motion. Sci Rep 7(1):9073. https://doi.org/10.1038/s41598-017-08290-6

Akeson M, Branton D, Kasianowicz JJ, Brandin E, Deamer DW (1999) Microsecond time-scale discrimination among polycytidylic acid, polyadenylic acid, and polyuridylic acid as homopolymers or as segments within single RNA molecules. Biophys J 77(6):3227–3233. https://doi.org/10.1016/S0006-3495(99)77153-5

Akhtarian S, Miri S, Doostmohammadi A, Brar SK, Rezai P (2021) Nanopore sensors for viral particle quantification: current progress and future prospects. Bioengineered 12(2):9189–9215. https://doi.org/10.1080/21655979.2021.1995991

Aquino RS, Lee ES, Park PW (2010) Diverse functions of glycosaminoglycans in infectious diseases. Progr Mol Biol Transl Sci 93:373–394. https://doi.org/10.1016/S1877-1173(10)93016-0

Ashkenasy N, Sánchez-Quesada J, Bayley H, Ghadiri MR (2005) Recognizing a single base in an individual DNA strand: a step toward DNA sequencing in nanopores. Angew Chem Int Ed 44(9):1401–1404. https://doi.org/10.1002/anie.200462114

Bacri L, Oukhaled A, Hémon E, Bassafoula FB, Auvray L, Daniel R (2011) Discrimination of neutral oligosaccharides through a nanopore. Biochem Biophys Res Commun 412(4):561–564. https://doi.org/10.1016/j.bbrc.2011.07.121

Bayat P, Rambaud C, Priem B, Bourderioux M, Bilong M, Poyer S, Pastoriza-Gallego M, Oukhaled A, Mathé J, Daniel R (2022) Comprehensive structural assignment of glycosaminoglycan oligo- and polysaccharides by protein nanopore. Nat Commun 13(1):5113. https://doi.org/10.1038/s41467-022-32800-4

Branton D, Deamer DW, Marziali A, Bayley H, Benner SA, Butler T, Di Ventra M, Garaj S, Hibbs A, Huang X, Jovanovich SB, Krstic PS, Lindsay S, Ling XS, Mastrangelo CH, Meller A, Oliver JS, Pershin YV, Ramsey JM, Riehn R, Soni GV, Tabard-Cossa V, Wanunu M, Wiggin M, Schloss JA (2008) The potential and challenges of nanopore sequencing. Nat Biotechnol 26(10):1146–1153. https://doi.org/10.1038/nbt.1495

Chi H, Shi L, Gan S, Fan G, Dong Y (2025) Innovative applications of nanopore technology in tumor screening: an exosome-centric approach. Biosensors (Basel) 15(4):199. https://doi.org/10.3390/bios15040199

Clark SJ, Keenan TDL, Fielder HL, Collinson LJ, Holley RJ, Merry CLR, van Kuppevelt TH, Day AJ, Bishop PN (2011) Mapping the differential distribution of glycosaminoglycans in the adult human retina, choroid, and sclera. Invest Ophthalmol Vis Sci 52(9):6511. https://doi.org/10.1167/iovs.11-7909

De Roeck A, De Coster W, Bossaerts L, Cacace R, De Pooter T, Van Dongen J, D'Hert S, De Rijk P, Strazisar M, Van Broeckhoven C, Sleegers K (2019) NanoSatellite: accurate characterization of expanded tandem repeat length and sequence through whole genome long-read sequencing on PromethION. Genome Biol 20(1):239. https://doi.org/10.1186/s13059-019-1856-3

Deamer D, Akeson M, Branton D (2016) Three decades of nanopore sequencing. Nat Biotechnol 34(5):518–524. https://doi.org/10.1038/nbt.3423

Don M (2003) The coulter principle: Foundation of an industry. JALA 8(6):72–81. https://doi.org/10.1016/s1535-5535(03)00023-6

Dorey A, Howorka S (2023) Unfolding the path to nanopore protein sequencing. Nat Nanotechnol 18(11):1259–1260. https://doi.org/10.1038/s41565-023-01480-6

Erxleben DA, Dodd RJ, Day AJ, Green DE, DeAngelis PL, Poddar S, Enghild JJ, Huebner JL, Kraus VB, Watkins AR, Reesink HL, Rahbar E, Hall AR (2024) Targeted analysis of the size distribution of heavy chain-modified hyaluronan with solid-state nanopores. Anal Chem 96(4):1606–1613. https://doi.org/10.1021/acs.analchem.3c04387

Euskirchen P, Bielle F, Labreche K, Kloosterman WP, Rosenberg S, Daniau M, Schmitt C, Masliah-Planchon J, Bourdeaut F, Dehais C, Marie Y, Delattre J-Y, Idbaih A (2017) Same-day genomic and epigenomic diagnosis of brain tumors using real-time nanopore sequencing. Acta Neuropathol 134(5):691–703. https://doi.org/10.1007/s00401-017-1743-5

Fennouri A, Przybylski C, Pastoriza-Gallego M, Bacri L, Auvray L, Daniel R (2012) Single molecule detection of glycosaminoglycan hyaluronic acid oligosaccharides and depolymerization enzyme activity using a protein nanopore. ACS Nano 6(11):9672–9678. https://doi.org/10.1021/nn3031047

Fennouri A, Ramiandrisoa J, Bacri L, Mathé J, Daniel R (2018) Comparative biosensing of glycosaminoglycan hyaluronic acid oligo- and polysaccharides using aerolysin and α-hemolysin nanopores. Eur Phys J E 41(10):127. https://doi.org/10.1140/epje/i2018-11733-5

Fried JP, Swett JL, Nadappuram BP, Mol JA, Edel JB, Ivanov AP, Yates JR (2021) *In situ* solid-state nanopore fabrication. Chem Soc Rev 50(8):4974–4992. https://doi.org/10.1039/D0CS00924E

Godec P, Pančur M, Ilenič N, Čopar A, Stražar M, Erjavec A, Pretnar A, Demšar J, Starič A, Toplak M, Žagar L, Hartman J, Wang H, Bellazzi R, Petrovič U, Garagna S, Zuccotti M, Park D, Shaulsky G, Zupan B (2019) Democratized image analytics by visual programming through integration of deep models and small-scale machine learning. Nat Commun 10(1):1–7. https://doi.org/10.1038/s41467-019-12397-x

Gorrie CL, Mirceta M, Wick RR, Judd LM, Wyres KL, Thomson NR, Strugnell RA, Pratt NF, Garlick JS, Watson KM, Hunter PC, McGloughlin SA, Spelman DW, Jenney AWJ, Holt KE (2018) Antimicrobial-resistant Klebsiella pneumoniae carriage and infection in specialized geriatric care wards linked to acquisition in the referring hospital. Clin Infect Dis 67(2): 161–170. https://doi.org/10.1093/cid/ciy027

Guerrini M, Beccati D, Shriver Z, Naggi A, Viswanathan K, Bisio A, Capila I, Lansing JC, Guglieri S, Fraser B, Al-Hakim A, Gunay NS, Zhang Z, Robinson L, Buhse L, Nasr M, Woodcock J, Langer R, Venkataraman G, Linhardt RJ, Casu B, Torri G, Sasisekharan R (2008) Oversulfated chondroitin sulfate is a contaminant in heparin associated with adverse clinical events. Nat Biotechnol 26(6):669–675. https://doi.org/10.1038/nbt1407

Gurnev PA, Nestorovich EM (2014) Channel-forming bacterial toxins in biosensing and macromolecule delivery. Toxins (Basel) 6(8):2483–2540. https://doi.org/10.3390/toxins6082483

He Y, Tsutsui M, Fan C, Taniguchi M, Kawai T (2011) Gate manipulation of DNA capture into nanopores. ACS Nano 5(10):8391–8397. https://doi.org/10.1021/nn203186c

He Y, Tsutsui M, Zhou Y, Miao X-S (2021) Solid-state nanopore systems: from materials to applications. NPG Asia Mater 13(1):48. https://doi.org/10.1038/s41427-021-00313-z

He S, Liu Y, Fang S, Li Y, Weng T, Tian R, Yin Y, Zhou D, Yin B, Wang Y, Liang L, Xie W, Wang D (2024) Solid-state nanopore DNA sequencing: advances, challenges and prospects. Coord Chem Rev 510:215816. https://doi.org/10.1016/j.ccr.2024.215816

Heerema SJ, Vicarelli L, Pud S, Schouten RN, Zandbergen HW, Dekker C (2018) Probing DNA translocations with inplane current signals in a graphene nanoribbon with a nanopore. ACS Nano 12(3):2623–2633. https://doi.org/10.1021/acsnano.7b08635

Im J, Lindsay S, Wang X, Zhang P (2019) Single molecule identification and quantification of glycosaminoglycans using solid-state nanopores. ACS Nano 13(6):6308–6318. https://doi.org/10.1021/acsnano.9b00618

Karawdeniya BI, Bandara YMNDY, Nichols JW, Chevalier RB, Dwyer JR (2018) Surveying silicon nitride nanopores for glycomics and heparin quality assurance. Nat Commun 9(1): 3278. https://doi.org/10.1038/s41467-018-05751-y

Kasianowicz JJ, Brandin E, Branton D, Deamer DW (1996) Characterization of individual polynucleotide molecules using a membrane channel. PNAS 93(24):13770–13773. https://doi.org/10.1073/pnas.93.24.13770

Kennedy E, Dong Z, Tennant C, Timp G (2016) Reading the primary structure of a protein with 0.07 Nm3 resolution using a subnanometre-diameter pore. Nat Nanotechnol 11(11):968–976. https://doi.org/10.1038/nnano.2016.120

Kim D, Lee JY, Yang JS, Kim JW, Kim VN, Chang H (2020) The architecture of SARS-CoV-2 transcriptome. Cell 181(4):914–921 e10. https://doi.org/10.1016/j.cell.2020.04.011

Kishimoto TK, Viswanathan K, Ganguly T, Elankumaran S, Smith S, Pelzer K, Lansing JC, Sriranganathan N, Zhao G, Galcheva-Gargova Z, Al-Hakim A, Bailey GS, Fraser B, Roy S, Rogers-Cotrone T, Buhse L, Whary M, Fox J, Nasr M, Dal Pan GJ, Shriver Z, Langer RS, Venkataraman G, Austen KF, Woodcock J, Sasisekharan R (2008) Contaminated heparin associated with adverse clinical events and activation of the contact system. N Engl J Med 358(23):2457–2467. https://doi.org/10.1056/NEJMoa0803200

Kono N, Arakawa K (2019) Nanopore sequencing: review of potential applications in functional genomics. Develop Growth Differ 61(5):316–326. https://doi.org/10.1111/dgd.12608

Kwok H, Briggs K, Tabard-Cossa V (2014) Nanopore fabrication by controlled dielectric breakdown. PLoS One 9(3):e92880. https://doi.org/10.1371/journal.pone.0092880

Leija-Salazar M, Sedlazeck FJ, Toffoli M, Mullin S, Mokretar K, Athanasopoulou M, Donald A, Sharma R, Hughes D, Schapira AHV, Proukakis C (2019) Evaluation of the detection of *GBA* missense mutations and other variants using the Oxford nanopore MinION. Mol Genet Genomic Med 7(3). https://doi.org/10.1002/mgg3.564

Li J, Stein D, McMullan C, Branton D, Aziz MJ, Golovchenko JA (2001) Ion-beam sculpting at nanometre length scales. Nature 412(6843):166–169. https://doi.org/10.1038/35084037

Li Q, Ying Y-L, Liu S-C, Lin Y, Long Y-T (2019) Detection of single proteins with a general nanopore sensor. ACS Sens 4(5):1185–1189. https://doi.org/10.1021/acssensors.9b00228

Li X, Lee KH, Shorkey S, Chen J, Chen M (2020) Different anomeric sugar bound states of maltose binding protein resolved by a cytolysin a nanopore tweezer. ACS Nano 14(2):1727–1737. https://doi.org/10.1021/acsnano.9b07385

Li M, Xiong Y, Cao Y, Zhang C, Li Y, Ning H, Liu F, Zhou H, Li X, Ye X, Pang Y, Zhang J, Liang X, Qing G (2023) Identification of tagged Glycans with a protein Nanopore. Nat Commun 14(1):1737. https://doi.org/10.1038/s41467-023-37348-5

Lieberman KR, Cherf GM, Doody MJ, Olasagasti F, Kolodji Y, Akeson M (2010) Processive replication of single DNA molecules in a nanopore catalyzed by Phi29 DNA polymerase. J Am Chem Soc 132(50):17961–17972. https://doi.org/10.1021/ja1087612

Lin K, Chen C, Wang C, Lian P, Wang Y, Xue S, Sha J, Chen Y (2022) Fabrication of solid-state nanopores. Nanotechnology 33(27):272003. https://doi.org/10.1088/1361-6528/ac622b

Liu Y, Zhang S, Wang Y, Wang L, Cao Z, Sun W, Fan P, Zhang P, Chen H-Y, Huang S (2022) Nanopore identification of alditol epimers and their application in rapid analysis of alditol-containing drinks and healthcare products. J Am Chem Soc 144(30):13717–13728. https://doi.org/10.1021/jacs.2c04595

Loman NJ, Quick J, Simpson JT (2015) A complete bacterial genome assembled de novo using only nanopore sequencing data. Nat Methods 12(8):733–735. https://doi.org/10.1038/nmeth.3444

Lu W, Zhao X, Li M, Li Y, Zhang C, Xiong Y, Li J, Zhou H, Ye X, Li X, Wang J, Liang X, Qing G (2024) Precise structural analysis of neutral glycans using aerolysin mutant T240R nanopore. ACS Nano 18(19):12412–12426. https://doi.org/10.1021/acsnano.4c01571

Lu C, Bonini A, Viel JH, Maglia G (2025) Toward single-molecule protein sequencing using nanopores. Nat Biotechnol 43(3):312–322. https://doi.org/10.1038/s41587-025-02587-y

Manrao EA, Derrington IM, Laszlo AH, Langford KW, Hopper MK, Gillgren N, Pavlenok M, Niederweis M, Gundlach JH (2012) Reading DNA at single-nucleotide resolution with a mutant MspA nanopore and Phi29 DNA polymerase. Nat Biotechnol 30(4):349–353. https://doi.org/10.1038/nbt.2171

Martin-Baniandres P, Lan W-H, Board S, Romero-Ruiz M, Garcia-Manyes S, Qing Y, Bayley H (2023) Enzyme-less nanopore detection of post-translational modifications within long polypeptides. Nat Nanotechnol 18(11):1335–1340. https://doi.org/10.1038/s41565-023-01462-8

Mayer SF, Cao C, Dal Peraro M (2022) Biological nanopores for single-molecule sensing. iScience 25(4):104145. https://doi.org/10.1016/j.isci.2022.104145

Meller A, Nivon L, Brandin E, Golovchenko J, Branton D (2000) Rapid nanopore discrimination between single polynucleotide molecules. Proc Natl Acad Sci 97(3):1079–1084. https://doi.org/10.1073/PNAS.97.3.1079

Moon J, Kim N, Kim T-J, Jun J-S, Lee HS, Shin H-R, Lee S-T, Jung K-H, Park K-I, Jung K-Y, Kim M, Lee SK, Chu K (2019) Rapid diagnosis of bacterial meningitis by nanopore 16S amplicon sequencing: a pilot study. Int J Med Microbiol 309(6):151338. https://doi.org/10.1016/j.ijmm.2019.151338

Nouri R, Jiang Y, Tang Z, Lian XL, Guan W (2021) Detection of SARS-CoV-2 with solid-state CRISPR-Cas12a-assisted nanopores. Nano Lett 21(19):8393–8400. https://doi.org/10.1021/acs.nanolett.1c02974

Orsini P, Minervini CF, Cumbo C, Anelli L, Zagaria A, Minervini A, Coccaro N, Tota G, Casieri P, Impera L, Parciante E, Brunetti C, Giordano A, Specchia G, Albano F (2018) Design and MinION testing of a nanopore targeted gene sequencing panel for chronic lymphocytic leukemia. Sci Rep 8(1):11798. https://doi.org/10.1038/s41598-018-30330-y

Ouldali H, Sarthak K, Ensslen T, Piguet F, Manivet P, Pelta J, Behrends JC, Aksimentiev A, Oukhaled A (2020) Electrical recognition of the twenty proteinogenic amino acids using an aerolysin nanopore. Nat Biotechnol 38(2):176–181. https://doi.org/10.1038/s41587-019-0345-2

Perez S, Makshakova O, Angulo J, Bedini E, Bisio A, de Paz JL, Fadda E, Guerrini M, Hricovini M, Hricovini M, Lisacek F, Nieto PM, Pagel K, Paiardi G, Richter R, Samsonov SA, Vivès RR, Nikitovic D, Ricard Blum S (2023) Glycosaminoglycans: what remains to be deciphered? JACS Au 3(3):628–656. https://doi.org/10.1021/jacsau.2c00569

Ramsay WJ, Bayley H (2018) Single-molecule determination of the isomers of <scp>d</Scp> -glucose and <scp>d</Scp> -fructose that bind to boronic acids. Angew Chem Int Ed 57(11): 2841–2845. https://doi.org/10.1002/anie.201712740

Ritmejeris J, Chen X, Dekker C (2024) Single-molecule protein sequencing with nanopores. Nat Rev Bioeng. https://doi.org/10.1038/s44222-024-00260-8

Rivas F, Zahid OK, Reesink HL, Peal BT, Nixon AJ, DeAngelis PL, Skardal A, Rahbar E, Hall AR (2018) Label-free analysis of physiological Hyaluronan size distribution with a solid-state Nanopore sensor. Nat Comun 9(1):1037. https://doi.org/10.1038/s41467-018-03439-x

Rivas F, DeAngelis PL, Rahbar E, Hall AR (2022) Optimizing the sensitivity and resolution of hyaluronan analysis with solid-state nanopores. Sci Rep 12(1):4469. https://doi.org/10.1038/s41598-022-08533-1

Shi W, Friedman AK, Baker LA (2017) Nanopore sensing. Anal Chem 89(1):157–188. https://doi.org/10.1021/acs.analchem.6b04260

Stoddart D, Heron AJ, Mikhailova E, Maglia G, Bayley H (2009) Single-nucleotide discrimination in immobilized DNA oligonucleotides with a biological nanopore. PNAS 106(19):7702–7707. https://doi.org/10.1073/pnas.0901054106

Suzuki A, Suzuki M, Mizushima-Sugano J, Frith MC, Makalowski W, Kohno T, Sugano S, Tsuchihara K, Suzuki Y (2017) Sequencing and phasing cancer mutations in lung cancers using a long-read portable sequencer. DNA Res 24(6):585–596. https://doi.org/10.1093/dnares/dsx027

Taniguchi M (2022) Challenges of the practical applications of solid-state nanopore platforms for sensing biomolecules. Appl Phys Express 15(7):070101. https://doi.org/10.35848/1882-0786/ac7bd4

Thyashan N, Ghimire ML, Lee S, Kim MJ (2024) Exploring single-molecule interactions: heparin and FGF-1 proteins through solid-state nanopores. Nanoscale 16(17):8352–8360. https://doi.org/10.1039/D4NR00274A

Varki A, Cummings RD, Esko JD, Stanley P, Hart GW, Aebi M, Mohnen D, Kinoshita T, Packer NH, Prestegard JH, Schnaar RL, Seeberger PH (2022) Essentials of glycobiology. Cold Spring Harbor Laboratory Press. https://doi.org/10.1101/9781621824213

Wang Y, Zhao Y, Bollas A, Wang Y, Au KF (2021) Nanopore sequencing technology, bioinformatics and applications. Nat Biotechnol 39(11):1348–1365. https://doi.org/10.1038/s41587-021-01108-x

Watkins A, Fasanello D, Stefanovski D, Schurer S, Caracappa K, D'Agostino A, Costello E, Freer H, Rollins A, Read C, Su J, Colville M, Paszek M, Wagner B, Reesink H (2021) Investigation of synovial fluid lubricants and inflammatory cytokines in the horse: a comparison of recombinant equine interleukin 1 beta-induced synovitis and joint lavage models. BMC Vet Res 17(1):189. https://doi.org/10.1186/s12917-021-02873-2

Waugh M, Briggs K, Gunn D, Gibeault M, King S, Ingram Q, Jimenez AM, Berryman S, Lomovtsev D, Andrzejewski L, Tabard-Cossa V (2020) Solid-state nanopore fabrication by automated controlled breakdown. Nat Protocols 15(1):122–143. https://doi.org/10.1038/s41596-019-0255-2

Xia K, Hagan JT, Fu L, Sheetz BS, Bhattacharya S, Zhang F, Dwyer JR, Linhardt RJ (2021) Synthetic Heparan sulfate standards and machine learning facilitate the development of solid-state Nanopore analysis. PNAS 118(11). https://doi.org/10.1073/pnas.2022806118

Xia B, Fang J, Ma S, Ma M, Yao G, Li T, Cheng X, Wen L, Gao Z (2023) Mapping the acetylamino and carboxyl groups on glycans by engineered α-hemolysin nanopores. J Am Chem Soc 145(34):18812–18824. https://doi.org/10.1021/jacs.3c03563

Xue L, Yamazaki H, Ren R, Wanunu M, Ivanov AP, Edel JB (2020) Solid-state nanopore sensors. Nat Rev Mater 5(12):931–951. https://doi.org/10.1038/s41578-020-0229-6

Yang J, Chi L (2017) Characterization of structural motifs for interactions between glycosaminoglycans and proteins. Carbohydr Res 452:54–63. https://doi.org/10.1016/j.carres.2017.10.008

Yao G, Ke W, Xia B, Gao Z (2024) Nanopore-based glycan sequencing: state of the art and future prospects. Chem Sci 15(17):6229–6243. https://doi.org/10.1039/D4SC01466A

Yao G, Xia B, Wei F, Wang J, Yang Y, Ma S, Ke W, Li T, Cheng X, Wen L, Long Y, Gao Z (2025) Glycan sequencing based on glycosidase-assisted Nanopore sensing. J Am Chem Soc 147(2): 1721–1731. https://doi.org/10.1021/jacs.4c12940

Yin B, Xie W, Fang S, He S, Ma W, Liang L, Yin Y, Zhou D, Wang Z, Wang D (2024) Research progress on saccharide molecule detection based on nanopores. Sensors *24*(16):5442. https://doi.org/10.3390/S24165442

Yusko EC, Johnson JM, Majd S, Prangkio P, Rollings RC, Li J, Yang J, Mayer M (2011) Controlling protein translocation through nanopores with bio-inspired fluid walls. Nat Nanotechnol 6(4):253–260. https://doi.org/10.1038/nnano.2011.12

Yusko EC, Bruhn BR, Eggenberger OM, Houghtaling J, Rollings RC, Walsh NC, Nandivada S, Pindrus M, Hall AR, Sept D, Li J, Kalonia DS, Mayer M (2017) Real-time shape approximation and fingerprinting of single proteins using a nanopore. Nat Nanotechnol 12(4):360–367. https://doi.org/10.1038/nnano.2016.267

Zhang F, Zhang Z, Linhardt RJ (2010) Glycosaminoglycans. In: Handbook of glycomics. Elsevier, pp 59–80. https://doi.org/10.1016/B978-0-12-373600-0.00003-2

Zhang S, Cao Z, Fan P, Wang Y, Jia W, Wang L, Wang K, Liu Y, Du X, Hu C, Zhang P, Chen H, Huang S (2022) A nanopore-based saccharide sensor. Angew Chem Int Ed 61(33). https://doi.org/10.1002/anie.202203769

Zhang Y, Yi Y, Li Z, Zhou K, Liu L, Wu HC (2024a) Peptide sequencing based on host–guest interaction-assisted nanopore sensing. Nat Methods 21(1):102–109. https://doi.org/10.1038/s41592-023-02095-4

Zhang S, Cao Z, Fan P, Sun W, Xiao Y, Zhang P, Wang Y, Huang S (2024b) Discrimination of disaccharide isomers of different glycosidic linkages using a modified MspA nanopore. Angew Chem Int Ed 63(8). https://doi.org/10.1002/anie.202316766

Zhao Y, Iarossi M, De Fazio AF, Huang J-A, De Angelis F (2022) Label-free optical analysis of biomolecules in solid-state nanopores: toward single-molecule protein sequencing. ACS Photon 9(3):730–742. https://doi.org/10.1021/acsphotonics.1c01825

Zhao Y, Su Z, Zhang X, Wu D, Wu Y, Li G (2024) Recent advances in Nanopore-based analysis for carbohydrates and Glycoconjugates. Anal Met 16(10):1454–1467. https://doi.org/10.1039/D3AY02040A

Zhou Y, Long X, Zhang Y, Zheng D, Jiang Y, Hu Y (2025) Advances and challenges in solid-state nanopores for DNA sequencing. Langmuir 41(9):5736–5761. https://doi.org/10.1021/acs.langmuir.4c04961

Decrypting Glycosaminoglycan "sulfation code" with Computational Approaches

Sergey A. Samsonov and Mateusz P. Marcisz

Contents

Abstract

Glycosaminoglycans (GAGs), linear anionic periodic polysaccharides, play pivotal roles in various biologically relevant processes within the extracellular matrix (ECM). These processes encompass cell development, proliferation, signaling, ECM assembly, coagulation, and angiogenesis. GAGs perform their functions through their interactions with specific protein partners, rendering them attractive targets for regenerative medicine and drug design. However, the molecular mechanisms governing protein-GAG interactions remain unclear. Classical structure determination techniques face significant challenges when dealing with protein-GAG complexes. This is due to GAGs' unique properties, including their extensive length, flexibility, periodicity, symmetry, multipose binding, and

S. A. Samsonov (✉) · M. P. Marcisz
Faculty of Chemistry, University of Gdańsk, Gdańsk, Poland
e-mail: sergey.samsonov@ug.edu.pl

F. Pfrengle et al. (eds.), *Complex Carbohydrates in Health and Disease*, Handbook of Experimental Pharmacology 288, https://doi.org/10.1007/164_2025_741

the high heterogeneity of their sulfation patterns constituting the "sulfation code." Consequently, only a limited number of experimental protein-GAG structures have been elucidated. Hence, theoretical approaches are particularly promising in deciphering the code for understanding the structure-function relationship of these complex molecules. In this chapter, we focus on the particularities, challenges, and advances of computational methods such as molecular docking, molecular dynamics, and free-energy calculations when applied to GAG-containing systems. These computational approaches offer valuable insights into the enigmatic world of protein-GAG interactions, paving the way for their enhanced understanding and potential therapeutic applications.

1 Glycosaminoglycans

1.1 Chemistry and Biology of Glycosaminoglycans

Glycosaminoglycans (GAGs) represent a class of linear anionic periodic polysaccharides that consist of disaccharide building blocks containing a uronic acid (glucuronic, GlcA, or iduronic, IdoA) and a hexosamine (N-acetylglucosamine, GlcNAc or N-acetylgalactososamine, GalNAc), which could be sulfated in different positions (Varki et al. 2022). Depending on their disaccharide unit composition, glycosidic linkage, and sulfation pattern, GAGs are divided into several classes: hyaluronic acid (HA), chondroitin sulfate/dermatan sulfate (CS/DS), heparin/heparan sulfate (HP/HS), and keratan sulfate (KS) (Fig. 1).

All GAGs constitute the glycan part of proteoglycans except HA, which is not attached to any protein and is the only unsulfated GAG. Forty-three proteoglycans with covalently bound GAG chains have been identified so far (Iozzo and Schaefer 2015). GAG sequences are chemically highly heterogeneous. For example, only for HS disaccharides (dp2, dp stands for the degree of polymerization), there are 48 theoretically possible combinations, 23 of which have been identified in mammals (Clerc et al. 2019). In total, out of 408 theoretically possible combinations for all GAG classes, 202 different GAG dp2 sequences have been detected in mammals (Clerc et al. 2019). GAGs are located in the extracellular matrix of the cell (ECM) where they interact with at least 827 different protein targets and so affect

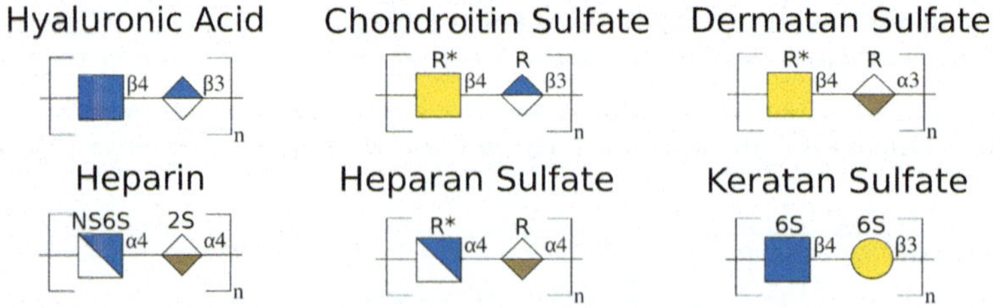

Fig. 1 GAG classes in SNFG representation. R* and R indicate heterogeneity of the sulfation within the GAG disaccharide units for GlcNAc/GalNAc and GlcA/IdoA monosaccharide blocks, respectively

the function of these proteins (Vallet et al. 2020). In particular, GAGs are crucial for processes such as cell proliferation, tissue regeneration, neuroplasticity, cell maturation, cell signaling, and angiogenesis. They are also actively involved in pathological processes such as cancer development, inflammatory response, infection, cardiovascular diseases, Parkinson's and Alzheimer's diseases, or viral infections including Sars-Cov-2. Additionally, by attracting numerous molecules in the connective tissues and thus allowing loadbearing capabilities, GAGs take part in the osmotic swelling response. Therefore, GAGs are very promising potential molecular targets for novel strategies in regenerative medicine and drug design (Abdelfadiel et al. 2023). At the same time, only one HS dp5 molecule, a specific high affinity binder of antithrombin, is proven as therapeutically active (Jin et al. 1997).

1.2 "Sulfation code" and Specificity of Interactions

The GAG-protein interactions are predominantly electrostatic driven and occur between sulfate and carboxyl groups of the GAG units and positively charged lysine, arginine, and sometimes protonated histidine residues on protein surfaces (Imberty et al. 2007). Whereas in some protein-GAG systems, the net charge of the GAG completely defines the strength of the interactions, in other systems the dependence of the specificity/selectivity of the interactions is dependent on the GAG sulfation pattern. The specificity/selectivity propensity of GAGs recognition by proteins is continuous. From a highly selective pattern of interaction (antithrombin with HS dp5) to the unselective one (TIMP-3 or osteoprogerin), there are intermediate levels of selectivity, including moderately selective and plastic interactions (Perez et al. 2022). The state of the art in GAG research is essentially based on the central concept that the amount (net sulfation) and particular positions of the individual sulfation groups (sulfation pattern) constitute a *"sulfation code"* (Habuchi et al. 2004), which ultimately determines the conformational, dynamic, thermodynamic, and recognition properties of GAGs, and, consequently, defines their biological activity in the key processes in ECM (Kjellén and Lindahl 2018). Thus, deciphering the "sulfation code" is essential for understanding GAG function, which is, in turn, of high importance for the development of novel medical approaches based on the exploitation of GAG molecular properties. Despite the apparent simplicity of the "sulfation code," it is still not decrypted. The nature of GAGs makes such deciphering challenging for many reasons as discussed in the next subsection.

1.3 Challenges in the GAG Research

Both experimental and theoretical studies of GAGs and protein-GAG interactions face several challenges that originate from the particular properties of this class of polysaccharides (Fig. 2) (Perez et al. 2022).

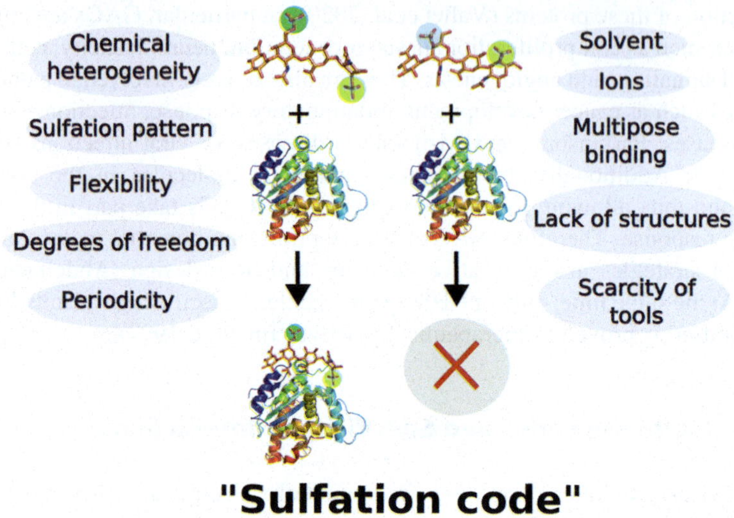

Fig. 2 "Sulfation code" concept and the challenges faced in GAG research to decrypt it

- *Chemical heterogeneity*. GAG molecules are not only highly variable in terms of dp2 unit potential combinations but could also be composed by different units within the same chain. In addition, a GAG chain length in vivo reaches up to 10^4 dp in the case of HA. This heterogeneity and complexity do not allow the experimental methods to deal exclusively with the homogeneous sequences yielding many properties of studied systems as ensemble averages per chains. In turn, for the computational approaches, which handle particular sequences and structures, the task of analyzing the whole statistical ensemble to obtain one-to-one correspondence with the experiment is unfeasible.
- *Sulfation pattern*. This particular example of chemical heterogeneity is especially important in terms of decrypting the "sulfation code." Unfortunately, often the experimental data only suggest how many sulfates per dp are in the sequence of the studied GAG but not the particular positions of the sulfate, which requires application of combinatorial approach from the in silico methods which leads to the dramatic increase of the calculations expenses even in case of relatively short GAG oligosaccharides.
- *Flexibility*. GAGs are highly flexible and dynamic molecules that imply difficulties in their crystallization which is the reason for the limited number of available experimental structures containing these molecules (Pérez et al. 2020). In computational analysis, this high flexibility needs to be taken into account, which makes theoretical approaches designed for rigid or semirigid ligand to be ineffective for GAG-containing systems. In one of the recent cutting-edge computational studies contributing to deciphering the "sulfation code" performed on a group of specific HS oligosaccharides, it was shown that the flexibility itself is one of the key variables that needs to be taken into account as a molecular

descriptor for understanding the "sulfation code" (Holmes et al. 2022). Besides the direct impact of the flexibility of the GAGs, their interacting partners are predominantly long side chains of Arg and Lys, which flexibility essentially contributes to the flexibility of entire protein-GAG systems.

- *Degrees of freedom*. GAG molecule's high number of degrees of freedom corresponds, first of all, to glycosidic linkages and ring puckering. The latter is particularly important to account for in the case of GlcA/IdoA monosaccharide units. In addition, the rotameric space of sulfate groups contributes to the challenge of conformational heterogeneity of GAGs encountered by the experimental techniques and to the computational expenses in the theoretical analysis.
- *Periodicity*. While the high flexibility of GAGs suggests that NMR can be a proper technique for studying these molecules, the fact that GAGs are periodic polymers leads to obstacles for the NMR to properly assign the spectrum. The same periodic nature of GAGs causes the challenges faced by the computational approaches: It is often difficult to propose a more favorable binding pose among several binding poses, which are shifted in relation to each other by one or several periodic units, and to properly rank the corresponding binding energies for them. Fortunately, the combination of NMR techniques and computational approaches can be essentially more effective than the application of either of the methods alone (Künze et al. 2021).
- *Solvent and ions*. Due to their charged nature GAGs maintain more stable contacts with water molecules in the binding interfaces than non-charged ligands, which is reflected by about 10 times more solvent molecules resolved by the X-ray spectroscopy in the interfaces of protein-GAG complexes in comparison to the representative dataset of protein-protein/peptide complexes (Teyra et al. 2011). This means that preferably high-resolution structures including interfacial water molecules should be obtained experimentally to gain sufficient insights into GAG recognition, in particular, to be efficiently used in GAG-based drug design development. Computationally, proper treatment of solvent is a general challenge for modeling techniques, while many solvent models potentially yield essentially different results for particular classes of molecules. Similarly to solvents, cations are very likely to affect GAG conformational and binding properties (Kogut et al. 2022a), rendering it important to consider specific ions in the experiment while using appropriate parameters for modeling protein-GAG-ion systems.
- *Multipose binding*. GAG recognition often represents "multipose binding" (Joseph et al. 2015), which implies that GAG molecules may bind to multiple regions of the protein, or they may exhibit multiple binding patterns/positions in the same region that correspond to hardly distinguishable binding affinities. This represents another difficulty for both the experimental and computational methods to analyze such system due to the additional number of binding poses that should not be disregarded when a protein-GAG complex is quantitatively described.
- *Lack of structures*. As a consequence of the challenges faced by the experimental approaches listed above, there are slightly more than 100 structures in the PDB representing protein-GAG complexes. About half of them are complexes between

GAGs and GAG-processing enzymes that are very similar to each other and in which GAGs are bound in pockets, and therefore, these structures offer limited insights into the GAG recognition (Samsonov and Pisabarro 2016). Other complexes include predominantly GAGs of dp2 (34 complexes). The available structures with longer GAGs are represented by 10 structures for dp3, 18 for dp4, 13 for dp5, 15 for dp6, 7 for dp7, 8 for dp8, and 1 for dp9, respectively (Pérez et al. 2020). Among these structures, less than 30 represent a nonredundant dataset with GAG ligands longer than dp3. This limitation imposes a serious constraint on the use of potentially powerful computational machine-learning (ML) approaches for protein-GAG systems since these approaches require essentially more data to be trained on.
- *Scarcity of GAG-specific tools*. Although the interest in GAG research essentially increased in recent decades, the number of computational tools and protocols specifically developed for GAGs is incomparably small when related to the ones existing for small molecules or peptides.

All these challenges leave the following persistent questions that should be answered to decipher the GAG "sulfation code." What are the physicochemical principles underlying the interactions of proteins/peptides with GAGs? What is the role of ions and solvent in GAG recognition? How specific is GAG recognition and which strategies should be used to effectively exploit this specificity for drug design? How to improve the predictive power of computational approaches that should be applied to GAG-containing systems? The main objective of the next section is to introduce the reader to the state of the art of GAG research performed with the use of theoretical approaches aiming to answer these questions and to provide several examples of their successful application. To get a more complete view on the topic, the reader is welcome to refer to numerous reviews in the field (Almond 2018; Sankaranarayanan et al. 2018; Nagarajan et al. 2019, 2020, 2022; Perez et al. 2022; Ricard-Blum and Perez 2022; Kogut et al. 2022b).

2 Computational Approaches for Protein-GAG Interactions

A general pipeline for computational analysis of protein-GAG interactions is established (Fig. 3). Within this pipeline, experimentally available or modeled structures of the protein receptor and GAG ligand are used for the molecular docking to the particular binding site/region on the receptor surface, the docked solutions are clustered, and representative structures from clusters are chosen for the further molecular dynamics (MD)-based analysis. As output, predicted structures of the complexes and corresponding binding energies including contributions of individual residues to the complex formation are obtained. All the data are compared, complemented, and filtered by the use of available experimental data. In this section, particular steps of the pipeline and their theoretical background are discussed.

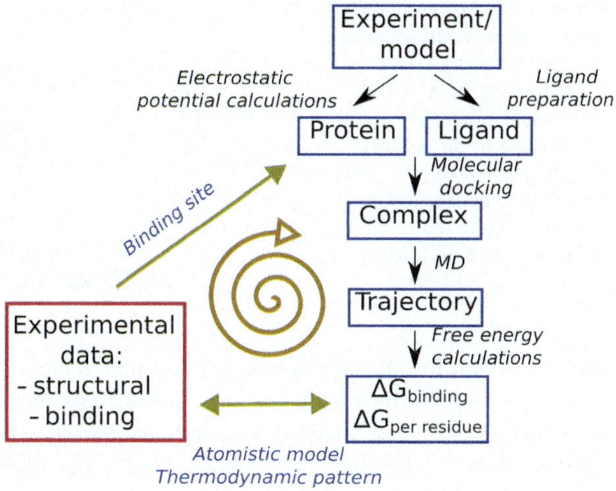

Fig. 3 A general pipeline for modeling protein-GAG interactions

2.1 Binding Site Predictions

As an initial step to computationally analyze protein-GAG interactions, when a corresponding experimental structure is not available, a GAG binding site on a protein surface should be predicted. The first attempt to identify principles for GAG binding site predictions was made in 1989, when Cardin and Weintraub compared sequences of the experimentally identified HP binding regions in a protein dataset and identified two sequence motifs that they proposed as consensus sequences for HP recognition. These sequence motifs are XBBXBX and XBBBXXBX, where B is a basic amino acid residue and X is a nonbasic residue (Cardin and Weintraub 1989). It was indeed proven that simply predicting protein motifs consisting of positively charged amino acids can help predict GAG binding sites (Hileman et al. 1998). Although this concept can be useful in the identification of the HP binding sites when the residues responsible for GAG binding are located within the same polypeptide chain, it is not capable of taking into account the cases when sequentially remote basic residues are close to each other in space. Such spatial proximity of several positively charged residues leads to the excess of the positive electrostatic potential in the corresponding volume, while GAGs involving complexes are dominated by electrostatic interactions (Nagarajan et al. 2022). Therefore, assessing protein surface charge distribution can be a very useful method for proposing GAG potential binding regions. More generally, it has been shown that calculations of excessive electrostatic potential formed by patches of protein residues are promising for the identification of binding regions for GAGs (Bitomsky and Wade 1999). One of the most recognized methods for calculating electrostatic potentials is the Poisson-Boltzmann surface area (PBSA) method, in which solute atoms containing point charges produce an electrostatic field, in both the solute and solvent region, which can be computed by solving the Poisson-Boltzmann equation (Warwicker and Watson 1982). This approach showed to be very successful in

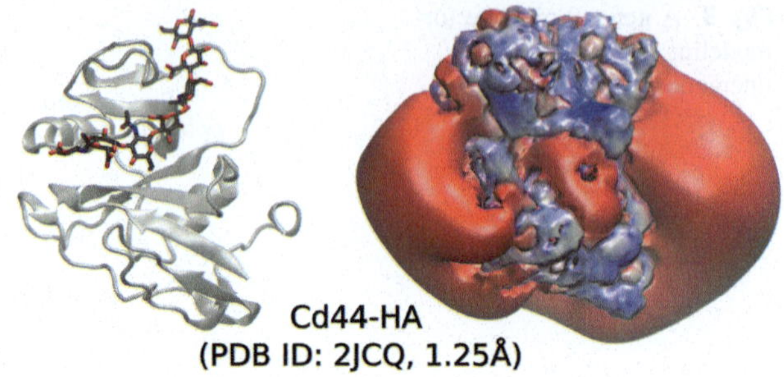

Fig. 4 Cd44-HA complex (left): Protein and ligand are in cartoon and licorice, respectively. PBSA electrostatic potential calculations (right) allow to propose protein surface areas favorable (blue isosurface) and unfavorable (red isosurface) to bind GAGs

predicting binding sites for GAGs as presented in the work of Samsonov et al., where it yielded true positive results for most of the experimental structures (Samsonov and Pisabarro 2016). The method not only enables the prediction of protein surface regions with a propensity for binding GAGs but also facilitates the exclusion of those regions that do not bind GAGs (Fig. 4).

2.2 Molecular Docking

When the binding site is known, the next step for the analysis is to obtain a binding pose of a ligand in the binding site. This can be done by the molecular docking approach, which predicts the preferred orientation of a molecule when it binds to another to create a stable complex and estimate the corresponding binding affinity. As a result, this technique has become crucial for the development of rational drug design protocols including structure-based virtual screening. There are two principal steps of molecular docking: (1) exploring of vast conformational space that represents diverse potential binding modes, including both prediction of the potential binding site and binding pose of the ligand (placement); (2) assessment of the interaction energy related to each of the predicted binding conformation (scoring). Some docking tools perform the so-called blind docking which means that the whole surface of the receptor is considered in the docking simulation and a priori knowledge about the binding site is not required. Other docking tools are limited to scanning only part of the receptor predefined by the user in advance (local docking). In case of GAGs the knowledge about a potential binding site can be obtained using, e.g., electrostatic potential calculations (see previous subsection). To successfully predict binding poses by finding the global energy minimum, search algorithms comprehensively scan the potential energy landscape. In the process of conformational search, ligand's and some protein's dihedral, translational, and rotational degrees of freedom can be sampled. All docking algorithms can be classified into two major groups based on the flexibility of ligands and receptors in the docking simulation: rigid docking and flexible docking. The rigid docking approach takes into account essential geometric complementarities between a receptor and a ligand but does not account for the flexibility of the interacting partners. The flexible docking accounts for multiple potential conformations of either a ligand, a receptor,

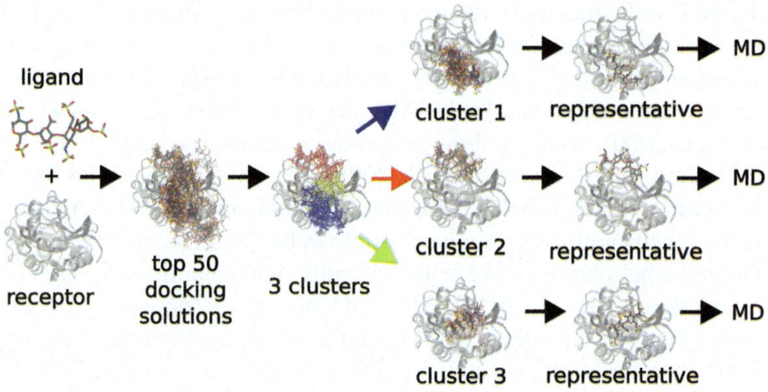

Fig. 5 Example of docking a GAG (HP dp4) to a protein (cathepsin O) with AutoDock 3 (Morris et al. 1998)

or both molecules. Due to the high flexibility of GAGs, the use of flexible docking is preferable. Docking employs diverse classes of algorithms for placement as well as for scoring of the docked poses. More details on the search algorithms and scoring schemes can be found in the review on this topic (Ferreira et al. 2015).

After obtaining and scoring the binding poses, the question arises regarding which and how many of these binding poses are representative and merit further analysis (Fig. 5). One of the ways to answer this question is to use structural clustering, which relies on the assumption that the structures belonging to clusters are representative. There are several clustering strategies with most of them simply rely on using the algorithm to find the structures with the largest number of neighbors containing a given number of shared atoms within a specified radius. To define a neighbor an appropriate metric should be introduced to calculate the differences between the structures. Traditionally such metric is RMSD (root mean square deviation), which compares the distances between all the equivalent atoms in the structures to be clustered. For GAGs, RMSatd (RMSD for atomic types) can be used to consider the periodicity of the molecules (Samsonov et al. 2014a).

The predictive power of GAG docking still essentially underperforms in comparison with that for small molecules and even protein-protein docking. One of the reasons is that many researchers still rely on docking programs that were not developed specifically for GAGs. Although diverse docking software is abundant, those optimized for other ligands usually do not perform satisfactorily when applied to GAG systems (Griffith et al. 2017). In the work of Uciechowska-Kaczmarzyk et al. (2019), only a few of fourteen conventional docking tools, which also included several commercial programs, yielded acceptable results. Among the tested software, a free-of-charge AutoDock 3 (AD3) (Morris et al. 1998) performs the best. Nevertheless, it is still limited in, for example, the constraint in the number of ligand torsional degrees of freedom. For the long and flexible GAG molecules allowed by the software, a maximum of 32 torsional degrees of freedom is not sufficient to sample the ligand conformational space efficiently, which forces the user to manually pick only the most relevant ones. This also implies the reliability of docking results obtained for GAG molecules longer than dp8 is questionable. Some docking programs particularly dedicated to GAG molecules have been based on AD including Vina-Carb (Nivedha et al. 2016) and GlycoTorch Vina (Boittier et al. 2020). Both of these programs outperformed AD Vina (Eberhardt et al. 2021) and Glide (Friesner et al. 2004) for the same protein-GAG dataset; however, neither of them took up the glove to challenge the AD3 and to directly compare the results with AD3 results. For researchers who are not experienced with computational approaches, there are user-friendly molecular docking servers such as ClusPro (Mottarella et al. 2014), HADDOCK (Van Zundert et al. 2016), or SwissDock (Aurélien Grosdidier 2011). Although their advantage is rather general accessibility than the strong specific performance since they usually do not offer all adjustable options as the regular docking software, they proved to be valuable tools. For example, ClusPro in the important study on SARS-CoV-2 helped in describing interactions between SARS-CoV-2 spike protein and HS (Clausen et al. 2020). HADDOCK showed its potential in a GAG-related study where interactions between HP and CXCL-8 were analyzed (Joseph et al. 2015). An alternative method to the use of docking software could be the manual placement of the given molecule near the anticipated binding site. Then the obtained structure could be refined by MD simulations to find potential binding poses (Vuorio et al. 2017).

Recently few novel approaches emerged to overcome the obstacles experienced by using conventional docking tools for GAGs. One of them is a method called Dynamic Molecular Docking (DMD) (Samsonov et al. 2014a) which involves using a steered MD approach to move a GAG molecule placed somewhere in the bulk solvent toward its binding site on the protein surface. DMD allows for full flexibility of both GAG and protein molecules independently of GAG length and receptor size. One significant drawback of this method is the need for a priori knowledge of the binding site, which is not always plausible. Additionally, the method is computationally expensive due to the size of the periodic boundary box required for the starting intermolecular configuration with the use of an explicit solvent model that is recommended with this approach. One more tool that is potentially useful for docking longer GAG molecules is the fragment-based approach (Samsonov et al. 2019). This method involves sampling the protein's surface by docking trimeric fragments of a GAG, which afterward are assembled into longer chains based on their overlaps. This approach allows for the docking of longer GAG molecules. Unfortunately, there is a major drawback in the case when a GAG binding site is located near negatively charged amino acid residues. Short trimeric fragments may be avoiding patches of negative electrostatic potential preventing a continuous assembly of a long GAG resulting in unsuccessful docking.

There are other tools that could potentially overcome the issues featured in the above-mentioned techniques. Repulsive Scaling Replica Exchange Molecular Dynamics (RS-REMD) is a novel MD-based method initially designed for protein-protein docking (Siebenmorgen and Zacharias 2020). In RS-REMD approach, the van der Waals radii increase in different replicas, which physically corresponds to the increase of the size of the atoms without affecting other types of interactions in

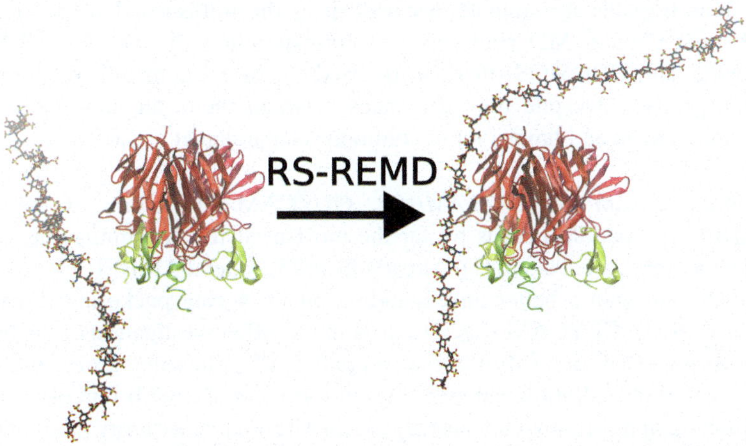

Fig. 6 The complex between trimeric APRIL protein (red cartoon), its receptors TACI (light green and yellow cartoon), and HP dp48 (licorice) obtained by RS-REMD docking approach

the system. This approach facilitates a thorough and robust search for binding sites and poses on the protein surface while allowing flexibility of the both docked molecule and the receptor sidechains. When applied to protein-GAG systems, it could potentially solve many difficulties characteristic for docking GAGs (Maszota-Zieleniak et al. 2021; Marcisz et al. 2022). In particular, this method does not decrease in performance when applied to long GAGs and can be effectively used with explicit solvent. Using this method, it was possible to dock HP dp48 to a tertiary complex of a trimeric APRIL protein bound to its three receptor molecules and to propose a molecular mechanism that explains the effect of HP in the signaling of the corresponding protein-receptor system (Marcisz et al. 2021b) (Fig. 6).

2.3 Molecular Dynamics

MD is a computational method to study how particles move over time by solving Newton's equations of motion. In MD, specific force fields (FFs) defining inter-atomic potentials are used to calculate potential energies and forces between any number of given molecules. FF parameters and corresponding potential energy functions may be derived from quantum chemical (QM) calculations and experimental data. Equations describing forces in the mentioned force fields are solved simultaneously in small time steps (typically in fs range) keeping, for example, pressure/volume and temperature in the required range. The product of solving those equations is the trajectory of the system defined as the coordinates of its particles as a function of time. The evaluation of the trajectory may provide macroscopic properties and a description of dynamical processes occurring in the system. All-atom (AA) MD simulations, in which every physical atom is explicitly represented by a particle, have proven that with recent advances in computational power, it is possible to simulate systems up to the millisecond range. There are several most popular MD packages and corresponding FF families (GLYCAM, CHARMM, and GROMOS) that can be used for MD simulations of the systems containing GAGs. Despite the differences between them, all they yield similar/comparable results and have a lot in common (Nagarajan et al. 2019).

- *Simulations with GLYCAM/AMBER.* GLYCAM FF family, a member of AMBER compatible FFs, is by far the most used for the simulations of GAG molecules (Paiardi et al. 2021). Already in 1993, general AMBER FF was used to simulate and gain a better understanding on IdoA ring puckering (Forster and Mulloy 1993). GLYCAM was also used in the MD investigation of the behavior of a library of HP dp6 (Muñoz-García et al. 2013). The same group investigated the effect of the sulfation pattern in the conformational equilibrium of IdoA in HP mimicking dp3 (Muñoz-García et al. 2012). The adiabatic energy maps of CS dp2 units were compared them with experimental data (Rodríguez-Carvajal et al. 2003). MD and MD-steered techniques showed the importance of explicit solvent in local molecular docking and investigated ligand and receptor flexibility in the GAG systems (Samsonov et al. 2014a). Furthermore, three different techniques, QM, MD, and NMR, were applied to characterize conformational properties of GAG monosaccharides. Among others, QM and MD results were compared for *gg/gt/tg* conformations distribution for GlcNAc and GalNAc rings (Samsonov et al. 2014b). In 2016 GLYCAM06 FF was evaluated in terms of parameters used for the modeling of GAGs. As an outcome, a new set of parameters has been added to GLYCAM for GAGs containing protonated GlcA and IdoA residues and sulfate groups. For eight HS dp6 containing differently sulfated GlcN residues, the impact of sulfation patterns on the conformation of neighboring IdoA residues was described (Hsieh et al. 2016). Lately, the potential influence of the GAG length on their binding to proteins was investigated (Potthoff et al. 2019). The MD simulations performed with GLYCAM06 helped to unravel the multipose binding phenomenon in a microsecond-scale investigation of HP interaction with FGF1 protein (Bojarski et al. 2019). The effect of HP binding on the Vascular Endothelial Growth Factor (VEGF) conformation was analyzed showing that HP stabilizes the structure of the VEGF protein in a particular conformation that interacts with the VEGF receptor, which is crucial for cell signaling (Uciechowska-Kaczmarzyk et al. 2018). In a recent MD-based study, the role of GAGs in the procathepsin B maturation mechanism was proposed (Bojarski et al. 2020).
- *Simulations with CHARMM.* CHARMM was used in 1991 to successfully investigate structural characteristics of the HA and CS molecules simulated in water and in vacuo (Scott et al. 1991). Later on, the conformational preferences of the HA molecule were analyzed using this force field (Almond et al. 1997, 1998a,b). In this work, two different CHARMM FFs were compared, and MD simulations were performed using the TIP3P water model. The authors also compared fiber diffraction data with their MD results. X-ray fiber diffraction data were also used

to study polar interactions of CS (Wiegel et al. 1999). The same group conducted a similar study on CS-A dp4 (Kaufmann et al. 1999). In 2000, an excellent study was conducted to rigorously describe the conformational preferences of CS and DS molecules, where similarities and differences between the data obtained from MD simulation in explicit solvent and from fiber diffraction were described (Almond and Sheehan 2000).

– *Simulations with GROMOS*. One of the first studies on HP using GROMOS was conducted in 2004 (Verli and Guimarães 2004). Another example of the use of GROMOS is the study aimed to reproduce the puckering of IdoA(2S) (Gandhi and Mancera 2010). The transition from the skewed boat to the chair conformation in the presence of SPC/E water was detected while not observing a reverse, chair to boat, transition. QM and MD simulations were performed to study configurations of CS and discovered that the use of standard parameters resulted in nonphysical behavior of sulfates such as disorientation and deformation of geometry. Luckily, reparametrization using charges derived from QM improved sulfate groups' behavior and enabled their MD simulations in SPC water (Cilpa et al. 2010).

2.4 Coarse-Grained Approaches

AA simulations can provide a high level of molecular details of any given system. However, for big systems on a long-time scale, AA simulations are not feasible due to the requirement of enormous computer resources. To overcome this challenge, one can apply a higher level of abstraction to the representation of investigated molecules. A group of atoms can be represented as a pseudoatom which leads to a decrease of the complexity of the system due to a lower number of interactions and degrees of freedom, which therefore allows more extensive sampling in a longer timescale at the same cost. FFs using such models are termed as coarse-grained (CG) FFs. CG models may be sufficient when atomistic precision is a lesser concern, and the main focus is the more global dynamic evolution of the studied system. Molinero and Goddard laid the foundations for the development of CG FFs for GAGs when they proposed their FF for the malto-oligosaccharides (Molinero and Goddard 2004). The first published GAG CG study was published by Bathe et al. one year later (Bathe et al. 2005). Authors proposed the CG model for the CS and HA molecules and analyzed their characteristics as a function of pH and ionic strength. Sattelle et al. in two studies further improved CG models by emphasizing the importance of ring puckering in CG models and implanting it into their model. They not only introduced this feature alongside inter-glycosidic torsions but also utilized two separate energy functions for puckering movements and inter-glycosidic connections (Sattelle et al. 2013). Another noteworthy approach has been proposed by Samsonov et al. where the authors introduced 28 pseudoatoms representing functional groups based on 17 distinct GAG residues implemented compatibly with AMBER AA FF (Samsonov et al. 2015). Kolesnikov and coauthors proposed a field-theoretic approach for CG modeling of GAGs in aqueous solutions with the addition of varying salt concentrations. This study utilized experimental osmotic pressure data to compute a degree of dissociation, static structure factor, radius of gyration, and persistence length for semiflexible polyelectrolyte chains in the salt solutions. By using adjustable parameters, this straightforward theoretical method demonstrated consistency with the experimental studies (Kolesnikov et al. 2014). In the recent work of Danielsson et al., the SUGRES-1P (Danielsson et al. 2023) model, which belongs to the family of UNICORN FFs (Liwo et al. 2014), was extended to HP using the parameters obtained in AMBER AA simulations. The model was evaluated for unbound HP of dp6 to dp68 in terms of its agreement with the X-ray diffraction data for HP. The SUGRES-1P FF was able to accurately predict the general shape and global characteristics of HP molecules. CG approaches can also be useful for docking longer GAGs, for which conventional docking approaches are not effective. In a study of Marcisz et al., the hybrid AA/CG docking approach was used to dock long GAGs (Marcisz et al. 2021c). In this model, the part of the GAG molecule directly contacting a protein was described with AA parameters, while for the GAG part that does not establish van der Waals contacts with the protein each monosaccharide unit was represented with a single pseudoatom.

2.5 Modeling Solvent and Ions

There are two principal classes of solvent computational models: explicit and implicit ones. In explicit solvent, each water molecule is represented by a particle. Explicit solvent models can be categorized depending on whether the model is flexible or rigid, by the number of the so-called sites which describe the number of interaction centers within a water molecule (varying from 2 to 6), by considering polarization effects. The implicit solvent models represent a continuous medium with the averaged solvent properties and are in general less accurate than explicit solvents. However, they are still useful for analyzing large systems where computational efficiency can be improved up to 100 times when implicit solvent is used.

TIP3P is the most popular explicit water model used not only in GAG computational studies but also in other MD-related studies, which is reflected in 37949 citations of the original work of Jorgensen et al. (1983). This 3-site water model is now established as a go-to solvent model in MD involving GAG systems because of its low computational cost when compared to the models as TIP4P (4-site) or TIP5P (5-site). The importance of the water model for GAG simulations has been shown in the study of Sarkar et al. (2016) where it was proven that water dominates specific protein-GAG interactions. Additionally, it was shown that the use of a proper solvent model is important in reproducing proper ring puckering in GAGs (Gandhi and Mancera 2010). In the study of Neamtu et al., influence of different water models on the properties of CS was shown (Neamtu et al. 2013). In that work, the TIP3P model favored intramolecular H-bonds, while TIP4P and TIP5P water models disfavored them. On the other hand, the TIP5P model yielded a higher number of water bridges along the CS4 dp8 than the other models. It was also shown that the second layer of hydration was better represented in the case of the TIP5P water model. Despite the impact of the water model on the reliability of computational studies, there has been very little discussion on this matter as there was only one attempt to partially cover this topic involving the GAG system (Verli and Guimarães 2004). Recently microsecond MD simulations were performed for HP, HA, and CS dp10 to systematically evaluate the effect of five implicit and six explicit solvent models on dynamic and structural properties of these GAGs (Marcisz and Samsonov 2023). It was shown that implicit solvent models fail to even qualitatively match experimental results regarding the HP ring puckering. While explicit water models well reproduced experimental findings regarding local parameters of GAGs (dihedral angles and ring puckering), many of them did not succeed in the case of global parameters (end-to-end distance, radius of gyration, volume, or the number of established hydrogen bonds). In the case of TIP3P, abnormal GAG chain curvature and end-to-end distance were observed. The question which solvent model is the best choice for modeling GAG-containing systems is still open.

GAGs interact with metal ions in the ECM, mediating countless biological processes in which they are involved. Interactions between HP and a wide range of metal ions were inspected experimentally using atomic absorption and spectrophotometry, and the overall trend for the HP-metal affinity was reported for 11 ion types (Stevic et al. 2011). Additionally, the conformation of IdoA(2S) is sensitive to the identity of adjacent residues and their substitution patterns in the fragment of HP and HS (Rudd et al. 2007). The glycosidic linkage geometry is also affected by an altered substitution pattern (Yates et al. 2000), affecting the electrostatic properties of the polysaccharides and, therefore, their interactions with ions. Dynamic properties of HP oligosaccharides were characterized by performing an extensive computational analysis of these HP molecules to investigate the impact of Ca^{2+} on the oligosaccharide conformation in a microsecond-timescale MD simulation taking into account also the parameters set used to model the ions (Kogut et al. 2022a). It was concluded that the flexibility of the monosaccharides, the glycosidic linkages, and ring puckering were not significantly affected by the presence of Ca^{2+} in contrast to H-bond propensities and the theoretically determined distribution of the radius of gyration of oligosaccharides' population. The findings indicated the crucial importance of the Ca^{2+} parameters for modeling protein-HP interactions involving Ca^{2+} and highlighted the limitation of the currently available parameters to model binary GAG-ion complexes. Subsequently, the essential role of metal ions in protein-GAG interaction was shown experimentally for a series of protein-GAG systems such as endostatin (Ricard-Blum et al. 2004) or annexin II (Shao et al. 2006) and many others. There are several experimental structures where Ca^{2+} ions were detected in the HP-protein interfaces (PDB ID: 1G5N, 2HYV, and 2HYU). In the computational study of Kogut et al. the effects of the impact of Ca^{2+} in annexin-Ca^{2+}-HP systems and the sensitivity of the used computational approaches to the presence of these ions were inspected (Kogut et al. 2021). The obtained results revealed the lack of Ca^{2+} notably affect the strength of protein-GAG interactions, the ligand's binding pose, and dissociation pathway.

2.6 Free-Energy Calculations

To estimate the strength of binding in the modeled protein-GAG complexes, free-energy calculations are performed. There are several strategies for calculating binding free energy. Relatively fast end-point techniques, such as empirical scoring functions, are mainly utilized for virtual screening purposes (e.g., in docking). Then, there are methods based on implicit solvent models such as molecular mechanics/Poisson-Boltzmann surface area (MM/PBSA) or molecular mechanics/generalized Born surface area (MM/GBSA) which are more accurate but are also more computationally expensive. Finally, there are also more rigorous methods like thermodynamic integration (TI) or free-energy perturbation (FEP). Some of the methods employ MD simulations with the explicit solvent and are based on rigorous postprocessing of the MD trajectories (Abel et al. 2017). Although powerful, FEP has not been widely used to study systems involving GAGs as only very few studies implemented this method (Sarkar et al. 2016; Babik et al. 2017).

MM/PBSA and its approximation MM/GBSA are arguably the most used in GAG-related studies due to their satisfactory accuracy and relatively low computational cost. The enthalpic term in the calculated free binding energy values consists of bonded, electrostatic, van der Waals and solvation contributions. The entropic term can be estimated by a normal-mode (NM) analysis of vibrational frequencies or quasi-harmonic QH analysis. However, consideration of the entropy contribution often increases the overall uncertainty in the calculated binding energies in general (Homeyer and Gohlke 2006) and in GAG complexes particularly (Gandhi and Mancera 2009). MM/PBSA and MM/GBSA were widely used to calculate binding free energies in the GAG complexes with a number of proteins (Kogut et al. 2022b). In addition, the application of these methods for free-energy decomposition per residue allows to establish which residues contribute most and least to binding. This analysis can be useful for direct comparison with NMR data (Künze et al. 2021) or for guiding rational mutagenesis design (Fig. 7) (Nordsieck et al. 2012).

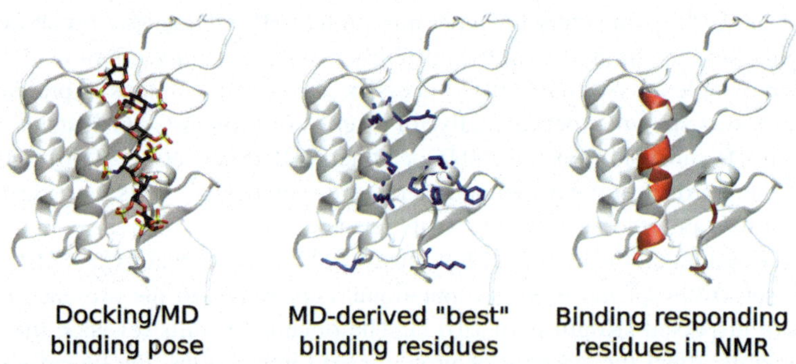

Fig. 7 Comparison of the MD-derived per residue free-energy decomposition data with the backbone NMR chemical shifts for a protein-GAG system. In each case, 10 top "binding" residues according to the calculation and the experiment are shown, respectively

An even faster end-point alternative method to MM/GBSA is Linear Interaction Energy (LIE), in which the van der Waals and scaled electrostatic energies are summed. Although LIE is computationally efficient, it has not gained the same popularity as MM/PBSA(GBSA) for studying GAG complexes. However, some studies reported its use as a helpful tool (Plazinski and Knys-Dziecіuch 2012). Both LIE and MM/PBSA methods have been rigorously studied in terms of the predictive power in the work of Marcisz et al., which suggested how many and how long MD simulations should be performed to effectively obtain binding affinities agree with the experiment. However, it is important to notice that free-energy calculation protocols could perform essentially differently for different molecular complexes (Marcisz et al. 2021a).

Another particularly rigorous approach is Potential of Mean Force PMF which could be determined by using umbrella sampling (US) or by employing the Jarzyński equality during steered MD simulations. PMF calculates the free-energy landscape of a system by integrating the potential energy of the system over all possible configurations and provides information on the relative stability of different conformations or states of the system. It was used, for example, for the free-energy analysis of a GAG dissociation from a protein (Kogut et al. 2021).

Although QM or hybrid methodologies that include QM approaches could be thought as alternative and potentially more accurate ways to calculate free energy of binding, they have not been used yet in GAG-related studies due to the high flexibility of the GAGs rendering QM-based calculations too computationally expensive.

3 Summary and Outlook

This chapter provides a general overview of some principal computational approaches as molecular docking, MD, and free-energy calculations to study GAGs, a particular class of biologically relevant and very challenging carbohydrate polymers. This chapter does not have the ambition to provide a full picture of the tremendously heterogeneous theoretical research of GAGs but rather to describe some aspects of this challenging but quickly advancing scientific field. In particular, there was no discussion about ML or artificial-intelligence-based approaches, which are especially common and effective to use nowadays. This is mainly because the number of the experimental structures of protein-GAG complexes is quite low to apply ML straightforward to these systems. The number of examples presented in this chapter is not exhaustive or evenly balanced. Instead, the authors have aimed to provide their interpretation of the current state of the topic, drawing primarily from their own experience and closely related literature. To conclude, GAG-related computational analysis is still complex, lags behind theoretical studies for other types of biomolecular systems, and has great room for improvement, which is expected to be filled with the application of more sophisticated novel methods including advanced MD techniques being more and more accessible with the growth of the available computational power. Another important need for success in GAG computational research is access to systematically obtained experimental data, especially structural and thermodynamic. A combination of the experiment with the computation will work more and more synergistically with the objective of the decryption of the GAG "sulfation code," which, in turn, will open up novel opportunities for drug design and regenerative medicine strategies based on the comprehension of the GAG function.

Funding This research was funded by National Science Centre of Poland, grant number 2023/49/B/ST4/00041.

References

Abdelfadiel EI, Gunta R, Villuri BK, Afosah DK, Sankaranarayanan NV, Desai UR (2023) Designing smaller, synthetic, functional mimetics of sulfated glycosaminoglycans as allosteric modulators of coagulation factors. J Med Chem 66(7):4503–4531. https://doi.org/10.1021/acs.jmedchem.3c00132

Abel R, Wang L, Harder ED, Berne BJ, Friesner RA (2017) Advancing drug discovery through enhanced free energy calculations. Account Chem Res. https://doi.org/10.1021/acs.accounts.7b00083

Almond A (2018) Multiscale modeling of glycosaminoglycan structure and dynamics: current methods and challenges. Current Opin Struct Biol. 50:58–64. https://doi.org/10.1016/j.sbi.2017.11.008

Almond A, Sheehan JK (2000) Glycosaminoglycan conformation: Do aqueous molecular dynamics simulations agree with x-ray fiber diffraction? Glycobiology. https://doi.org/10.1093/glycob/10.3.329

Almond A, Sheehan JK, Brass A (1997) Molecular dynamics simulations of the two disaccharides of hyaluronan in aqueous solution. Glycobiology. https://doi.org/10.1093/glycob/7.5.597

Almond A, Brass A, Sheehan JK (1998a) Deducing polymeric structure from aqueous molecular dynamics simulations of oligosaccharides: Predictions from simulations of hyaluronan tetrasaccharides compared with hydrodynamic and X-ray fibre diffraction data. J Mol Biol. https://doi.org/10.1006/jmbi.1998.2245

Almond A, Brass A, Sheehan JK (1998b) Dynamic exchange between stabilized conformations predicted for hyaluronan tetrasaccharides: Comparison of molecular dynamics simulations with available NMR data. Glycobiology. https://doi.org/10.1093/glycob/8.10.973

Aurélien Grosdidier OM Vincent Zoete (2011) SwissDock, a protein-small molecule docking web service based on EADock DSS. Nucl Acids Res 39(Web Server issue). https://doi.org/10.1093/NAR/GKR366. publisher: Nucleic Acids Res

Babik S, Samsonov SA, Pisabarro MT (2017) Computational drill down on FGF1-heparin interactions through methodological evaluation. Glycoconjugate J 34(3):427–440. https://doi.org/10.1007/s10719-016-9745-4

Bathe M, Rutledge GC, Grodzinsky AJ, Tidor B (2005) A coarse-grained molecular model for glycosaminoglycans: application to chondroitin, chondroitin sulfate, and hyaluronic acid. Biophys J 88(6):3870–3887. https://doi.org/10.1529/biophysj.104.058800

Bitomsky W, Wade RC (1999) Docking of glycosaminoglycans to heparin-binding proteins: Validation for aFGF, bFGF, and antithrombin and application to IL-8. J Am Chem Soc. https://doi.org/10.1021/ja983319g

Boittier ED, Burns JM, Gandhi NS, Ferro V (2020) GlycoTorch Vina: Docking designed and tested for glycosaminoglycans. J Chem Inf Model. https://doi.org/10.1021/acs.jcim.0c00373

Bojarski KK, Sieradzan AK, Samsonov SA (2019) Molecular dynamics insights into protein-glycosaminoglycan systems from microsecond-scale simulations. Biopolymers 110(7): e23252. https://doi.org/10.1002/bip.23252

Bojarski KK, Karczyńska AS, Samsonov SA (2020) Role of glycosaminoglycans in procathepsin B maturation: molecular mechanism elucidated by a computational study. J Chem Inf Model 60(4):2247–2256. https://doi.org/10.1021/acs.jcim.0c00023

Cardin AD, Weintraub HJ (1989) Molecular modeling of protein-glycosaminoglycan interactions. Arteriosclerosis. https://doi.org/10.1161/01.atv.9.1.21

Cilpa G, Hyvönen MT, Koivuniemi A, Riekkola ML (2010) Atomistic insight into chondroitin-6-sulfate glycosaminoglycan chain through quantum mechanics calculations and molecular dynamics simulation. J Comput Chem 31(8):1670–1680. https://doi.org/10.1002/jcc.21453

Clausen TM, Sandoval DR, Spliid CB, Pihl J, Perrett HR, Painter CD, Narayanan A, Majowicz SA, Kwong EM, McVicar RN, Thacker BE, Glass CA, Yang Z, Torres JL, Golden GJ, Bartels PL, Porell RN, Garretson AF, Laubach L, Feldman J, Yin X, Pu Y, Hauser BM, Caradonna TM, Kellman BP, Martino C, Gordts PLSM, Chanda SK, Schmidt AG, Godula K, Leibel SL, Jose J, Corbett KD, Ward AB, Carlin AF, Esko JD (2020) SARS-CoV-2 infection depends on cellular heparan sulfate and ACE2. Cell 183(4):1043–1057.e15. https://doi.org/10.1016/j.cell.2020.09.033

Clerc O, Mariethoz J, Rivet A, Lisacek F, Pérez S, Ricard-Blum S (2019) A pipeline to translate glycosaminoglycan sequences into 3D models. Application to the exploration of glycosaminoglycan conformational space. Glycobiology 29(1):36–44. https://doi.org/10.1093/glycob/cwy084

Danielsson A, Samsonov SA, Liwo A, Sieradzan AK (2023) Extension of the SUGRES-1P coarse-grained model of polysaccharides to heparin. J Chem Theory Comput 19(17):6023–6036. https://doi.org/10.1021/acs.jctc.3c00511

Eberhardt J, Santos-Martins D, Tillack AF, Forli S (2021) AutoDock Vina 1.2.0: new docking methods, expanded force field, and Python bindings. J Chem Inf Model. https://doi.org/10.1021/acs.jcim.1c00203

Ferreira LG, Dos Santos RN, Oliva G, Andricopulo AD (2015) Molecular docking and structure-based drug design strategies. Molecules. https://doi.org/10.3390/molecules200713384. ISSN: 14203049.

Forster MJ, Mulloy B (1993) Molecular dynamics study of iduronate ring conformation. Biopolymers. https://doi.org/10.1002/bip.360330407

Friesner RA, Banks JL, Murphy RB, Halgren TA, Klicic JJ, Mainz DT, Repasky MP, Knoll EH, Shelley M, Perry JK, Shaw DE, Francis P, Shenkin PS (2004) Glide: A new approach for rapid, accurate docking and scoring. 1. Method and assessment of docking accuracy. J Med Chem. https://doi.org/10.1021/jm0306430

Gandhi NS, Mancera RL (2009) Free energy calculations of glycosaminoglycan - Protein interactions. Glycobiology. https://doi.org/10.1093/glycob/cwp101

Gandhi NS, Mancera RL (2010) Can current force fields reproduce ring puckering in 2-O-sulfo–l-iduronic acid? A molecular dynamics simulation study. Carbohydrate Res. https://doi.org/10.1016/j.carres.2009.12.020

Griffith AR, Rogers CJ, Miller GM, Abrol R, Hsieh-Wilson LC, Goddard WA (2017) Predicting glycosaminoglycan surface protein interactions and implications for studying axonal growth. Pro Natl Acad Sci USA. https://doi.org/10.1073/pnas.1715093115

Habuchi H, Habuchi O, Kimata K (2004) Sulfation pattern in glycosaminoglycan: does it have a code? Glycoconjugate J. 21(1-2):47–52. https://doi.org/10.1023/B:GLYC.0000043747.87325.5e

Hileman RE, Fromm JR, Weiler JM, Linhardt RJ (1998) Glycosaminoglycan-protein interactions: Definition of consensus sites in glycosaminoglycan binding proteins. BioEssays. https://doi.org/10.1002/(SICI)1521-1878(199802)20:2textless156::AID-BIES8textgreater3.0.CO;2-R. ISSN:02659247.

Holmes SG, Nagarajan B, Desai UR (2022) 3-O-Sulfation induces sequence-specific compact topologies in heparan sulfate that encode a dynamic sulfation code. Comput Struct Biotechnol J 20:3884–3898. https://doi.org/10.1016/j.csbj.2022.07.013

Homeyer N, Gohlke H (2006) Free energy calculations by the molecular mechanics PoissonBoltzmann surface area method. Mol Inf 31(2):114–122. https://doi.org/10.1002/minf.201100135

Hsieh PH, Thieker DF, Guerrini M, Woods RJ, Liu J (2016) Uncovering the relationship between sulphation patterns and conformation of iduronic acid in heparan sulphate. Sci Rep 6:29602. https://doi.org/10.1038/srep29602

Imberty A, Lortat-Jacob H, Pérez S (2007) Structural view of glycosaminoglycan-protein interactions. Carbohydrate Res. https://doi.org/10.1016/j.carres.2006.12.019. ISSN: 00086215

Iozzo RV, Schaefer L (2015) Proteoglycan form and function: A comprehensive nomenclature of proteoglycans. Matrix Biol 42:11–55. https://doi.org/10.1016/j.matbio.2015.02.003

Jin L, Abrahams JP, Skinner R, Petitou M, Pike RN, Carrell RW (1997) The anticoagulant activation of antithrombin by heparin. Proc Natl Acad Sci 94(26):14683–14688. https://doi.org/10.1073/pnas.94.26.14683

Jorgensen WL, Chandrasekhar J, Madura JD, Impey RW, Klein ML (1983) Comparison of simple potential functions for simulating liquid water. J Chem Phys 79(2):926–935. https://doi.org/10.1063/1.445869. Publisher: American Institute of Physics

Joseph PRB, Mosier PD, Desai UR, Rajarathnam K (2015) Solution NMR characterization of chemokine CXCL8/IL-8 monomer and dimer binding to glycosaminoglycans: structural plasticity mediates differential binding interactions. Biochem J 472(1):121–133. https://doi.org/10.1042/BJ20150059

Kaufmann J, Möhle K, Hofmann HJ, Arnold K (1999) Molecular dynamics of a tetrasaccharide subunit of chondroitin 4-sulfate in water. Carbohydrate Res 318(1-4):1–9. https://doi.org/10.1016/s0008-6215(99)00091-9

Kjellén L, Lindahl U (2018) Specificity of glycosaminoglycan-protein interactions. Current Opinion Struct Biol 50:101–108. https://doi.org/10.1016/j.sbi.2017.12.011

Kogut MM, Maszota-Zieleniak M, Marcisz M, Samsonov SA (2021) Computational insights into the role of calcium ions in protein-glycosaminoglycan systems. Phys Chem Chem Phys PCCP 23(5):3519–3530. https://doi.org/10.1039/d0cp05438k

Kogut MM, Danielsson A, Ricard-Blum S, Samsonov SA (2022a) Impact of calcium ions on the structural and dynamic properties of heparin oligosaccharides by computational analysis. Comput Biol Chem 99:107727. https://doi.org/10.1016/j.compbiolchem.2022.107727

Kogut MM, Marcisz M, Samsonov SA (2022b) Modeling glycosaminoglycan-protein complexes. Current Opinion Struct Biol 73:102332. https://doi.org/10.1016/j.sbi.2022.102332

Kolesnikov AL, Budkov YA, Nogovitsyn EA (2014) Coarse-grained model of glycosaminoglycans in aqueous salt solutions. A field-theoretical approach. J Phys Chem B 118(46):13037–13049. https://doi.org/10.1021/jp503749a

Künze G, Huster D, Samsonov SA (2021) Investigation of the structure of regulatory proteins interacting with glycosaminoglycans by combining NMR spectroscopy and molecular modeling - the beginning of a wonderful friendship. Biol Chem 402(11):1337–1355. https://doi.org/10.1515/hsz-2021-0119

Liwo A, Baranowski M, Czaplewski C, Gołaś E, He Y, Jagieła D, Krupa P, Maciejczyk M, Makowski M, Mozolewska MA, Niadzvedtski A, Ołdziej S, Scheraga HA, Sieradzan AK, Slusarz R, Wirecki T, Yin Y, Zaborowski B (2014) A unified coarse-grained model of biological macromolecules based on mean-field multipole-multipole interactions. J Mol Model 20(8):2306. https://doi.org/10.1007/s00894-014-2306-5

Marcisz M, Samsonov SA (2023) Solvent model benchmark for molecular dynamics of glycosaminoglycans. J Chem Inf Model 63(7):2147–2157. https://doi.org/10.1021/acs.jcim.2c01472

Marcisz M, Huard B, Lipska AG, Samsonov SA (2021a) Further analyses of APRIL/APRIL-receptor/glycosaminoglycan interactions by biochemical assays linked to computational studies. Glycobiology 31(7):772–786. https://doi.org/10.1093/glycob/cwab016

Marcisz M, Maszota-Zieleniak M, Huard B, Samsonov SA (2021b) Advanced molecular dynamics approaches to model a tertiary complex APRIL/TACI with long glycosaminoglycans. Biomolecules 11(9):1349. https://doi.org/10.3390/biom11091349

Marcisz M, Zacharias M, Samsonov SA (2021c) Modeling protein-glycosaminoglycan complexes: does the size matter? J Chem Inf Model 61(9):4475–4485. https://doi.org/10.1021/acs.jcim.1c00664

Marcisz M, Gaardløs M, Bojarski KK, Siebenmorgen T, Zacharias M, Samsonov SA (2022) Explicit solvent repulsive scaling replica exchange molecular dynamics (RS-REMD) in molecular modeling of protein-glycosaminoglycan complexes. J Comput Chem 43(24):1633–1640. https://doi.org/10.1002/jcc.26965

Maszota-Zieleniak M, Marcisz M, Kogut MM, Siebenmorgen T, Zacharias M, Samsonov SA (2021) Evaluation of replica exchange with repulsive scaling approach for docking glycosaminoglycans. J Comput Chem 42(15):1040–1053. https://doi.org/10.1002/jcc.26496

Molinero V, Goddard WA (2004) M3B: A coarse grain force field for molecular simulations of malto-oligosaccharides and their water mixtures. J Phys Chem B. https://doi.org/10.1021/jp0354752

Morris GM, Goodsell DS, Halliday RS, Huey R, Hart WE, Belew RK, Olson AJ (1998) Automated docking using a Lamarckian genetic algorithm and an empirical binding free energy function. J Comput Chem. https://doi.org/10.1002/(SICI)1096-987X(19981115)19:14%3C1639::AID-JCC10%3E3.0.CO;2-B

Mottarella SE, Beglov D, Beglova N, Nugent MA, Kozakov D, Vajda S (2014) Docking server for the identification of heparin binding sites on proteins. J Chem Inf Model 54(7):2068–2078. https://doi.org/10.1021/CI500115J. Publisher: American Chemical Society

Muñoz-García JC, Lõpez-Prados J, Angulo J, Díaz-Contreras I, Reichardt N, De Paz JL, Martñ-Lomas M, Nieto PM (2012) Effect of the substituents of the neighboring ring in the conformational equilibrium of iduronate in heparin-like trisaccharides. Chem Eur J https://doi.org/10.1002/chem.201202770

Muñoz-García JC, Solera C, Carrero P, De Paz JL, Angulo J, Nieto PM (2013) 3D structure of a heparin mimetic analogue of a FGF-1 activator. A NMR and molecular modelling study. Organic Biomol Chem. https://doi.org/10.1039/c3ob41789a

Nagarajan B, Sankaranarayanan NV, Desai UR (2019) Perspective on computational simulations of glycosaminoglycans. Wiley Interdiscip Rev Comput Mol Sci. https://doi.org/10.1002/wcms.1388. ISSN: 17590884

Nagarajan B, Sankaranarayanan NV, Desai UR (2020) Rigorous analysis of free solution glycosaminoglycan dynamics using simple, new tools. Glycobiology 30(8):516–527. https://doi.org/10.1093/glycob/cwaa015

Nagarajan B, Holmes SG, Sankaranarayanan NV, Desai UR (2022) Molecular dynamics simulations to understand glycosaminoglycan interactions in the free- and protein-bound states. Current Opin Struct Biol 74:102356. https://doi.org/10.1016/j.sbi.2022.102356

Neamtu A, Tamba B, Patras X (2013) Cellulose chemistry and technology molecular dynamics simulations of chondroitin sulfate in explicit solvent: Point charge water models compared. Cellulose Chem Technol 47(3):191–202

Nivedha AK, Thieker DF, Makeneni S, Hu H, Woods RJ (2016) Vina-Carb: Improving glycosidic angles during carbohydrate docking. J Chem Theory Comput. https://doi.org/10.1021/acs.jctc.5b00834

Nordsieck K, Pichert A, Samsonov SA, Thomas L, Berger C, Pisabarro MT, Huster D, Beck-Sickinger AG (2012) Residue 75 of interleukin-8 is crucial for its interactions with glycosaminoglycans. Chembiochem Eur J Chem Biol 13(17):2558–2566. https://doi.org/10.1002/cbic.201200467

Paiardi G, Milanesi M, Wade RC, D'ursi P, Rusnati M (2021) A bittersweet computational journey among glycosaminoglycans. Biomolecules. https://doi.org/10.3390/biom11050739. ISSN: 2218273X

Perez S, Makshakova O, Angulo J, Bedini E, Bisio A, de Paz JL, Fadda E, Guerrini M, Hricovini M, Hricovini M, Lisacek F, Nieto PM, Pagel K, Pairardi G, Richter R, Samsonov SA, Vivès RA, Nikitovic D, Ricard Blum S (2022) Glycosaminoglycans: what remains to be deciphered? JACS Au https://doi.org/10.1021/jacsau.2c00569

Plazinski W, Knys-Dzieciuch A (2012) Interactions between CD44 protein and hyaluronan: Insights from the computational study. Mol BioSyst. https://doi.org/10.1039/c2mb05399c

Potthoff J, Bojarski KK, Kohut G, Lipska AG, Liwo A, Kessler E, Ricard-Blum S, Samsonov SA (2019) Analysis of procollagen C-proteinase enhancer-1/glycosaminoglycan binding sites and of the potential role of calcium ions in the interaction. Int J Mol Sci 20(20):5021. https://doi.org/10.3390/ijms20205021

Pérez S, Bonnardel F, Lisacek F, Imberty A, Ricard Blum S, Makshakova O (2020) GAG-DB, the new interface of the three-dimensional landscape of glycosaminoglycans. Biomolecules 10(12): 1660. https://doi.org/10.3390/biom10121660

Ricard-Blum S, Perez S (2022) Glycosaminoglycan interaction networks and databases. Current Opin Struct Biol 74:102355. https://doi.org/10.1016/j.sbi.2022.102355

Ricard-Blum S, Féraud O, Lortat-Jacob H, Rencurosi A, Fukai N, Dkhissi F, Vittet D, Imberty A, Olsen BR, van der Rest M (2004) Characterization of endostatin binding to heparin and heparan sulfate by surface plasmon resonance and molecular modeling: role of divalent cations. J Biol Chem 279(4):2927–2936. https://doi.org/10.1074/jbc.M309868200

Rodríguez-Carvajal MA, Imberty A, Pérez S (2003) Conformational behavior of chondroitin and chondroitin sulfate in relation to their physical properties as inferred by molecular modeling. Biopolymers. https://doi.org/10.1002/bip.10304

Rudd TR, Guimond SE, Skidmore MA, Duchesne L, Guerrini M, Torri G, Cosentino C, Brown A, Clarke DT, Turnbull JE, Fernig DG, Yates EA (2007) Influence of substitution pattern and cation binding on conformation and activity in heparin derivatives. Glycobiology 17(9): 983–993. https://doi.org/10.1093/glycob/cwm062

Samsonov SA, Pisabarro MT (2016) Computational analysis of interactions in structurally available protein-glycosaminoglycan complexes. Glycobiology 26(8):850–861. https://doi.org/10.1093/glycob/cww055

Samsonov SA, Gehrcke JP, Pisabarro MT (2014a) Flexibility and explicit solvent in molecular-dynamics-based docking of protein-glycosaminoglycan systems. J Chem Inf Model 54(2): 582–592. https://doi.org/10.1021/ci4006047

Samsonov SA, Theisgen S, Riemer T, Huster D, Pisabarro MT (2014b) Glycosaminoglycan monosaccharide blocks analysis by quantum mechanics, molecular dynamics, and nuclear magnetic resonance. BioMed Res Int 2014:808071. https://doi.org/10.1155/2014/808071

Samsonov SA, Bichmann L, Pisabarro MT (2015) Coarse-grained model of glycosaminoglycans. J Chem Inf Model 55(1):114–124. https://doi.org/10.1021/ci500669w

Samsonov SA, Zacharias M, Chauvot de Beauchene I (2019) Modeling large protein-glycosaminoglycan complexes using a fragment-based approach. J Comput Chem 40(14): 1429–1439. https://doi.org/10.1002/jcc.25797

Sankaranarayanan NV, Nagarajan B, Desai UR (2018) So you think computational approaches to understanding glycosaminoglycan-protein interactions are too dry and too rigid? Think again! Current Opin Struct Biol 50:91–100. https://doi.org/10.1016/j.sbi.2017.12.004

Sarkar A, Yu W, Desai UR, Mackerell AD, Mosier PD (2016) Estimating glycosaminoglycan-protein interaction affinity: Water dominates the specific antithrombin-heparin interaction. Glycobiology. https://doi.org/10.1093/glycob/cww073

Sattelle BM, Shakeri J, Almond A (2013) Does microsecond sugar ring flexing encode 3D-shape and bioactivity in the heparanome? Biomacromolecules 14(4):1149–1159. https://doi.org/10.1021/bm400067g

Scott JE, Cummings C, Brass A, Chen Y (1991) Secondary and tertiary structures of hyaluronan in aqueous solution, investigated by rotary shadowing-electron microscopy and computer simulation. Hyaluronan is a very efficient network-forming polymer. Biochem J. https://doi.org/10.1042/bj2740699

Shao C, Zhang F, Kemp MM, Linhardt RJ, Waisman DM, Head JF, Seaton BA (2006) Crystallographic analysis of calcium-dependent heparin binding to annexin A2. J Biol Chem. https://doi.org/10.1074/jbc.M604502200

Siebenmorgen T, Zacharias M (2020) Efficient refinement and free energy scoring of predicted protein-protein complexes using replica exchange with repulsive scaling. J Chem Inf Model. https://doi.org/10.1021/acs.jcim.0c00853

Stevic I, Parmar N, Paredes N, Berry LR, Chan AKC (2011) Binding of heparin to metals. Cell Biochem Biophys 59(3):171–178. https://doi.org/10.1007/s12013-010-9129-5

Teyra J, Samsonov SA, Schreiber S, Pisabarro MT (2011) SCOWLP update: 3D classification of protein-protein, -peptide, -saccharide and -nucleic acid interactions, and structure-based binding inferences across folds. BMC Bioinf 12:398. https://doi.org/10.1186/1471-2105-12-398

Uciechowska-Kaczmarzyk U, Babik S, Zsila F, Bojarski KK, Beke-Somfai T, Samsonov SA (2018) Molecular dynamics-based model of VEGF-A and its heparin interactions. J Mol Graph Model 82:157–166. https://doi.org/10.1016/j.jmgm.2018.04.015

Uciechowska-Kaczmarzyk U, Chauvot de Beauchene I, Samsonov SA (2019) Docking software performance in protein-glycosaminoglycan systems. J Mol Graph Model 90:42–50. https://doi.org/10.1016/j.jmgm.2019.04.001

Vallet SD, Clerc O, Ricard-Blum S (2020) Glycosaminoglycan–protein interactions: The first draft of the glycosaminoglycan interactome. J Histochem Cytochem. https://doi.org/10.1369/0022155420946403. ISSN: 15515044

Van Zundert GC, Rodrigues JP, Trellet M, Schmitz C, Kastritis PL, Karaca E, Melquiond AS, Van Dijk M, De Vries SJ, Bonvin AM (2016) The HADDOCK2.2 web server: user-friendly integrative modeling of biomolecular complexes. J Mol Biol 428(4):720–725. https://doi.org/10.1016/J.JMB.2015.09.014. Publisher: Academic Press

Varki A, Cummings RD, Esko JD, Stanley P, Hart GW, Aebi M, Mohnen D, Kinoshita T, Packer NH, Prestegard JH, Schnaar RL, Seeberger PH (eds) (2022) Essentials of glycobiology, 4th edn. Cold Spring Harbor Laboratory Press, Cold Spring Harbor (NY)

Verli H, Guimarães JA (2004) Molecular dynamics simulation of a decasaccharide fragment of heparin in aqueous solution. Carbohydrate Res. https://doi.org/10.1016/j.carres.2003.09.026

Vuorio J, Vattulainen I, Martinez-Seara H (2017) Atomistic fingerprint of hyaluronan–CD44 binding. PLOS Comput Biol 13(7):e1005663. https://doi.org/10.1371/journal.pcbi.1005663

Warwicker J, Watson HC (1982) Calculation of the electric potential in the active site cleft due to -helix dipoles. J Mol Biol. https://doi.org/10.1016/0022-2836(82)90505-8

Wiegel D, Kaufmann J, Arnold K (1999) Polar interactions of chondroitinsulfate: Surface free energy and molecular dynamics simulations. Colloids Surf B Biointerfaces. https://doi.org/10.1016/S0927-7765(98)00115-5

Yates EA, Santini F, De Cristofano B, Payre N, Cosentino C, Guerrini M, Naggi A, Torri G, Hricovini M (2000) Effect of substitution pattern on 1H, 13C NMR chemical shifts and 1JCH coupling constants in heparin derivatives. Carbohydrate Res 329(1):239–247. https://doi.org/10.1016/S0008-6215(00)00144-0

Part III

Biological Functions & Medical Relevance

Glycosylation in Stem Cell Biology

Chika Ogura and Shoko Nishihara

Contents

C. Ogura
Laboratory of Cell Biology, Department of Science and Engineering for Sustainable Innovation, Faculty of Science and Engineering, Soka University, Tokyo, Japan

S. Nishihara (✉)
Laboratory of Cell Biology, Department of Science and Engineering for Sustainable Innovation, Faculty of Science and Engineering, Soka University, Tokyo, Japan

Glycan & Life System Integration Center (GaLSIC), Soka University, Tokyo, Japan
e-mail: shoko@soka.ac.jp

F. Pfrengle et al. (eds.), *Complex Carbohydrates in Health and Disease*, Handbook of Experimental Pharmacology 288, https://doi.org/10.1007/164_2025_748

Abstract

Embryonic stem cells are pluripotent stem cells originally derived from the inner cell mass of blastocysts and have the essential characteristics of pluripotency and self-renewal. Pluripotent stem cells can differentiate into all of the cell types constituting the adult body. Our current understanding is that pluripotent stem cells transition through three stages: a naïve state, a formative state, and a primed state. The stemness and differentiation of pluripotent stem cells depend on cell-surface glycans, which work as essential modulators in ligand–receptor interactions, cell–cell interactions, and cell–extracellular matrix interactions. Cell-surface glycans bind to various signal ligands, including Wnt, fibroblast growth factors, and bone morphogenetic proteins, and are tissue-specific and developmentally regulated. In addition, intracellular *O*-linked *N*-acetylglucosamine, a modification found on only nuclear or cytoplasmic proteins, regulates core transcription factors involved in stemness, phosphorylation of downstream signal components, epigenetics, and liquid–liquid phase separation. Thus, various kinds of glycans regulate each stem cell status; furthermore, different glycan structures at each stage are simultaneously epigenetically regulated by the polycomb repressive complex PRC2. Understanding the functions of glycans in stemness and differentiation is increasingly important for both innovative clinical applications and basic research. This chapter focuses on the roles of glycans in mouse and human pluripotent stem cells.

Keywords

Embryogenesis · Embryonic stem cell · Glycan · Glycosylation · Glycosyltransferase · Pluripotent stem cell

1 Introduction

Embryonic stem cells (ESCs), originally derived from the inner cell mass (ICM) of blastocysts, are pluripotent stem cells that have the essential characteristics of pluripotency and self-renewal. Pluripotent stem cells have the ability to differentiate into every type of cell making up the adult body. The first ES cell lines were established from mouse pre-implantation embryos in 1981 (Evans and Kaufman

1981; Martin 1981) and were used to generate gene knockout (KO) mice (Thompson et al. 1989). The establishment of human ESCs from human blastocysts in 1998 (Thomson et al. 1998) paved the way for research into regenerative medicine. The momentum for regenerative medicine was augmented by the discovery of induced pluripotent stem cells (iPSCs), which can be established by the forced expression of several transcription factors in somatic cells such as fibroblasts (Takahashi and Yamanaka 2006; Takahashi et al. 2007; Yu et al. 2007). Extrinsic signals such as leukemia inhibitory factor (LIF), bone morphogenetic protein 4 (BMP4), Wnt, and fibroblast growth factor (FGF), in addition to intrinsic transcription factors including Oct3/4, Nanog, and Sox2, also determine stemness and differentiation.

Most cell-surface proteins and secreted proteins are glycosylated. The glycan structures present on glycoproteins and glycolipids on the cell surface are developmentally and tissue-specifically regulated and change widely during development and disease. The majority of glycosylation is carried out in the lumens of the endoplasmic reticulum (ER) and Golgi apparatus by almost 200 glycosyltransferases, which are also developmentally and tissue-specifically regulated. Therefore, glycans are widely used as cell markers because they reflect the state of the cell. Some surface glycoproteins and glycolipids are used as markers of ESCs and iPSCs; for example, stage-specific embryonic antigen-1 (SSEA-1) is used to mark mouse ESCs/iPSCs; and SSEA-3, SSEA-5, TRA-1-60, and TRA-1-81 antigens are used to mark human ESCs/iPSCs (Muramatsu and Muramatsu 2004; Adewumi et al. 2007; Atwood et al. 2008; Tang et al. 2011; Nishihara 2017, 2018). On the other hand, *O*-GlcNAc transferase (OGT), the only glycosyltransferase in the cytosol and nucleus, catalyzes the addition of a single *O*-linked β-*N*-acetylglucosamine (*O*-GlcNAc) residue from a donor uridine diphosphatide GlcNAc (UDP-GlcNAc) to a serine (Ser) or threonine (Thr) residue of cytosolic and nuclear proteins such as transcription factors, epigenetic factors, and signaling molecules (Haltiwanger et al. 1992).

Moreover, glycans function as essential modulators of cell–cell interactions, cell–extracellular matrix interactions, and, more importantly, ligand–receptor interactions. Sulfated glycans such as heparan sulfate (HS) and chondroitin sulfate (CS) bind to signal ligands including Wnt, BMP, FGF, and Hedgehog, functioning as co-receptors or stabilizing factors in various cell types (Sasaki et al. 2008; Nishihara 2010, 2020; Nakato and Li 2016). Because Wnt, FGF, and BMP signals are essential signals for pluripotency, self-renewal, and cell fate decisions in stem cells, these glycans are also considered to regulate pluripotency, self-renewal, and cell fate decisions in stem cells. The aim of this chapter is to introduce these glycan functions, with a focus on mouse and human pluripotent stem cells, and provide an overview of the roles of glycans in stem cells.

2 Glycosylation in Mouse and Human

Glycosyltransferases transfer a sugar moiety from a sugar-nucleotide donor substrate to an acceptor substrate. Most glycosylation of glycoproteins and glycolipids is performed in the lumens of the endoplasmic reticulum and Golgi apparatus by more than 200 glycosyltransferases (Narimatsu 2004), the majority of which are type-II membrane proteins comprising an *N*-terminal cytoplasmic region, a transmembrane region, a stem region, and the *C*-terminal catalytic region. Thus, the glycoproteins and glycolipids that are secreted from cells or associated with the cell-surface membrane are glycosylated sequentially by various glycosyltransferases as they pass through the endoplasmic reticulum and Golgi apparatus. Because glycosyltransferases are responsible for synthesizing the huge diversity of complex oligosaccharides, the glycan structures on proteins and lipids in different species can be estimated by an analysis of their glycosyltransferases.

Another type of glycosylation is carried out in the cytosol and nucleus by OGT, which is the only glycosyltransferase found in these structures (Haltiwanger et al. 1992). OGT transfers a single *O*-GlcNAc residue from UDP-GlcNAc to the serine and threonine residues of cytosolic and nuclear proteins, while *O*-GlcNAcase (OGA) removes it from proteins.

During evolution from yeast to mammals, organisms have acquired glycosyltransferases with divergent substrate specificities, with the result that humans and mice have almost 200 of these enzymes. A molecular evolutionary study has shown that the fundamental gene families of glycosyltransferases were established before the divergence of deuterostomes and protostomes, including *Caenorhabditis elegans* and *Drosophila* (Kaneko et al. 2001). Subsequently, early in the history of the vertebrate lineage, gene duplication led to an increase in intrafamilial genes, leading to the acquisition of a variety of substrate specificities for the synthesis of more complex glycan structures. Therefore, although the number of *Drosophila* glycosyltransferases is almost half that of human ones (Yamamoto-Hino et al. 2015), many fundamental glycan structures in humans can be found in *Drosophila*, such as HS, CS, *O*-GlcNAc, *O*-α-fucose (*O*-Fuc), *O*-β-glucose (*O*-Glc), *O*-linked mannose (*O*-Man), mucin-type *O*-glycans such as T antigen (Galβ1,3GalNAcα1-*O*-Ser/Thr) and Tn antigen (GalNAcα1-*O*-Ser/Thr), and some *N*-glycans (Toyoda et al. 2000; Altmann et al. 2001; Kanie et al. 2009; Aoki and Tiemeyer 2010). Glycan structures that are conserved across species may have important functions in mammalian development, including the status of pluripotent stem cells.

3 Mouse and Human Embryogenesis

After passing through the 2-cell, 4-cell, 8-cell, and morula stages, the mouse zygote forms a structure called the blastocyst in which the ICM is surrounded by a trophoblast at embryonic day 3.5–4.5 (E3.5–4.5). The blastocyst implants and becomes an egg cylinder at E4.5–5.5, after which important developmental events occur, such as the formation of an anterior–posterior axis and gastrulation (E6.5) (Fig. 1). The ICM differentiates into epiblasts at E5.5, and into three germ layers consisting of endoderm, mesoderm, and ectoderm after E6.5 (Rossant and Tam 2009).

Similarly, the human zygote forms a blastocyst at E5.0–6.0 after passing through the 2-cell, 4-cell, 8-cell, and morula stages. At E7.0–12, the post-implanted human embryo becomes the bilaminar embryonic disc in which the ICM is divided into two layers, the epiblast and hypoblast, which differs from mouse embryo. During gastrulation, the primitive streak arises in the epiblast at E13–14, and a portion of the epiblast cells invaginates into hypoblast (Fig. 1) (Shahbazi 2020). At E15–17, the epiblast cells differentiate into three germ layers: the remaining epiblast cells differentiate to ectoderm; the expanding cells in the internal two layers differentiate to mesoderm; and the cells incorporated in the hypoblast differentiate to endoderm.

Mouse ESCs are established from the ICM at E3.5–4.5, while human ESCs are established at E6.0–7.0 (Evans and Kaufman 1981; Martin 1981). Human iPSCs can be established by the transfection of four genes, *OCT3/4*, *SOX2*, *KLF4*, and *c-MYC*, into human fibroblasts (Takahashi and Yamanaka 2006; Takahashi et al. 2007; Yu et al. 2007). These cells are called pluripotent stem cells: they have both self-renewal

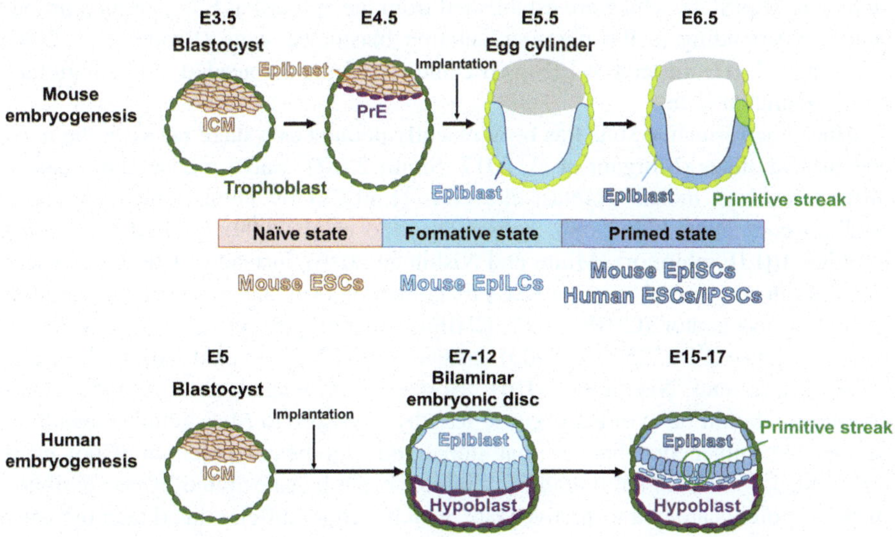

Fig. 1 Mouse and human embryogenesis. Schematic diagram of mouse and human embryogenesis. (Upper) Mouse ICM in the blastocyst differentiates to epiblast and primitive endoderm (PrE) at E3.5–4.5; the blastocyst then implants and forms an egg cylinder at E4.5–5.5. Gastrulation occurs at E6.5. Mouse embryonic stem cells (ESCs) correspond to the naïve state, mouse epiblast-like cells (EpiLCs) to the formative state, and mouse EpiSCs and human ESCs/induced pluripotent stem cells (iPSCs) to the primed state. (Lower) The human blastocyst implants at E5.0–7.0 and becomes the bilaminar embryonic disc at E7.0–12. The primitive streak arises in the epiblast at E13–14, and a portion of the epiblast cells invaginates into hypoblast

ability, proliferating and maintaining the undifferentiated state, and also pluripotency to differentiate into all cell types constituting the whole body of embryo.

4 Mouse and Human Pluripotent Stem Cells

Mouse and human pluripotent stem cells, including ESCs and iPSCs, exist in three different states, known as naïve, formative, and primed. The naïve state corresponds to the pre-implantation blastocyst stage of the embryo, the formative state corresponds to the early post-implantation blastocyst stage, and the primed state corresponds to the late post-implantation blastocyst stage before gastrulation (Table 1, Fig. 1) (Weinberger et al. 2016; Morgani et al. 2017; Nishihara 2017, 2018; Smith 2017). Pluripotent stem cells in each state show different characteristics. First, naïve state mouse ESCs are derived from the ICM of E3.5 embryos (Evans and Kaufman 1981; Martin 1981). Their developmental stage corresponds to the pre-implantation blastocyst; single-cell dissociation is possible, and their proliferation is fast. Pluripotency and self-renewal are maintained by adding LIF to the culture medium (Boeuf et al. 1997; Niwa et al. 2009). Second, mouse epiblast stem cells (EpiSCs), which are established from the epiblast at E5.5, are in a primed state corresponding to the post-implantation blastocyst stage (Brons et al. 2007; Tesar et al. 2007). Instead of LIF, FGF2 and Activin A are needed to maintain their undifferentiated state.

Third, the formative state has been recently defined as a stage between the naïve and primed states (Morgani et al. 2017; Smith 2017). Naïve mouse ESCs can be differentiated to mouse epiblast-like cells (EpiLCs) by culture and passage in medium containing FGF2 and Activin A (Guo et al. 2009), or FGF2 and JAK inhibitor 1 (LIF inhibitor) (Miura and Nishihara 2016), instead of LIF. Conversely, EpiLCs can be reverted to naïve ESCs by culture and passage in medium containing LIF, Wnt accelerator (CHIR99021, CHIR), and mitogen-activated protein kinase (Mek) inhibitor (PD0325901, PD03), instead of FGF2 (Zhou et al. 2010; Illich et al. 2016; Miura and Nishihara 2016). However, prolonged culture under these conditions should be avoided because lengthy exposure to Mek inhibitor results in irreversible epigenetic changes that impair developmental potential (Choi et al. 2017; Yagi et al. 2017) although Mek inhibitor can be substituted by Src inhibitor. In short, EpiLCs are in a formative state, which is more differentiated than the naive state but has not yet reached the primed state.

Mouse EpiSCs share a similar gene expression pattern and status of epigenetic regulation with conventional human ESCs, although their pattern and status are distinct from those of mouse ESCs. Thus, the status of conventional human ESCs is defined as the primed state (Tesar et al. 2007; Nichols and Smith 2009; Smith et al. 2014). Primed state pluripotent stem cells proliferate slowly and cannot be dissociated without RhoA/Rho-associated coiled-coil containing protein kinase (ROCK) inhibitor, and their gene transfer efficiency is low. Mouse and conventional human iPSCs are also defined as, respectively, naïve and primed pluripotent stem

Table 1 Characteristic features of naïve and primed states of pluripotent stem cells

Feature	Naïve state	Formative state	Primed state
Corresponding embryonic stage	Pre-implantation blastocyst	Early post-implantation blastocyst	Late post-implantation blastocyst
Established stem cells	Mouse ESCs/iPSCs	Mouse EpiLCs	Mouse EpiSCs, human ESCs/iPSCs
Required growth factors	LIF	Low FGF2, low Activin	FGF2, Activin
Pluripotency factors	Oct3/4, Sox2, Nanog, Klf2, Klf4	Oct3/4	Oct3/4, Sox2, Nanog
Specific markers	Rex1, Nr0b1, Klf4	Otx2	Fgf5, T
Glycan surface markers	SSEA-1[a] Gb3[b]	SSEA-1, 3, 4, 5	SSEA-3,4,5[c] Tra-1-60[c] Tra-1-81[c]
DNA methylation	Low	Intermediate	High
X chromosome inactivation	XaXa[d]	XaXa or XaXi	XaXi[e]
Metabolism	Oxidative phosphorylation	Oxidative phosphorylation, glycolysis	Glycolysis
Mitochondrial activity	High	Low	Low
Teratoma formation	Possible	Possible	Possible
Differentiation potential	Pluripotency	Pluripotency	Pluripotency (except for germ cells)
Blastocyst chimera	High	Partial	Very low
Morphology of cell colonies	Round	Flattened/rossette	Flattened
Single cell dissociation	Possible	Impossible[f]	Impossible[f]
Growth rate	Fast	Fast	Slow

ES embryonic stem, *iPS* induced pluripotent stem, *LIF* leukemia inhibitory factor, *FGF* fibroblast growth factor, *SSEA* stage-specific embryonic antigen
[a]glycan marker for mouse ESCs/iPSCs
[b]glycan marker for supposed naïve human ESCs/iPSCs
[c]glycan marker for conventional human ESCs/iPSCs
[d]Xa, activated X chromosome
[e]Xi, inactivated X chromosome
[f]impossible without ROCK inhibitor

cells because of characteristic features similar to those of mouse and human ESCs, respectively.

For functional analysis and application to regenerative medicine, an easy and reproducible method for establishing naïve human ESCs/iPSCs is desirable. Many methods for inducing naïve human ESCs/iPSCs have been developed, compared, and evaluated by means of gene expression patterns, methylation landscapes, and cell-surface markers (Zhou et al. 2023), although the naïve pluripotent state captured in mouse ESCs might be very transient in human embryos. The induction and maintenance of naïve human ESCs/iPSCs require mouse embryonic fibroblast cells and several inhibitors. The characteristics of conventional human ESCs/iPSCs differ depending on the cell line; therefore, the culture conditions are an important consideration.

5 Blastoid, Gastruloid, and Organoid Generation Using Pluripotent Stem Cells

While the mechanisms underlying early developmental processes in mice have been studied both in vivo and in vitro, research into human early development in vivo has lagged behind because of ethical issues surrounding the use of human embryos. In recent years, however, several methods have been developed to generate human early development models, including blastoids that mimic the human early blastocyst, gastruloids that mimic the human gastrula, and organoids that mimic human organs (Fig. 2) (Sullivan and Santos 2023; Yamanaka et al. 2023; Zhou et al. 2023).

Human blastoids are generated from naïve human ESCs/iPSCs and replicate the early blastocyst. They can be assembled by different methods from naïve human ESCs/iPSCs either alone or with other stem cells such as trophoblast stem cells and

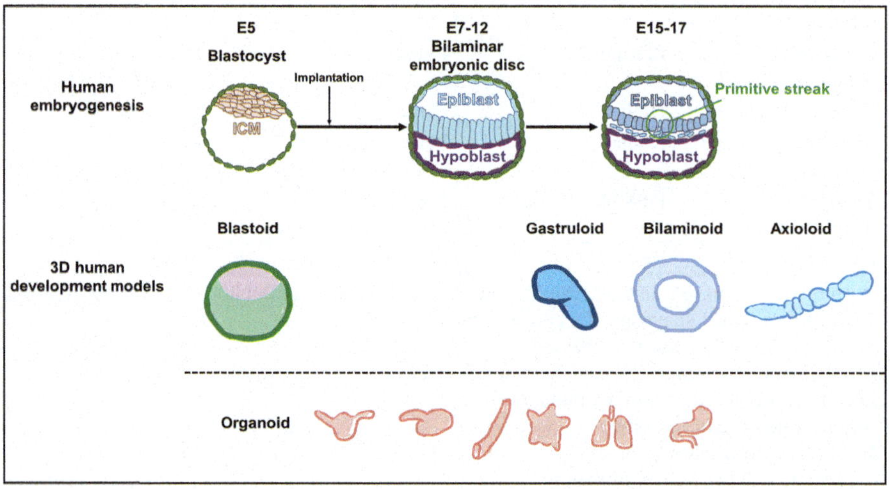

Fig. 2 Overview of the available models of human early development. Each 3D culture model mimics a stage of human early development. Blastoids, representing the human early blastocyst, are assembled from naïve human embryonic stem cells (ESCs)/induced pluripotent stem cells (iPSCs) either alone or with other stem cells such as trophoblast stem cells and extraembryonic endoderm stem cells. Gastruloids, representing the human gastrula, bilaminoids, representing the pregastrula stage, and axioloids, representing the somitogenesis stage, are developed from primed human ESCs/iPSCs. Organoids, representing various human organs, are generated from primed human ESCs/iPSCs or tissue stem cells

extraembryonic endoderm stem cells (Zhou et al. 2023). Blastoids are cultured in suspension during their generation; they are then transferred to an attachment culture to induce post-implantation differentiation. They are expected to be used as a pre/post-implantation model for understanding human early embryogenesis and the application of fertility treatment.

Gastrulation is a very important process in development when specification of the three germ layers, axis formation (e.g., anterior–posterior and left–right), and somitogenesis occur. Again, due to ethics and the 14-day rule, research on human gastrulation has lagged behind, but human gastruloids, which mimic the early gastrula (Sullivan and Santos 2023), human bilaminoids, which mimic the embryo at the pregastrula stage (Okubo et al. 2024), and human axioloids, which mimic the embryo at the somitogenesis stage (Yamanaka et al. 2023), have been developed from primed human ESCs/iPSCs and are expected to find use as models of the human gastrula.

Several methods have been reported for generating human organoids that mimic human organs and tissues. Currently, there are many types of organoids, including brain, heart, intestinal, and blood vessel organoids (Haniffa et al. 2021). A discussion of each organoid is beyond the scope of this chapter, but organoid technology is expected to have applications in disease research, drug assessment, and regenerative medicine.

6 Cell Markers for Mouse and Human Pluripotent Stem Cells

Many of the markers for mouse and human pluripotent stem cells, such as ESCs/iPSCs are glycans (Table 1) (Muramatsu and Muramatsu 2004; Adewumi et al. 2007; Atwood et al. 2008; Tang et al. 2011; Nishihara 2017, 2018). Mouse ESCs express SSEA-1, also called Lewis X carbohydrate antigen (LeX; Galβ1,4(Fucα1,3)GlcNAc), on both glycoproteins and glycolipids (Muramatsu and Muramatsu 2004; Atwood et al. 2008). Human ESCs do not express SSEA-1, but human ESCs/iPSCs commonly express SSEA-3, SSEA-4, SSEA-5, TRA-1-60, TRA-1-81, and R-10G antigens and Globo H, which are used as markers for identification (Adewumi et al. 2007; Liang et al. 2010; Natunen et al. 2011; Tang et al. 2011; Tateno et al. 2011; Fujitani et al. 2013; Kawabe et al. 2013; Matsumoto et al. 2015). SSEA-3 (R-3GalNAcβ1,3Galα1,4R′) and SSEA-4 (NeuAcα2,3Galβ1,3GalNAcβ1,3Galα1,4R′) are carried on globo-series glycolipids (Kannagi et al. 1983). SSEA-5 antigen is an H type 1 structure (Fucα1,2Galβ1,3GlcNAc) on a glycolipid or a mucin-type *O*-glycan. Both TRA-1-60 and TRA-1-81 antigens contain type 1 lactosamine (Galβ1,3GlcNAcβ1,3Galβ1,4GlcNAc) on a mucin-type *O*-glycan (Natunen et al. 2011). R-10G antibody recognizes keratan sulfate with low-sulfated structures (Kawabe et al. 2013). The Globo H structure is Fucα1,2Galβ1,3GalNAcβ1,3Galα1,4R′, while the H type 3 structure (Fucα1,2Galβ1,3GalNAc), which is included in Globo H, is also detected on glycolipids and glycoproteins and used as a human ESC/iPSC marker (Tateno et al. 2011). Thus, multiple carbohydrate markers are used for the identification of ESCs/iPSCs. In addition, CD77, which recognizes globotriaosylceramide Gb3, is a marker of the human blastocyst and naïve human ESCs prepared under two different conditions (Collier et al. 2017) (Table 1).

As mentioned above, many glycan structures form the basis of cell-surface markers for mouse and human pluripotent stem cells; however, their biological functions are not well understood. An RNA interference (RNAi) screen of more than 100 glycosyltransferases in mouse ESCs has been used to identify, via alkaline phosphatase activity, cell-surface glycans that are essential for naïve ESC self-renewal (Sasaki et al. 2011). In the next sections, our current knowledge of the roles of glycans in mouse and human pluripotent stem cells, including the results from RNAi screening of glycosyltransferases, will be described.

7 The Role of Glycans in Naïve Mouse Pluripotent Stem Cells

In naïve mouse ESCs, the pluripotency-associated transcription factor network comprises Oct3/4, Sox2, Nanog, Tbx3, Klf4, Klf2, Tfcp2l1, and Esrrb, which modulate each other and collaborate to regulate downstream gene expression to maintain self-renewal and pluripotency (Fig. 3). The pluripotency-associated transcription factor networks function downstream of exogenous signals including LIF, BMP4, and Wnt; these signals are essential for maintaining the naïve state. By contrast, FGF4 signal operates both in lineage commitment to primitive endoderm and as the trigger to differentiate from the naïve state (Kang et al. 2012, 2017; Molotkov et al. 2017). In the naïve state of pluripotent stem cells, glycans contribute to these four major signaling pathways: LIF, BMP4, Wnt, and FGF.

7.1 LacdiNAc Modulates LIF Signaling

LIF must be added to the culture medium of naïve pluripotent stem cells (Table 1) because it induces the expression of the above-mentioned pluripotency-associated transcription factors through activation of the Jak/Stat3 and PI3K/Akt pathways (Boeuf et al. 1997; Niwa et al. 2009; Martello et al. 2012, 2013; Nicola and Babon 2015; Ohtsuka et al. 2015). The Jak/Stat3 pathway directly activates Klf4 and Tfcp2l1; the PI3K/Akt pathway directly activates Tbx3 (Niwa et al. 2009; Martello et al. 2012, 2013).

An RNAi screen for glycosyltransferases identified β1,4 *N*-acetylgalactosaminyltransferase 3 (β1,4GalNAc-T3), which synthesizes the LacdiNAc (GalNAcβ1,4GlcNAc) structure, as essential for self-renewal and pluripotency (Sasaki et al. 2011). LacdiNAc, which is a unique terminal structure in the outer chain moieties of *N*- or *O*-glycans, is conserved from *Drosophila* to humans (Aoki and Tiemeyer 2010). It is highly expressed on naïve mouse ESCs and decreases during differentiation. LacdiNAc on LIF receptor (LIFR) and glycoprotein 130 (gp130) is required to maintain the naïve state via LIF/Stat3 signaling (Figs. 3 and 4a). Many ligands, receptors, and Src family kinases are assembled in lipid rafts:

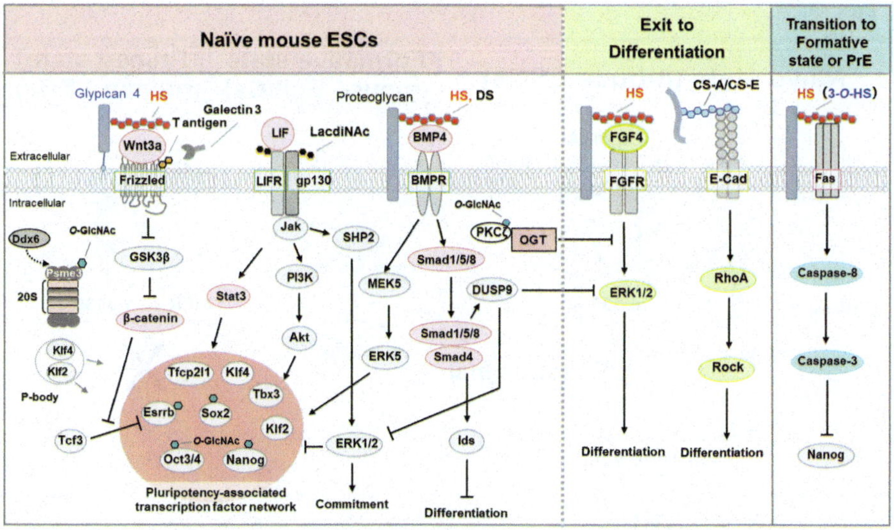

Fig. 3 Glycans and signals that function in naïve mouse embryonic stem cells and during exit and transition from the naïve state. In naïve mouse embryonic stem cells (ESCs), glycans contribute to the four major signals that are essential for maintaining the naïve state or triggering differentiation: leukemia inhibitory factor (LIF), bone morphogenetic protein 4 (BMP4), Wnt, and fibroblast growth factor 4 (FGF4). The LacdiNAc glycan structure contributes to maintaining the naïve state through LIF/STAT signaling. Heparan sulfate (HS) contributes to maintaining the naïve state through Wnt/β-catenin and BMP/Smad signaling and to exiting to differentiation through FGF4/extracellular signal-regulated kinase (ERK) signaling. Glypican 4, an HS proteoglycan, specifically regulates Wnt/β-catenin signaling. Through Fas signaling, the 3-*O*-sulfated heparan sulfate (3-*O*-HS) structure contributes to the transition from naïve state to formative state, which is followed by epiblast and three germ layer differentiation, or differentiation to primitive endoderm (PrE). Chondroitin sulfate A (CS-A) and CS-E contribute to the exit to differentiation through E-cadherin/Ras homolog family member A (RhoA) signaling. In the naïve state, *O*-GlcNAc on protein kinase Cζ (PKCζ) inhibits the phosphorylation and activation of PKCζ, which in turn inhibits FGF4/ERK signaling. *O*-GlcNAc on Oct3/4, Esrrb, and Nanog also contributes to maintaining the naïve state, while *O*-GlcNAc on Sox2 has inhibitory effect. *O*-GlcNAcylation of proteasome activator subunit 3 (Psme3) protein promotes Ddx6 degradation, which is essential for assembly of processing body (P-body), leading to the release and translation of mRNAs of Klf4 and Klf2, pluripotent transcription factors. Lastly, the endocytosis of Frizzled-5 is directly regulated by galectin-3, which binds to the T antigen on Frizzled-5

microdomains within the lipid bilayer of the cellular membrane enriched in glycosphingolipids (GSLs), sphingomyelin, and cholesterol, with a subset of transmembrane or glycosylphosphatidylinositol (GPI)-anchored proteins (Lingwood and Simons 2010). In the naïve state, LacdiNAc structures on LIFR and on gp130 contribute to interactions with caveolin-1 complex components of caveolae (a specific type of lipid raft incorporating caveolin-1); this interaction stabilizes LIFR and gp130 localization to the raft (Fig. 4a). The localization of receptors to caveolae is required to induce sufficiently strong transduction of LIF signaling to maintain self-renewal and the naïve state.

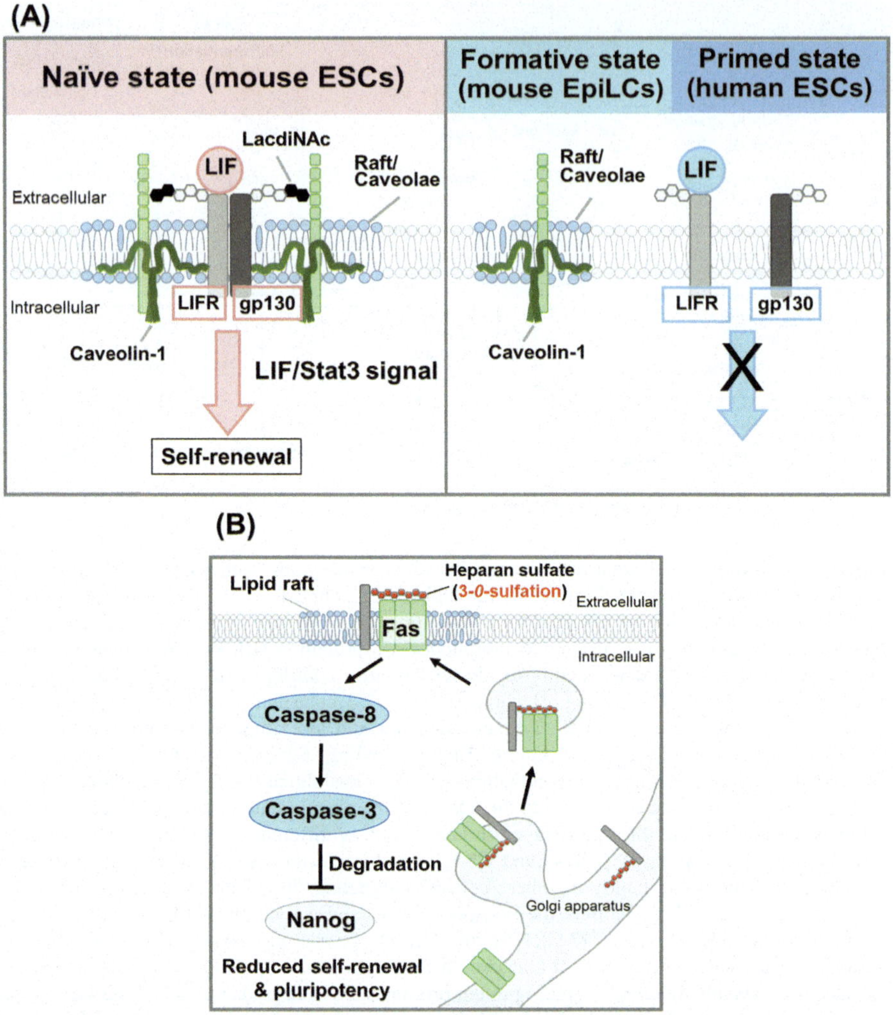

Fig. 4 The localization of ligand receptors to lipid rafts via glycans regulates pluripotent and differentiation signaling. (**a**) LacdiNAc modifications on leukemia inhibitory factor (LIF) receptor (LIFR) and glycoprotein 130 (gp130) are required for inducing and maintaining the naïve state. In the naïve state (mouse embryonic stem cells; ESCs), LacdiNAc structures on LIFR and gp130 contribute to the interaction of these proteins with caveolin-1 complex components and stabilize their localization to the caveolae (a type of lipid raft). This localization is vital for inducing sufficiently strong LIF signaling to maintain self-renewal and the naïve state. In the formative state (mouse epiblast-like cells; EpiLCs) or primed state (human induced pluripotent stem cells; iPSCs), LIFR and gp130 do not have LacdiNAc structures and cannot localize to the caveolae; therefore, the cells are no longer able to induce strong LIF signaling. (**b**) Transition of mouse pluripotent stem cells from the naïve to formative state or differentiation to primitive endoderm requires Fas signaling through 3-*O*-sulfated heparan sulfate (3-*O*-HS). During mouse ESCs transition from the naïve state, 3-*O*-HS structures are upregulated and bind to Fas, facilitating translocation of intracellular Fas into lipid rafts on the cell surface. This accumulation of Fas in the lipid rafts activates Fas/caspase-8/caspase-3 signaling; in turn, activated caspase-3 degrades Nanog, a key transcription factor for self-renewal and pluripotency of the naïve state

7.2 Heparan Sulfate and Dermatan Sulfate Modulate BMP4 and Wnt Signaling

In addition to LIF, several extrinsic signaling pathways, including BMP4/mothers against decapentaplegic homolog (Smad) and Wnt/β-catenin, are involved in self-renewal and pluripotency in naïve pluripotent stem cells (Fig. 3). BMP4 signaling via Smad1/5/8 with Smad4 induces Inhibitor of differentiation (Id; a negative regulator of basic helix–loop–helix transcription factors), which suppresses neural determination (Ying et al. 2003) and also upregulates Dual-specificity phosphatase 9 (DUSP9) to specifically reduce extracellular signal-regulated kinase (ERK) activity, which is required to exit the naïve state (Li et al. 2012). BMP4 also uses the MEK5/ERK5 pathway, which induces Klf2, a pluripotency factor of the naïve state (Morikawa et al. 2016). Synergistically with LIF, Wnt signaling promotes the maintenance of self-renewal via Esrrb activation (Niwa 2014), although Wnt signaling alone is insufficient to maintain self-renewal of naïve mouse ESCs (Ogawa et al. 2006).

The RNAi screen for glycosyltransferases also identified exostosin glycosyltransferase 1 (EXT1), which synthesizes disaccharide repeats of HS (Sasaki et al. 2008). In naïve mouse ESCs, HS is present at high levels (Nairn et al. 2007) and contributes to BMP4/Smad signaling in naïve mouse ESCs (Fig. 3). In *Ext1* KO mouse ESCs, phosphorylation of Smad1/5/8 in response to BMP4 is reduced (Kraushaar et al. 2012). BMP4/Smad signaling is also impaired in mouse ESCs that have knockdown (KD) of the genes encoding 3′-phosphoadenosine 5′--phosphosulfate transporters 1 (PAPST1) and PAPST2 or *N*-deacetylase/*N*-sulfotransferase 1 (NDST1) and NDST2, and these ESCs show reduced HS sulfation (Sasaki et al. 2009). PAPST1 and PAPST2 transport PAPS, a donor substrate of sulfotransferases, from the cytosol to the Golgi apparatus where sulfation takes place, while NDST1 and NDST2 catalyze the first step of HS sulfation. Moreover, BMP4 binds directly to heparin ($K_D = 69.4$ nM) (Sasaki et al. 2009). In this case, HS works as a co-receptor or stabilizer of BMP similar to its function for the *Drosophila* BMP homolog in germline stem cells (Guo and Wang 2009; Hayashi et al. 2009; Nakato and Li 2016; Nishihara 2020). Dermatan sulfate (DS), another type of glycosaminoglycan (GAG), also contributes to self-renewal and pluripotency in the naïve state through BMP4 (Ogura and Nishihara 2021), probably by the same mechanism as HS (Fig. 3). Dermatan-4-*O*-sulfotransferase-1 (D4ST1) transfers sulfate to the C-4 hydroxyl group of GalNAc in the iduronic acid (IdoA)-GalNAc repeating disaccharide structure of DS. The activity of the ESC self-renewal marker is reduced in *D4ST1* KD mouse ESCs and conversely increased in *D4ST1* overexpressing mouse ESCs.

The observation of strong binding of heparin to Wnt3a ($K_D = 26.0$ nM) (Sasaki et al. 2008) and *Drosophila* mutant analysis (Tabata and Takei 2004; Nishihara 2010, 2020) has led to the prediction that HS contributes to Wnt signaling in naïve mouse ESCs. However, the situation in mouse ESCs is complex and not yet resolved. In response to Wnt3a, Wnt signaling is upregulated in *Ext1* KO mouse ESCs, which completely lack HS (Kraushaar et al. 2012), but downregulated both in ESCs of the null mutant *Glypican 4* (Fico et al. 2012), which lacks an HS proteoglycan, and in mouse ESCs with transient *Extl* KD (Sasaki et al. 2008). The contributions of HS to Wnt signaling might depend on core proteins: HS on proteoglycans other than Glypican 4 might act as a reservoir, inhibiting the binding of Wnt to cognate receptors, whereas HS on Glypican 4 might act as a co-receptor or stabilizer.

7.3 T Antigen Modulates Wnt Receptor Endocytosis

The RNAi screen identified a fourth glycosyltransferase, core 1 β1,3-galactosyltransferase (C1GalT1), that synthesizes T antigen (Galβ1,3GalNAc), one of most abundant mucin-type *O*-glycans, as essential for self-renewal and pluripotency (Pecori et al. 2020). The elongation pathway via C1GalT1 is the most prominent among mucin-type *O*-glycosylation modifications in naïve mouse ESCs. *C1GalT1* KD results in the loss of ESC pluripotency and in differentiation to trophectoderm. Moreover, mucin-type *O*-glycosylation on Frizzled-5, a Wnt receptor, directly regulates its endocytosis via galectin-3 binding to T antigen on Frizzled-5; and a reduction in T antigen results in mouse ESCs exiting the pluripotent state via excess activation of canonical Wnt signaling (Fig. 5). This seems to be a novel regulatory mechanism that modulates Wnt signaling, and, consequently, ESC pluripotency.

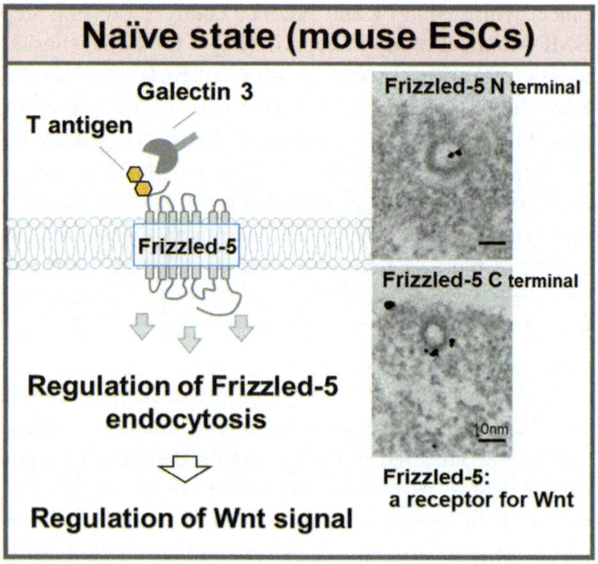

Fig. 5 T antigen, a mucin-type *O*-glycan, regulates pluripotency in naïve mouse embryonic stem cells (ESCs) via endocytosis of Frizzled-5. The Wnt receptor, Frizzled-5, is modified with T antigen (Galβ1,43GalNAc). Galectin-3 binds to T antigen, causing endocytosis of Frizzled-5, and thereby regulates Wnt signaling. The photos on the right show vesicles containing Frizzled-5 (*N*- and *C*-terminal labels) taken up by endocytosis

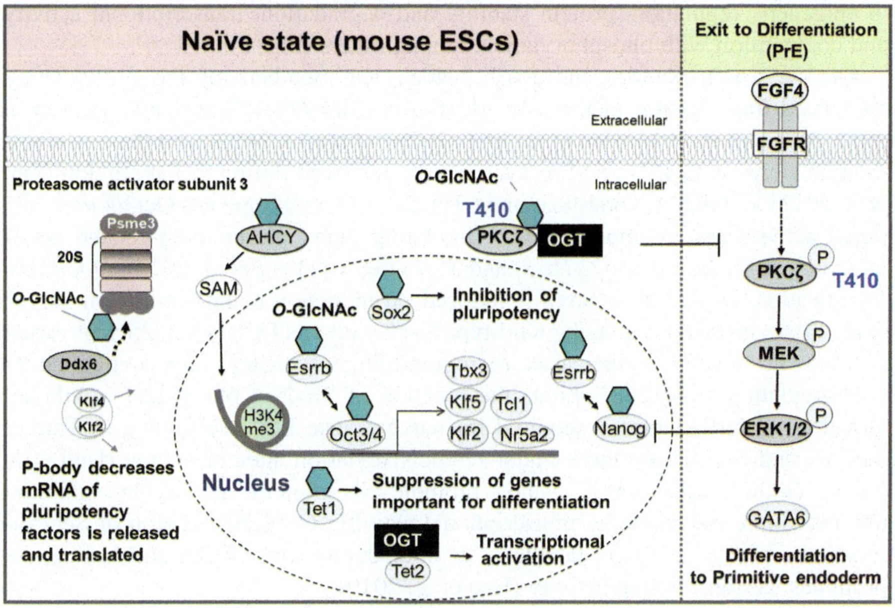

Fig. 6 *O*-GlcNAc is required for maintaining the naïve state of mouse embryonic stem cells via several pathways. *O*-GlcNAc modification of Thr228 of Oct3/4 regulates the transcriptional activity of Oct3/4 and induces genes essential for maintaining naïve status, including *Klf2*, *Klf5*, *Nr5a2*, *Tbx3*, and *Tcl1*. *O*-GlcNAcylation of Nanog promotes the stability and transcriptional activity of Nanog. *O*-GlcNAcylation of Ser25 of Essrb similarly stabilizes Essrb, facilitating its interactions with Oct3/4 and Nanog, and promoting its transcription activity. By contrast, *O*-GlcNAcylation of Ser248 of Sox2 has an inhibitory effect on the maintenance and establishment of pluripotency. In addition, *O*-GlcNAcylation of ten-eleven translocation methylcytosine dioxygenase 1 (Tet1) stabilizes Tet1 protein and represses developmentally important genes. Tet2 directly interacts with OGT and facilitates transcriptional activation. *O*-GlcNAcylation of Thr136 of adenosylhomocysteinase (AHCY), which regulates intracellular S-adenosylmethionine (SAM) through the methionine cycle, increases its activity and is important for histone H3 lysine 4 (H3K4me3) trimethylation to maintain pluripotency. On the other hand, *O*-GlcNAcylation of PKCζ on Thr410 inhibits the FGF4–PKCζ–MEK–ERK1/2 pathway via inhibition of PKCζ phosphorylation at this residue. Moreover, *O*-GlcNAcylation of Ser111 of Proteasome activator subunit 3 (Psme3) mediates the interaction of Psme3 with DEAD box polypeptide 6 (Ddx6), which is essential for assembly of processing body (P-body), and promotes Ddx6 degradation, leading to a reduction in P-body, and the release and translation of Klf4 and Klf2 mRNAs

7.4 *O*-GlcNAc Is also Required for the Naïve State

The RNAi screen for glycosyltransferases also identified OGT (Miura et al. 2018), the only known glycosyltransferase in the cytosol and nucleus. As mentioned above, OGT transfers GlcNAc from UDP-GlcNAc to serine and threonine residues of core proteins, whereas OGA removes it (Hart 2014). GlcNAc is required for the viability of naïve mouse ESCs, and the *Ogt* KO mouse dies during embryogenesis (Shafi et al. 2000; Hart et al. 2011). *O*-GlcNAc has many functions in naïve mouse ESCs related to epigenetic regulation, protein stability and degradation, transcriptional activity, and competition with phosphorylation, among others (Fig. 6).

Oct3/4, Sox2, Nanog, and Esrrb, which are members of the pluripotency-associated transcription factor network (Fig. 3), are *O*-GlcNAcylated (Jang et al. 2012; Myers et al. 2016; Hao et al. 2019, 2023; Abo et al. 2022), and the *O*-GlcNAc modifications on Oct3/4 and Sox2 are rapidly removed during differentiation (Jang et al. 2012). *O*-GlcNAc modification of Thr228 in Oct3/4 regulates Oct3/4 transcriptional activity and is important for stimulating many pluripotency-related genes, including *Klf2*, *Klf5*, *Nr5a2*, *Tbx3*, and *Tcl1* (Fig. 6) (Jang et al. 2012). *O*-GlcNAc modification of Ser248 in Sox2 is reduced on induction of differentiation (Myers et al. 2016); however, replacing wild-type Sox2 with an *O*-GlcNAc-deficient variant (Ser248Ala) results in increased reprogramming efficiency, suggesting that *O*-GlcNAcylation at Ser248 inhibits the function of Sox2 in pluripotency, although global *O*-GlcNAcylation is required for naïve mouse ESCs self-renewal. Thr278, Ser267, and Ser269 are three major *O*-GlcNAcylation sites of Nanog (Hao et al. 2023). *O*-GlcNAcylation of Nanog promotes its protein stability by inhibiting ubiquitination and enhances transcriptional activity. *O*-GlcNAcylation of Ser25 of Essrb also stabilizes Essrb, facilitates its interactions with Oct3/4 and Nanog, and promotes its transcription activity (Hao et al. 2019).

Ten-eleven translocation methylcytosine dioxygenase 1 (Tet1) and Tet2, which are expressed in naïve mouse ESCs, belong to the Tet family of proteins that convert 5-methylcytosine to 5-hydroxymethylcytosine, an intermediate in the DNA demethylation pathway (Vella et al. 2013). Tet1 is stabilized by *O*-GlcNAcylation and required for maintaining naïve ESCs through repression of developmentally important genes by associating with the chromatin repression complexes Sin3A and NuRD (Fig. 6) (Shi et al. 2013). Tet2 directly interacts with OGT and facilitates transcriptional activation (Chen et al. 2013), possibly because the Tet2–OGT complex positively regulates gene transcription through histone H2B GlcNAcylation at Ser112. Adenosylhomocysteinase, a key metabolic enzyme in the methionine cycle, regulates intracellular *S*-adenosylmethionine levels and thereby the methylation status of histones and DNA. In naïve mouse ESCs, *O*-GlcNAcylation of Thr136 on adenosylhomocysteinase increases its activity and is important for maintaining trimethylation of histone H3 lysine 4 (H3K4me3) to continue mouse ESC pluripotency (Zhu et al. 2020).

Proteasome activator subunit 3 (Psme3; also known as PA28g or Reg g) is a nuclear regulatory component that binds to the 20S proteasome (Li et al. 2005). Ser111 *O*-GlcNAcylation of Psme3 mediates its interaction with DEAD box polypeptide 6 (Ddx6), which is essential for assembly of processing body (P-body), and promotes degradation of Ddx6 (Fig. 6). This leads to reduced amounts of P-body, and the release and translation of Klf4 and Klf2 mRNAs belonging to the pluripotency-associated transcription factor network (Pecori et al. 2021a).

In addition, *O*-GlcNAc modification of protein kinase C-ζ (PKCζ) on Thr410, a phosphorylation site, inhibits the FGF4–PKCζ–MEK–ERK1/2 pathway in naïve mouse ESCs via inhibition of PKCζ phosphorylation, and consequently, it inhibits differentiation to primitive endoderm to maintain the naïve state (Fig. 6) (Miura et al. 2018).

In summary, LacdiNAc, HS, DS, T antigen (Galβ1,3GalNAc), and *O*-GlcNAc have essential functions in the naïve state of mouse pluripotent stem cells. Notably, these glycan structures are conserved from *Drosophila* to mammals.

8 The Role of Glycans in Differentiation from the Naïve Pluripotent State

Overall levels of HS, CS/DS, and hyaluronic acid (HA) increase during the differentiation of mouse ESCs (Nairn et al. 2007). Both HS and CS have been shown to contribute to the transition from the naïve pluripotent state.

8.1 *N*-sulfation of HS, CS-A, and CS-E Is Required for Exit from the Naïve State

FGF4 signaling has dual roles during pre-implantation development of the mouse embryo (Molotkov et al. 2017). FGF4 is the key signal for specification of primitive endoderm versus pluripotent epiblast within the ICM (Kang et al. 2012, 2017; Molotkov et al. 2017). Naïve mouse ESCs secrete FGF4, and FGF4/FGF receptor 1 (FGFR1) signaling is essential both for exit from naïve state pluripotency (Fig. 3) and for attaining the primed state (Molotkov et al. 2017). In this pathway, HS is required for binding of FGF4 to its cognate receptor, activation of downstream cascades, and exit from the naïve state (Fig. 3) (Johnson et al. 2007; Kraushaar et al. 2012). In *Ext1* KO mouse embryoid bodies, high expression of Nanog and Oct3/4 is maintained, and no lineage markers are expressed (Kraushaar et al. 2012). *Ndst1* and *Ndst2* KO mouse ESCs retain high Nanog and Oct3/4 expression, and fail to induce the primitive ectoderm marker FGF5 and the primitive endoderm marker GATA6 (Lanner et al. 2010). Moreover, *Ndst1* and *Ndst2* KO mouse ESCs show reduced ERK1/2 phosphorylation in response to FGF4 but not in response to FGF2, demonstrating that either *N*-sulfation is selectively required for FGF4, or that FGF4 is more sensitive than FGF2 to the reduction of HS.

CS also contributes to exit from the naïve pluripotent state (Izumikawa et al. 2014). CS-A and CS-E, which share the common structure of a sulfated hydroxyl group at positions 4′ of GalNAc, bind to E-cadherin on naïve mouse ESCs and stimulate differentiation through Rock signaling (Fig. 3).

8.2 3-*O*-sulfation of HS Is Required for Transition to the Formative State

As mentioned above, naïve mouse ESCs can be converted to formative EpiLCs by culture and passage in medium containing FGF2 and Activin A (Guo et al. 2009), or

FGF2 and JAK inhibitor 1 (Miura and Nishihara 2016), instead of LIF. During the transition from naïve to formative state or the differentiation of primitive endoderm, HS 3-*O*-sulfotransferase 5 (HS3OST-5) is upregulated, leading to an increase in 3-*O*-sulfated HS (Figs. 3 and 4b) (Hirano et al. 2012, 2013). 3-*O*-sulfated HS binds to the heparin-binding domain of the first apoptosis signal receptor (Fas), which belongs to the tumor necrosis factor receptor family, causing translocation of Fas into lipid rafts on the surface of mouse ESCs (Fig. 4b). The accumulation of Fas in lipid rafts activates the Fas/caspase-8/caspase-3 signaling cascade; in turn, the activated caspase-3 degrades Nanog, a key transcription factor for maintaining self-renewal and pluripotency of mouse ESCs.

9 The Role of Glycans in the Formative or Primed State of Pluripotent Stem Cells

As outlined above, mouse EpiLCs correspond to the formative state (Smith 2017), while mouse EpiSCs and conventional human ESCs/iPSCs, which have similar gene expression patterns and epigenetic regulation status, correspond to the primed state of pluripotent stem cells (Tesar et al. 2007; Nichols and Smith 2009; Smith 2017).

9.1 OGT and Sialylated Podocalyxin Are Required for the Formative State

During transition, from the naïve to formative state in mouse ESCs, expression of nucleocytoplasmic OGT does not change, while that of cytosolic OGA is significantly increased (Miura and Nishihara 2016). OGT is required for the survival of formative mouse EpiLCs but is not involved in maintaining the undifferentiated formative state, whereas OGA is required for neither survival nor maintenance, although the *O*-GlcNAcylation pattern in formative mouse EpiLCs is distinct from that in naïve ESCs.

The role of OGT in the primed state has not been well clarified. Excess *O*-GlcNAc modification does not affect the pluripotency of human ESCs; instead, it restricts the differentiation of hPSCs towards specific cell lineages (Maury et al. 2013). In human ESCs, Thr250/Ser251 and Ser278 in ring finger protein 1 (RING1B), which is the catalytic subunit of polycomb repressive complex 1 (PRC1; a major epigenetic repressor of pluripotency and differentiation), are *O*-GlcNAcylated and the extent of *O*-GlcNAcylation decreases during differentiation (Maury et al. 2015). In chromatin immunoprecipitation (ChIP)-sequencing analysis, RING1B lacking *O*-GlcNAc is found to be enriched near cell cycle genes, whereas *O*-GlcNAcylated RING1B is preferentially enriched near neuronal genes, suggesting that this modification may influence the interaction of PRC1 with DNA and may be an essential regulator of stem cell growth and differentiation.

During peri-implantation development of the mouse embryo, a cluster of EpiSCs undergo epithelialization, creating an apical surface in contact with a cavity

(Bedzhov and Zernicka-Goetz 2014). At E5.0–5.25, the lumen is formed by hollowing, which results from charge repulsion between apical membranes coated by anti-adhesive glycoproteins such as the sialylated mucin podocalyxin (Podxl). *Podxl* KO mouse ESCs cultured in Matrigel as a model system for embryogenesis show defective lumenogenesis (Shahbazi et al. 2017). In the human embryo and human ESCs, the conditions for inducing exit from naïve pluripotency and lumenogenesis lead to both rosette and lumen formation. Human PODXL also coats lumen, indicating that the sialylation of this mucin is important for lumenogenesis in both mouse and human.

9.2 ST6GAL1, HA-Mediated Motility, and HS Are Required for the Primed State

Glycan function in the primed state has begun to be clarified using conventional human ESCs/iPSCs. In human ESCs, β-galactoside *a*2,6-sialyltransferase 1 (ST6GAL1) and its product, α2,6 sialylated glycans, are preferentially expressed and play an essential role in maintaining the primed state by regulating OCT3/4 and SOX2 (Wang et al. 2015). Moreover, inhibition of ST6GAL1 activity impairs the generation of human iPSCs. These observations indicate that ST6GAL1 functions in the regulation of pluripotency and differentiation in primed human ESCs/iPSCs.

Hyaluronan synthase 2 (HAS2) and receptor for HA-mediated motility (RHAMM), an HA-binding protein, are expressed in human ESCs, as well as in all stages of early pre-implantation human embryos (Choudhary et al. 2007). RHAMM expression is downregulated during the differentiation of human ESCs, while HAS2 is upregulated. KD analysis of each gene shows that RHAMM is important for maintaining human ESCs pluripotency, viability, and cell cycle control; however, HAS2 plays a role in the differentiation of human ESCs, and the addition of exogenous HA to the medium enhances their differentiation to mesodermal and cardiac lineages.

GAGs, including HS and CS, are most highly expressed on human ESCs/iPSCs (Fujitani et al. 2013), although their functions in primed human ESCs/iPSCs have not been clarified as yet. Nevertheless, serum-replacement media containing FGF2 and heparin have been developed for maintaining their self-renewal and pluripotency (Furue et al. 2008). Murine embryonic fibroblasts, which are used as feeder cells for human ESCs and iPSCs, secrete HS proteoglycans that bind and stabilize FGF2 (Levenstein et al. 2008; Stelling et al. 2013), and are probably required for FGFR dimerization during the activation of FGF signaling. Supplementation of culture medium with HS is important for cost-saving in the mass production process of human iPSCs for regenerative therapy (Horiguchi et al. 2018). Recently, HS-deficient human ESCs have been shown to maintain self-renewal and pluripotency under standard culture conditions with high levels of FGF2 (Syangtan et al. 2025). Although HA is not sulfated, artificially sulfating this glycan can regulate its binding to FGF2, and highly sulfated HA (at a level of 3.63 for the maximum 4 hydroxyl groups per disaccharide unit) binds FGF2 more strongly than heparin (Miura et al. 2019). Adding sulfated HA to the medium facilitates the culture of pluripotent human iPSCs under feeder cell-free conditions without needing both FGF2 and serum by enhancing the signal derived from the FGF secreted by human iPSCs themselves.

Several markers of human ESCs/iPSCs incorporate Fucα1,2Gal, such as the Globo H structure (Fucα1,2Galβ1,3GalNAcβ1,3Galα1,4R'), H type 1 structure (Fucα1,2Galβ1,3GlcNAc) recognized as SSEA-5 antigen, and H type 3 structure (Fucα1,2Galβ1,3GalNAc) (Tang et al. 2011). The function of these glycan structures remains unknown. It has been shown, however, that iPSCs with no defects can be established from an individual with Bombay blood-type carrying a nonsense mutation 826C to T (276 Gln to Ter) in the *fucosyltransferase 1* (*FUT1*) gene and a missense mutation 739G to A (247 Gly to Ser) in the *FUT2* gene, leading to a deficiency in H antigen (Fucα1,2Galβ1,4GlcNAc) on red blood cells (Seifinejad et al. 2010). Both FUT1 and FUT2 are involved in synthesis of the Fucα1,2 structure in the above markers, suggesting that these Fucα1,2 structures may not have essential functions in the human primed state.

In summary, OGT and sialylated Podxl have important functions in the formative state, while OGT, ST6GAL1, RHAMM, and HS have important functions in the primed state. Among these factors, OGT, ST6GAL1, and HS are conserved from *Drosophila* to mammals. However, *Drosophila* does not have sialomucin Podxl, HA, or the HA-binding protein RHAMM.

10 The Role of Glycans in Transition from the Formative or Primed State to Naïve State

As mentioned above, primed mouse EpiSCs or formative mouse EpiLCs can be reverted to naïve ESC-like cells by culture in medium containing LIF, Wnt accelerator, and Mek inhibitor (Zhou et al. 2010; Illich et al. 2016; Miura and Nishihara 2016). Furthermore, naïve human ESCs can be induced from conventional human ESCs in a primed state under either "PXGL" conditions, where P is PD0325901 (Mek inhibitor), X is XAV939 (Wnt inhibitor), G is Gö6963 (PKC inhibitor), and L is LIF (Bredenkamp et al. 2019); or "5iLA" conditions, where 5i is Mek inhibitor, Wnt accelerator, RAF inhibitor, SRC inhibitor, and ROCK inhibitor; L is LIF; and A is Activin A (Theunissen et al. 2016).

10.1 LacdiNAc, OGT, and OGA Are Required for the Formative to Naïve Transition

Both LacdiNAc expression (Sasaki et al. 2011) and LIF/Stat3 signaling (Bao et al. 2009; Yang et al. 2010) contribute to reversion from the formative to the naïve state in mice (Fig. 4a). The rate of reversion in mouse EpiLCs with KD of the gene encoding β4GalNAc-T3, which synthesizes LacdiNAc, is significantly lower than that in control cells, indicating that upregulation of LacdiNAc is required to transition from the formative state back to the naïve state (Sasaki et al. 2011). In addition, the efficiency of reversion from the formative to naïve state is significantly lower in culture in the presence of an OGT or OGA inhibitor, indicating that both OGT and OGA are required for mouse EpiLCs to transition to naïve ESC-like cells (Miura and Nishihara 2016).

10.2 Knockout of B3GALT5 Promotes the Primed to Naïve Transition in Human ESCs

Similar to naïve mouse ESCs, naïve human ESCs express neolacto-series GSLs, including SSEA-1 (Galβ1,4(Fucα1,3)GlcNAc) and H type 2 antigen (Fucα1,2Galβ1,4GlcNAc) (Lin et al. 2020). During the transition from the primed to naïve state, the expression pattern switches from globo-series GSLs (e.g., SSEA-3, SSEA-4, and Globo H) and lacto-series GSLs (e.g., fucosyl-Lc4Cer) to neolacto-series GSLs (e.g., SSEA-1 and H type 2 antigen), along with significant downregulation of β-1,3-galactosyltransferase (B3GALT5), the key glycosyltransferase for the biosynthesis of globo- and lacto-series GSLs, including conventional human ESCs markers, in the primed state. In addition, *B3GALT5* KO in conventional primed human ESCs changes the GSL profile to expression of neolacto-series and facilitates the transition to naïve state.

11 The Role of Glycans in Trophoblast Stem Cells

After fertilization, every cell in the embryo has totipotency: that is, the ability to produce both the embryo itself and the cells that will become the placenta and other tissues that nourish the embryo. In mice, the first cell lineage commitment occurs in the embryo around E3.5 after fertilization, forming a blastocyst with the trophoblast (the cells that will form the placenta) on the outside and the ICM (the multipotent cell population that will form the embryonic body) on the inside. Trophoblast stem cells (TSCs) can be established from the blastocyst or trophoblast and have the ability to differentiate to the three major types of trophoblast cell lineage: cytotrophoblast, extravillous cytotrophoblast, and syncytiotrophoblast (Tanaka et al. 1998; Okae et al. 2018).

11.1 PIG-A Is Required for Human Trophoblast Formation

Phosphatidyl-inositol-glycan class A (PIG-A) is required for the first step in the synthesis of GPI (Kinoshita et al. 1997). *PIG-A*-deficient human ESCs, which lack GPI-anchored cell-surface proteins, can form embryoid bodies and initiate differentiation into the three embryonic germ layers. However, they fail to form trophoblast cells because of the lack of GPI-anchored BMP co-receptors, resulting in impaired activation of BMP4 signaling (Chen et al. 2008).

11.2 GALNT3 Is Required for the Epithelial State in TSCs

Multipotent TSCs isolated from pre-implantation blastocysts can be cultured indefinitely in the presence of FGF4 (Tanaka et al. 1998). TSCs have epithelial properties with tight cell–cell adhesion and apical–basal polarity. During implantation, the first developmental epithelial-to-mesenchymal transition occurs in TSCs, beginning with the disruption of epithelial cell–cell adhesion. Polypeptide *N*-acetylgalactosaminyltransferase 3 (GALNT3) belongs to the glycosyltransferase family that transfers *N*-acetylgalactosamin (GalNAc) to proteins in the first step of mucin-type *O*-glycosylation. By promoting mucin-type *O*-glycosylation of serine/threonine residues in proteins such as E-cadherin, GALNT3 is essential for the epithelial state in TSCs (Raghu et al. 2019). Loss of GALNT3 results in the retention of E-cadherin in the Golgi apparatus, disruption of linear adherens junction assembly, and induction of the epithelial-to-mesenchymal transition.

12 Regulation of Glycosylation Through Glycosyltransferases in Stem Cells

12.1 PRC2 Controls More Than 40% of Glycosyltransferase Expression in the Naïve to Formative Transition

As described above, mouse ESCs and mouse EpiLCs reflect two distinct pluripotent states known as the naïve state and the formative state, respectively; together, they provide a useful in vitro model system to examine the pluripotent state transition that occurs at implantation (Smith 2017). Total glycome analysis, which determines the absolute amounts of all glycan structures such as *N*-glycans, *O*-glycans, GAGs, GSLs, and free oligosaccharides, shows that all types of glycan undergo dramatic changes during the transition from naïve to formative state (Fig. 7a) (Pecori et al. 2021b). The identification of novel specific glycan structures in each cell has also provided additional markers to distinguish between the naïve and formative pluripotent states. As compared with other types of cells, such as human ESCs and normal/carcinoma somatic cell lines (Fujitani et al. 2013), both mouse ESCs and mouse EpiLCs show a typical comprehensive glycomic profile: there is greater expression of GSLs than of other types of glycans, although GSL expression decreases during the transition from mouse ESCs to mouse EpiLCs (Fig. 7a). The amount of free oligosaccharides also decreases during the transition, whereas *N*-glycans, *O*-glycans, and GAGs increase in content.

Quantitative analysis of the gene expression of all glycosyltransferases in the samples used for total glycome analysis has correspondingly shown that all types of glycosyltransferase undergo marked changes during transition. An in-depth analysis of sequencing datasets in the ChIP-Atlas database (https://chip-atlas.org) (Oki et al. 2018), coupled with experiments using mouse ESCs treated with a PRC2 inhibitor, has identified PRC2, which promotes gene repression via trimethylation of histone H3 lysine 27 (H3K27me3) (Oki et al. 2018), as a key component in the network

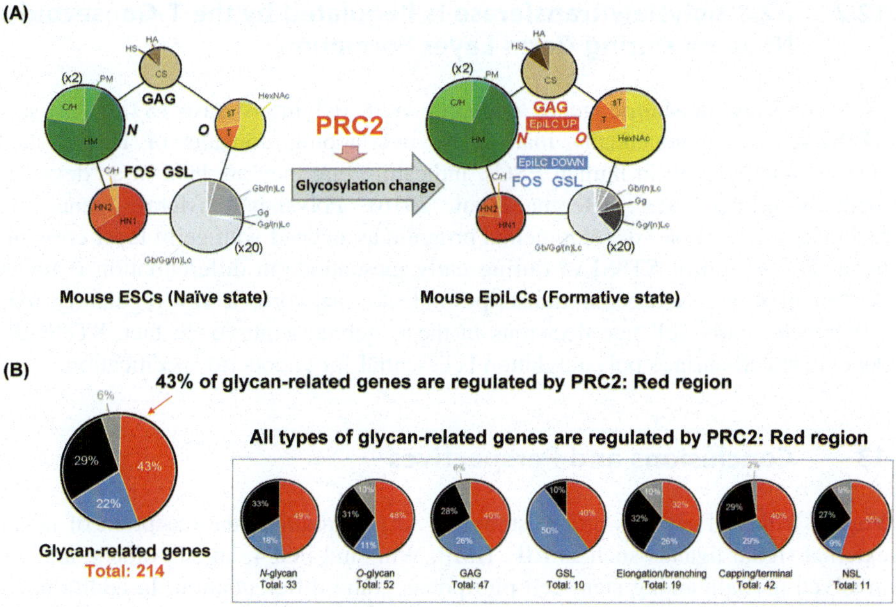

Fig. 7 PRC2 contributes to total glycosylation dynamics and the regulatory network during the transition from naïve mouse embryonic stem cells to formative epiblast-like cells. (**a**) Changes in the absolute amount of structures encompassing all types of glycan during the transition from mouse embryonic stem cells (ESCs) to epiblast-like cells (EpiLCs). The size of each pie chart reflects the absolute mean quantity of glycans (pmol/100 μg of protein). For visualization, the *N*-linked and GSL pie chart sizes have been scaled down by a factor of 2 and 20, respectively. The identification of novel specific structures of all glycosylation types has provided additional markers to distinguish between the naïve and formative pluripotent states. (**b**) Number of glycan-related genes with expression regulated by PRC2 during the transition from mouse ESCs (naïve state) to EpiLCs (formative state). PRC2 regulates 43% of the expression of glycan-related genes such as glycosyltransferases (left), contributing to changes in every type of glycan structure during the transition (right). Red segment, glycan-related genes showing a similar expression pattern during EpiLC transition and PRC2 inhibitor treatment; blue segment, genes showing opposite patterns; black segment, genes showing changes in EpiLCs but not upon PRC2 inhibitor treatment; grey segment, genes with unchanged expression

orchestrating the dramatic glycosylation changes during the mouse ESC to EpiLC transition. Overall, the expression of more than 40% of glycan-related genes, including glycosyltransferases is regulated simultaneously by PRC2; these glycan-related genes encompass every type of the glycan such as *N*- and *O*-glycans, GAGs, and GSLs (Fig. 7b). Therefore, PRC2 integrally regulates multiple glycosylation changes by simultaneously controlling the expression of all types of glycosyltransferase during transition from the naive to formative state.

12.2 α2,8-polysialyltransferase Is Regulated by the T-Goosecoid Network During Germ Layer Formation

Two polysialyltransferase genes encoding *α*2,8-sialyltransferase (ST8SIA2) and ST8SIA4, are epigenetically marked by overlapping domains of H3K27 and H3K4 trimethylation in human ESCs, indicating that they are held in a "developmentally primed" state (Berger et al. 2016). Polysialyltransferase genes are recognized as part of a developmental program associated with germ layer commitment. Expression of ST8SIA4 during early mesendoderm differentiation is under control of the T-Goosecoid transcription factor network, a key regulatory axis required for early cell fate decisions in the vertebrate embryo. In fact, ST8SIA4-dependent cell-surface polysialylation is essential for endoderm specification.

13 Conclusions and Perspectives

Various kinds of glycans on the stem cell surface regulate the reception of major external signal ligands such as LIF, BMP, Wnt, and FGF4, among others, and are involved in maintaining stem cell pluripotency and differentiation. In addition, *O*-GlcNAc, the only intracellular glycan modification, regulates the stability and activation of pluripotency-associated transcription factors, the status of epigenetic DNA and chromatin, and the phosphorylation of downstream signaling molecules involved in maintaining stem cell pluripotency and differentiation. Many of the glycans and glycosyltransferases that regulate stem cells and differentiation are conserved from *Drosophila* to mammals, suggesting that there is a shared fundamental mechanism underlying stem cell regulation. More importantly, however, the various glycosyltransferases that synthesize almost every type of glycan involved in the control of stem cell status are regulated simultaneously by epigenetic factors such as PRC2 during the naïve to formative state transition, providing a model of early embryogenesis. It is possible that different kinds of glycan structures, which previously have been thought to be unrelated to each other, are simultaneously expressed and cooperate to regulate the developmental process. The complex changes in the glycome that occur during early developmental transitions are integrally controlled by defined regulatory networks.

As described in this chapter, glycans function in various aspects of early embryogenesis. Although not mentioned in detail, they also have an important role in organ development as components of the extracellular matrix. Furthermore, glycans are thought to function in implantation and other processes. Incorporating their functions into regenerative medicine will open up new possibilities and perspectives for cell therapy, as well as for the development of culture media and scaffolding materials.

Acknowledgment The author's work cited in this review was supported in part by grants from JST, JSPS, MEXT, and PMAC to S.N. S.N. is supported by the "Human Glycome Atlas Project" as one of the "Large-Scale Academic Frontier Promotion Projects" of the Ministry of Education,

Culture, Sports, Science and Technology (MEXT). GaLSIC is supported by J-GlycoNet, one of the official Joint Usage/Research Centers of Japan by MEXT.

References

Abo H, Kume M, Pecori F et al (2022) Disaccharide-tag for highly sensitive identification of *O*-GlcNAc-modified proteins in mammalian cells. PLoS One 17:e0267804. https://doi.org/10.1371/journal.pone.0267804

Adewumi O, Aflatoonian B, Ahrlund-Richter L et al (2007) Characterization of human embryonic stem cell lines by the international stem cell initiative. Nat Biotechnol 25:803–816. https://doi.org/10.1038/nbt1318

Altmann F, Fabini G, Ahorn H, Wilson IBH (2001) Genetic model organisms in the study of *N*-glycans. Biochimie 83:703–712. https://doi.org/10.1016/s0300-9084(01)01297-4

Aoki K, Tiemeyer M (2010) The glycomics of glycan glucuronylation in *Drosophila melanogaster*. Methods Enzymol 480:297–321. https://doi.org/10.1016/s0076-6879(10)80014-x

Atwood JA, Cheng L, Alvarez-Manilla G et al (2008) Quantitation by Isobaric Labeling: applications to Glycomics. J Proteome Res 7:367–374. https://doi.org/10.1021/pr070476i

Bao S, Tang F, Li X et al (2009) Epigenetic reversion of post-implantation epiblast to pluripotent embryonic stem cells. Nature 461:1292–1295. https://doi.org/10.1038/nature08534

Bedzhov I, Zernicka-Goetz M (2014) Self-organizing properties of mouse pluripotent cells initiate morphogenesis upon implantation. Cell 156:1032–1044. https://doi.org/10.1016/j.cell.2014.01.023

Berger RP, Sun YH, Kulik M et al (2016) ST8SIA4-dependent polysialylation is part of a developmental program required for germ layer formation from human pluripotent stem cells. Stem Cells 34:1742–1752. https://doi.org/10.1002/stem.2379

Boeuf H, Hauss C, Graeve FD et al (1997) Leukemia inhibitory factor–dependent transcriptional activation in embryonic stem cells. J Cell Biol 138:1207–1217. https://doi.org/10.1083/jcb.138.6.1207

Bredenkamp N, Stirparo GG, Nichols J et al (2019) The cell-surface marker sushi containing domain 2 facilitates establishment of human naive pluripotent stem cells. Stem Cell Rep 12:1212–1222. https://doi.org/10.1016/j.stemcr.2019.03.014

Brons IGM, Smithers LE, Trotter MWB et al (2007) Derivation of pluripotent epiblast stem cells from mammalian embryos. Nature 448:191–195. https://doi.org/10.1038/nature05950

Chen G, Ye Z, Yu X et al (2008) Trophoblast differentiation defect in human embryonic stem cells lacking PIG-A and GPI-anchored cell-surface proteins. Cell Stem Cell 2:345–355. https://doi.org/10.1016/j.stem.2008.02.004

Chen Q, Chen Y, Bian C et al (2013) TET2 promotes histone *O*-GlcNAcylation during gene transcription. Nature 493:561–564. https://doi.org/10.1038/nature11742

Choi J, Huebner AJ, Clement K et al (2017) Prolonged Mek1/2 suppression impairs the developmental potential of embryonic stem cells. Nature 548:219–223. https://doi.org/10.1038/nature23274

Choudhary M, Zhang X, Stojković P et al (2007) Putative role of hyaluronan and its related genes, HAS2 and RHAMM, in human early preimplantation embryogenesis and embryonic stem cell characterization. Stem Cells 25:3045–3057. https://doi.org/10.1634/stemcells.2007-0296

Collier AJ, Panula SP, Schell JP et al (2017) Comprehensive cell surface protein profiling identifies specific markers of human naive and primed pluripotent states. Cell Stem Cell 20:874–890.e7. https://doi.org/10.1016/j.stem.2017.02.014

Evans MJ, Kaufman MH (1981) Establishment in culture of pluripotential cells from mouse embryos. Nature 292:154–156. https://doi.org/10.1038/292154a0

Fico A, Chevigny AD, Egea J et al (2012) Modulating glypican4 suppresses tumorigenicity of embryonic stem cells while preserving self-renewal and pluripotency. Stem Cells 30:1863–1874. https://doi.org/10.1002/stem.1165

Fujitani N, Furukawa J, Araki K et al (2013) Total cellular glycomics allows characterizing cells and streamlining the discovery process for cellular biomarkers. Proc Natl Acad Sci 110:2105–2110. https://doi.org/10.1073/pnas.1214233110

Furue MK, Na J, Jackson JP et al (2008) Heparin promotes the growth of human embryonic stem cells in a defined serum-free medium. Proc Natl Acad Sci 105:13409–13414. https://doi.org/10.1073/pnas.0806136105

Guo Z, Wang Z (2009) The glypican Dally is required in the niche for the maintenance of germline stem cells and short-range BMP signaling in the *Drosophila* ovary. Development 136:3627–3635. https://doi.org/10.1242/dev.036939

Guo G, Yang J, Nichols J et al (2009) Klf4 reverts developmentally programmed restriction of ground state pluripotency. Development 136:1063–1069. https://doi.org/10.1242/dev.030957

Haltiwanger RS, Blomberg MA, Hart GW (1992) Glycosylation of nuclear and cytoplasmic proteins. Purification and characterization of a uridine diphospho-*N*-acetylglucosamine:polypeptide beta-N-acetylglucosaminyltransferase. J Biol Chem 267:9005–9013

Haniffa M, Taylor D, Linnarsson S et al (2021) A roadmap for the human developmental cell atlas. Nature 597:196–205. https://doi.org/10.1038/s41586-021-03620-1

Hao Y, Fan X, Shi Y et al (2019) Next-generation unnatural monosaccharides reveal that Esrrb *O*-GlcNAcylation regulates pluripotency of mouse embryonic stem cells. Nat Commun 10:4065. https://doi.org/10.1038/s41467-019-11942-y

Hao Y, Li X, Qin K et al (2023) Chemoproteomic and transcriptomic analysis reveals that *O*-GlcNAc regulates mouse embryonic stem cell fate through the pluripotency network. Angew Chem Int Ed 62:e202300500. https://doi.org/10.1002/anie.202300500

Hart GW (2014) Minireview series on the thirtieth anniversary of research on *O*-GlcNAcylation of nuclear and cytoplasmic proteins: nutrient regulation of cellular metabolism and physiology by *O*-GlcNAcylation. J Biol Chem 289:34422–34423. https://doi.org/10.1074/jbc.r114.609776

Hart GW, Slawson C, Ramirez-Correa G, Lagerlof O (2011) Cross talk between *O*-GlcNAcylation and phosphorylation: roles in signaling, transcription, and chronic disease. Annu Rev Biochem 80:825–858. https://doi.org/10.1146/annurev-biochem-060608-102511

Hayashi Y, Kobayashi S, Nakato H (2009) *Drosophila* glypicans regulate the germline stem cell niche. J Cell Biol 187:473–480. https://doi.org/10.1083/jcb.200904118

Hirano K, Sasaki N, Ichimiya T et al (2012) 3-*O*-sulfated heparan sulfate recognized by the antibody HS4C3 contribute to the differentiation of mouse embryonic stem cells via Fas signaling. PLoS One 7:e43440. https://doi.org/10.1371/journal.pone.0043440

Hirano K, Kuppevelt THV, Nishihara S (2013) The transition of mouse pluripotent stem cells from the naïve to the primed state requires Fas signaling through 3-*O* sulfated heparan sulfate structures recognized by the HS4C3 antibody. Biochem Biophys Res Commun 430:1175–1181. https://doi.org/10.1016/j.bbrc.2012.12.005

Horiguchi I, Urabe Y, Kimura K, Sakai Y (2018) Effects of glucose, lactate and basic FGF as limiting factors on the expansion of human induced pluripotent stem cells. J Biosci Bioeng 125:111–115. https://doi.org/10.1016/j.jbiosc.2017.08.004

Illich DJ, Zhang M, Ursu A et al (2016) Distinct signaling requirements for the establishment of ESC pluripotency in late-stage EpiSCs. Cell Rep 15:787–800. https://doi.org/10.1016/j.celrep.2016.03.073

Izumikawa T, Sato B, Kitagawa H (2014) Chondroitin sulfate is indispensable for pluripotency and differentiation of mouse embryonic stem cells. Sci Rep 4:3701. https://doi.org/10.1038/srep03701

Jang H, Kim TW, Yoon S et al (2012) *O*-glcnac regulates pluripotency and reprogramming by directly acting on core components of the pluripotency network. Cell Stem Cell 11:62–74. https://doi.org/10.1016/j.stem.2012.03.001

Johnson CE, Crawford BE, Stavridis M et al (2007) Essential alterations of heparan sulfate during the differentiation of embryonic stem cells to Sox1-enhanced green fluorescent protein-expressing neural progenitor cells. Stem Cells 25:2389–2389. https://doi.org/10.1634/stemcells.2006-0445erratum

Kaneko M, Nishihara S, Narimatsu H, Saitou N (2001) The evolutionary history of glycosyltransferase genes. Trends Glycosci Glycotechnol 13:147–155. https://doi.org/10.4052/tigg.13.147

Kang M, Piliszek A, Artus J, Hadjantonakis A-K (2012) FGF4 is required for lineage restriction and salt-and-pepper distribution of primitive endoderm factors but not their initial expression in the mouse. Development 140:267–279. https://doi.org/10.1242/dev.084996

Kang M, Garg V, Hadjantonakis A-K (2017) Lineage establishment and progression within the inner cell mass of the mouse blastocyst requires FGFR1 and FGFR2. Dev Cell 41:496–510.e5. https://doi.org/10.1016/j.devcel.2017.05.003

Kanie Y, Yamamoto-Hino M, Karino Y et al (2009) Insight into the regulation of glycan synthesis in *Drosophila* chaoptin based on mass spectrometry. PLoS One 4:e5434. https://doi.org/10.1371/journal.pone.0005434

Kannagi R, Cochran NA, Ishigami F et al (1983) Stage-specific embryonic antigens (SSEA-3 and -4) are epitopes of a unique globo-series ganglioside isolated from human teratocarcinoma cells. EMBO J 2:2355–2361. https://doi.org/10.1002/j.1460-2075.1983.tb01746.x

Kawabe K, Tateyama D, Toyoda H et al (2013) A novel antibody for human induced pluripotent stem cells and embryonic stem cells recognizes a type of keratan sulfate lacking oversulfated structures. Glycobiology 23:322–336. https://doi.org/10.1093/glycob/cws159

Kinoshita T, Ohishi K, Takeda J (1997) GPI-anchor synthesis in mammalian cells: genes, their products, and a deficiency. J Biochem 122:251–257. https://doi.org/10.1093/oxfordjournals.jbchem.a021746

Kraushaar DC, Rai S, Condac E et al (2012) Heparan sulfate facilitates fgf and bmp signaling to drive mesoderm differentiation of mouse embryonic stem cells. J Biol Chem 287:22691–22700. https://doi.org/10.1074/jbc.m112.368241

Lanner F, Lee KL, Sohl M et al (2010) Heparan sulfation–dependent fibroblast growth factor signaling maintains embryonic stem cells primed for differentiation in a heterogeneous state. Stem Cells 28:191–200. https://doi.org/10.1002/stem.265

Levenstein ME, Berggren WT, Lee JE et al (2008) Secreted proteoglycans directly mediate human embryonic stem cell-basic fibroblast growth factor 2 interactions critical for proliferation. Stem Cells 26:3099–3107. https://doi.org/10.1634/stemcells.2007-1056

Li X, Lonard DM, Jung SY et al (2005) The SRC-3/AIB1 coactivator is degraded in a ubiquitin- and ATP-independent manner by the REGgamma proteasome. Cell 124:381–392. https://doi.org/10.1016/j.cell.2005.11.037

Li Z, Fei T, Zhang J et al (2012) BMP4 signaling acts via dual-specificity phosphatase 9 to control erk activity in mouse embryonic stem cells. Cell Stem Cell 10:171–182. https://doi.org/10.1016/j.stem.2011.12.016

Liang Y-J, Kuo H-H, Lin C-H et al (2010) Switching of the core structures of glycosphingolipids from globo- and lacto- to ganglio-series upon human embryonic stem cell differentiation. Proc Natl Acad Sci 107:22564–22569. https://doi.org/10.1073/pnas.1007290108

Lin R-J, Kuo M-W, Yang B-C et al (2020) B3GALT5 knockout alters gycosphingolipid profile and facilitates transition to human naïve pluripotency. Proc Natl Acad Sci USA 117:27435–27444. https://doi.org/10.1073/pnas.2003155117

Lingwood D, Simons K (2010) Lipid rafts as a membrane-organizing principle. Science 327:46–50. https://doi.org/10.1126/science.1174621

Martello G, Sugimoto T, Diamanti E et al (2012) Esrrb Is a pivotal target of the Gsk3/Tcf3 axis regulating embryonic stem cell self-renewal. Cell Stem Cell 11:491–504. https://doi.org/10.1016/j.stem.2012.06.008

Martello G, Bertone P, Smith A (2013) Identification of the missing pluripotency mediator downstream of leukaemia inhibitory factor. EMBO J 32:2561–2574. https://doi.org/10.1038/emboj.2013.177

Martin GR (1981) Isolation of a pluripotent cell line from early mouse embryos cultured in medium conditioned by teratocarcinoma stem cells. Proc Natl Acad Sci 78:7634–7638. https://doi.org/10.1073/pnas.78.12.7634

Matsumoto S, Nakao H, Kawabe K et al (2015) A cytotoxic antibody recognizing Lacto-N-fucopentaose I (LNFP I) on human induced pluripotent stem (hiPS) Cells. J Biol Chem 290: 20071–20085. https://doi.org/10.1074/jbc.m115.657692

Maury JJP, Chan KK-K, Zheng L et al (2013) Excess of *O*-linked *N*-acetylglucosamine modifies human pluripotent stem cell differentiation. Stem Cell Res 11:926–937. https://doi.org/10.1016/j.scr.2013.06.004

Maury JJP, Farran CAE, Ng D et al (2015) RING1B *O*-GlcNAcylation regulates gene targeting of polycomb repressive complex 1 in human embryonic stem cells. Stem Cell Res 15:182–189. https://doi.org/10.1016/j.scr.2015.06.007

Miura T, Nishihara S (2016) *O*-GlcNAc is required for the survival of primed pluripotent stem cells and their reversion to the naïve state. Biochem Biophys Res Commun 480:655–661. https://doi.org/10.1016/j.bbrc.2016.10.111

Miura T, Kume M, Kawamura T et al (2018) *O*-GlcNAc on PKCζ inhibits the FGF4-PKC-ζ-MEK-ERK1/2 pathway via inhibition of PKCζ phosphorylation in mouse embryonic stem cells. Stem Cell Rep 10:272–286. https://doi.org/10.1016/j.stemcr.2017.11.007

Miura T, Yuasa N, Ota H et al (2019) Highly sulfated hyaluronic acid maintains human induced pluripotent stem cells under feeder-free and bFGF-free conditions. Biochem Biophys Res Commun 518:506–512. https://doi.org/10.1016/j.bbrc.2019.08.082

Molotkov A, Mazot P, Brewer JR et al (2017) Distinct requirements for FGFR1 and FGFR2 in primitive endoderm development and exit from pluripotency. Dev Cell 41:511–526.e4. https://doi.org/10.1016/j.devcel.2017.05.004

Morgani S, Nichols J, Hadjantonakis A-K (2017) The many faces of pluripotency: in vitro adaptations of a continuum of in vivo states. BMC Dev Biol 17:7. https://doi.org/10.1186/s12861-017-0150-4

Morikawa M, Koinuma D, Mizutani A et al (2016) BMP sustains embryonic stem cell self-renewal through distinct functions of different krüppel-like factors. Stem Cell Rep 6:64–73. https://doi.org/10.1016/j.stemcr.2015.12.004

Muramatsu T, Muramatsu H (2004) Carbohydrate antigens expressed on stem cells and early embryonic cells. Glycoconj J 21:41–45. https://doi.org/10.1023/b:glyc.0000043746.77504.28

Myers SA, Peddada S, Chatterjee N et al (2016) SOX2 *O*-GlcNAcylation alters its protein-protein interactions and genomic occupancy to modulate gene expression in pluripotent cells. elife 5: e10647. https://doi.org/10.7554/elife.10647

Nairn AV, Kinoshita-Toyoda A, Toyoda H et al (2007) Glycomics of proteoglycan biosynthesis in murine embryonic stem cell differentiation. J Proteome Res 6:4374–4387. https://doi.org/10.1021/pr070446f

Nakato H, Li J-P (2016) Functions of heparan sulfate proteoglycans in development: insights from *Drosophila* models. Int Rev Cell Mol Biol 325:275–293. https://doi.org/10.1016/bs.ircmb.2016.02.008

Narimatsu H (2004) Construction of a human glycogene library and comprehensive functional analysis. Glycoconj J 21:17–24. https://doi.org/10.1023/b:glyc.0000043742.99482.01

Natunen S, Satomaa T, Pitkänen V et al (2011) The binding specificity of the marker antibodies Tra-1-60 and Tra-1-81 reveals a novel pluripotency-associated type 1 lactosamine epitope. Glycobiology 21:1125–1130. https://doi.org/10.1093/glycob/cwq209

Nichols J, Smith A (2009) Naive and primed pluripotent states. Cell Stem Cell 4:487–492. https://doi.org/10.1016/j.stem.2009.05.015

Nicola NA, Babon JJ (2015) Leukemia inhibitory factor (LIF). Cytokine Growth Factor Rev 26: 533–544. https://doi.org/10.1016/j.cytogfr.2015.07.001

Nishihara S (2010) Glycosyltransferases and transporters that contribute to proteoglycan synthesis in *Drosophila*: identification and functional analyses using the heritable and inducible RNAi system. Methods Enzymol 480:323–351. https://doi.org/10.1016/s0076-6879(10)80015-1

Nishihara S (2017) Glycans define the stemness of naïve and primed pluripotent stem cells. Glycoconj J 34:737–747. https://doi.org/10.1007/s10719-016-9740-9

Nishihara S (2018) Glycans in stem cell regulation: from *Drosophila* tissue stem cells to mammalian pluripotent stem cells. FEBS Lett 592:3773–3790. https://doi.org/10.1002/1873-3468.13167

Nishihara S (2020) Functional analysis of glycosylation using *Drosophila melanogaster*. Glycoconj J 37:1–14. https://doi.org/10.1007/s10719-019-09892-0

Niwa H (2014) The pluripotency transcription factor network at work in reprogramming. Curr Opin Genet Dev 28:25–31. https://doi.org/10.1016/j.gde.2014.08.004

Niwa H, Ogawa K, Shimosato D, Adachi K (2009) A parallel circuit of LIF signalling pathways maintains pluripotency of mouse ES cells. Nature 460:118–122. https://doi.org/10.1038/nature08113

Ogawa K, Nishinakamura R, Iwamatsu Y et al (2006) Synergistic action of Wnt and LIF in maintaining pluripotency of mouse ES cells. Biochem Biophys Res Commun 343:159–166. https://doi.org/10.1016/j.bbrc.2006.02.127

Ogura C, Nishihara S (2021) Dermatan-4-*O*-Sulfotransferase-1 contributes to the undifferentiated state of mouse embryonic stem cells. Front Cell Dev Biol 9:733964. https://doi.org/10.3389/fcell.2021.733964

Ohtsuka S, Nakai-Futatsugi Y, Niwa H (2015) LIF signal in mouse embryonic stem cells. JAK-STAT 4:e1086520. https://doi.org/10.1080/21623996.2015.1086520

Okae H, Toh H, Sato T et al (2018) Derivation of human trophoblast stem cells. Cell Stem Cell 22: 50–63.e6. https://doi.org/10.1016/j.stem.2017.11.004

Oki S, Ohta T, Shioi G et al (2018) ChIP-Atlas: a data-mining suite powered by full integration of public ChIP-seq data. EMBO Rep 19:e46255. https://doi.org/10.15252/embr.201846255

Okubo T, Rivron N, Kabata M et al (2024) Hypoblast from human pluripotent stem cells regulates epiblast development. Nature 626(7998):357–366. https://doi.org/10.1038/s41586-023-06871-2

Pecori F, Akimoto Y, Hanamatsu H et al (2020) Mucin-type *O*-glycosylation controls pluripotency in mouse embryonic stem cells via Wnt receptor endocytosis. J Cell Sci 133:jcs245845. https://doi.org/10.1242/jcs.245845

Pecori F, Kondo N, Ogura C et al (2021a) Site-specific *O*-GlcNAcylation of Psme3 maintains mouse stem cell pluripotency by impairing P-body homeostasis. Cell Rep 36:109361. https://doi.org/10.1016/j.celrep.2021.109361

Pecori F, Yokota I, Hanamatsu H et al (2021b) A defined glycosylation regulatory network modulates total glycome dynamics during pluripotency state transition. Sci Rep 11:1276. https://doi.org/10.1038/s41598-020-79666-4

Raghu D, Mobley RJ, Shendy NAM et al (2019) GALNT3 maintains the epithelial state in trophoblast stem cells. Cell Rep 26:3684–3697.e7. https://doi.org/10.1016/j.celrep.2019.02.093

Rossant J, Tam PPL (2009) Blastocyst lineage formation, early embryonic asymmetries and axis patterning in the mouse. Development 136:701–713. https://doi.org/10.1242/dev.017178

Sasaki N, Okishio K, Ui-Tei K et al (2008) Heparan sulfate regulates self-renewal and pluripotency of embryonic stem cells. J Biol Chem 283:3594–3606. https://doi.org/10.1074/jbc.m705621200

Sasaki N, Hirano T, Ichimiya T et al (2009) The 3′-phosphoadenosine 5′-phosphosulfate transporters, papst1 and 2, contribute to the maintenance and differentiation of mouse embryonic stem cells. PLoS One 4:e8262. https://doi.org/10.1371/journal.pone.0008262

Sasaki N, Shinomi M, Hirano K et al (2011) LacdiNAc (GalNAcβ1-4GlcNAc) contributes to self-renewal of mouse embryonic stem cells by regulating leukemia inhibitory factor/STAT3 Signaling. Stem Cells 29:641–650. https://doi.org/10.1002/stem.615

Seifinejad A, Taei A, Totonchi M et al (2010) Generation of human induced pluripotent stem cells from a Bombay individual: Moving towards "universal-donor" red blood cells. Biochem Biophys Res Commun 391:329–334. https://doi.org/10.1016/j.bbrc.2009.11.058

Shafi R, Iyer SPN, Ellies LG et al (2000) The *O*-GlcNAc transferase gene resides on the X chromosome and is essential for embryonic stem cell viability and mouse ontogeny. Proc Natl Acad Sci 97:5735–5739. https://doi.org/10.1073/pnas.100471497

Shahbazi MN (2020) Mechanisms of human embryo development: from cell fate to tissue shape and back. Development 147:dev190629. https://doi.org/10.1242/dev.190629

Shahbazi MN, Scialdone A, Skorupska N et al (2017) Pluripotent state transitions coordinate morphogenesis in mouse and human embryos. Nature 552:239–243. https://doi.org/10.1038/nature24675

Shi F-T, Kim H, Lu W et al (2013) Ten-eleven translocation 1 (Tet1) is regulated by *O*-linked *N*-acetylglucosamine transferase (Ogt) for target gene repression in mouse embryonic stem cells. J Biol Chem 288:20776–20784. https://doi.org/10.1074/jbc.m113.460386

Smith A (2017) Formative pluripotency: the executive phase in a developmental continuum. Development 144:365–373. https://doi.org/10.1242/dev.142679

Smith ZD, Chan MM, Humm KC et al (2014) DNA methylation dynamics of the human preimplantation embryo. Nature 511:611–615. https://doi.org/10.1038/nature13581

Stelling MP, Lages YM, Tovar AM et al (2013) Matrix-bound heparan sulfate is essential for the growth and pluripotency of human embryonic stem cells. Glycobiology 23:337–345. https://doi.org/10.1093/glycob/cws133

Sullivan AE, Santos SD (2023) The ever-growing world of gastruloids: autogenous models of mammalian embryogenesis. Curr Opin Genet Dev 82:102102. https://doi.org/10.1016/j.gde.2023.102102

Syangtan D, Mahbuba DA, Masuko S et al (2025) Heparan sulfate regulates the fate decisions of human pluripotent stem cells. Stem Cell Reports 20(1):102384. https://doi.org/10.1016/j.stemcr.2024.11.014

Tabata T, Takei Y (2004) Morphogens, their identification and regulation. Development 131:703–712. https://doi.org/10.1242/dev.01043

Takahashi K, Yamanaka S (2006) Induction of pluripotent stem cells from mouse embryonic and adult fibroblast cultures by defined factors. Cell 126:663–676. https://doi.org/10.1016/j.cell.2006.07.024

Takahashi K, Tanabe K, Ohnuki M et al (2007) Induction of pluripotent stem cells from adult human fibroblasts by defined factors. Cell 131:861–872. https://doi.org/10.1016/j.cell.2007.11.019

Tanaka S, Kunath T, Hadjantonakis A-K et al (1998) Promotion of trophoblast stem cell proliferation by FGF4. Science 282:2072–2075. https://doi.org/10.1126/science.282.5396.2072

Tang C, Lee AS, Volkmer J-P et al (2011) An antibody against SSEA-5 glycan on human pluripotent stem cells enables removal of teratoma-forming cells. Nat Biotechnol 29:829–834. https://doi.org/10.1038/nbt.1947

Tateno H, Toyota M, Saito S et al (2011) Glycome diagnosis of human induced pluripotent stem cells using lectin microarray. J Biol Chem 286:20345–20353. https://doi.org/10.1074/jbc.m111.231274

Tesar PJ, Chenoweth JG, Brook FA et al (2007) New cell lines from mouse epiblast share defining features with human embryonic stem cells. Nature 448:196–199. https://doi.org/10.1038/nature05972

Theunissen TW, Friedli M, He Y et al (2016) Molecular criteria for defining the naive human pluripotent state. Cell Stem Cell 19:502–515. https://doi.org/10.1016/j.stem.2016.06.011

Thompson S, Clarke AR, Pow AM et al (1989) Germ line transmission and expression of a corrected HPRT gene produced by gene targeting in embryonic stem cells. Cell 56:313–321. https://doi.org/10.1016/0092-8674(89)90905-7

Thomson JA, Itskovitz-Eldor J, Shapiro SS et al (1998) Embryonic stem cell lines derived from human blastocysts. Science 282:1145–1147. https://doi.org/10.1126/science.282.5391.1145

Toyoda H, Kinoshita-Toyoda A, Selleck SB (2000) Structural analysis of glycosaminoglycans in *Drosophila* and caenorhabditis elegans and demonstration that tout-velu, a *Drosophila* Gene related to EXT tumor suppressors, affects heparan sulfate in vivo. J Biol Chem 275:2269–2275. https://doi.org/10.1074/jbc.275.4.2269

Vella P, Scelfo A, Jammula S et al (2013) Tet proteins connect the *O*-linked *N*-acetylglucosamine transferase Ogt to chromatin in embryonic stem cells. Mol Cell 49:645–656. https://doi.org/10.1016/j.molcel.2012.12.019

Wang Y-C, Stein JW, Lynch CL et al (2015) Glycosyltransferase ST6GAL1 contributes to the regulation of pluripotency in human pluripotent stem cells. Sci Rep 5:13317. https://doi.org/10.1038/srep13317

Weinberger L, Ayyash M, Novershtern N, Hanna JH (2016) Dynamic stem cell states: naive to primed pluripotency in rodents and humans. Nat Rev Mol Cell Biol 17:155–169. https://doi.org/10.1038/nrm.2015.28

Yagi M, Kishigami S, Tanaka A et al (2017) Derivation of ground-state female ES cells maintaining gamete-derived DNA methylation. Nature 548:224–227. https://doi.org/10.1038/nature23286

Yamamoto-Hino M, Yoshida H, Ichimiya T et al (2015) Phenotype-based clustering of glycosylation-related genes by RNAi-mediated gene silencing. Genes Cells 20:521–542. https://doi.org/10.1111/gtc.12246

Yamanaka Y, Hamidi S, Yoshioka-Kobayashi K et al (2023) Reconstituting human somitogenesis in vitro. Nature 614:509–520. https://doi.org/10.1038/s41586-022-05649-2

Yang J, van Oosten AL, Theunissen TW et al (2010) Stat3 activation is limiting for reprogramming to ground state pluripotency. Cell Stem Cell 7:319–328. https://doi.org/10.1016/j.stem.2010.06.022

Ying Q-L, Nichols J, Chambers I, Smith A (2003) BMP induction of Id proteins suppresses differentiation and sustains embryonic stem cell self-renewal in collaboration with STAT3. Cell 115:281–292. https://doi.org/10.1016/s0092-8674(03)00847-x

Yu J, Vodyanik MA, Smuga-Otto K et al (2007) Induced pluripotent stem cell lines derived from human somatic cells. Science 318:1917–1920. https://doi.org/10.1126/science.1151526

Zhou H, Li W, Zhu S et al (2010) Conversion of mouse epiblast stem cells to an earlier pluripotency state by small molecules. J Biol Chem 285:29676–29680. https://doi.org/10.1074/jbc.c110.150599

Zhou J, Hu J, Wang Y, Gao S (2023) Induction and application of human naive pluripotency. Cell Rep 42:112379. https://doi.org/10.1016/j.celrep.2023.112379

Zhu Q, Cheng X, Cheng Y et al (2020) *O*-GlcNAcylation regulates the methionine cycle to promote pluripotency of stem cells. Proc Natl Acad Sci 117:7755–7763. https://doi.org/10.1073/pnas.1915582117

Role of Antibody Glycosylation in Health, Disease, and Therapy

Falk Nimmerjahn

Contents

Abstract

Immunoglobulin G (IgG) antibodies are an essential component of humoral immunity protecting the host from recurrent infections. Among all antibody isotypes, IgG antibodies have a uniquely long half-life, can basically reach any tissue in the body, and have the ability to kill opsonized target cells, which has made them the molecule of choice for therapeutic interventions in cancer and autoimmunity. Moreover, IgG antibodies in the form of pooled serum IgG preparations from healthy donors are used to treat chronic inflammatory and autoimmune diseases, providing evidence that serum IgG antibodies can have an active immunomodulatory activity. Research over the last two decades has established that the single sugar moiety attached to each IgG heavy chain plays a very important role in modulating the pro- and anti-inflammatory activities of

F. Nimmerjahn (✉)
Institute of Genetics, Department of Biology, University of Erlangen-Nuremberg, Erlangen, Germany
e-mail: falk.nimmerjahn@fau.de

F. Pfrengle et al. (eds.), *Complex Carbohydrates in Health and Disease*, Handbook of Experimental Pharmacology 288, https://doi.org/10.1007/164_2025_744

IgG. Moreover, specific sugar moieties such as sialic acid and galactose residues can serve as highly specific biomarkers for ongoing inflammatory processes. This chapter will summarize how different sugar residues in the IgG sugar moiety change upon inflammation and how such changes may translate to altered IgG function and hence maybe useful for optimizing or modulating the function of therapeutic antibodies.

Keywords

Disease · Fc receptors · Glycosylation · IgG · Therapy

1 Introduction

Immunoglobulin G (IgG) molecules serve as a crucial link between the innate and adaptive immune systems. IgG molecules consist of two heavy and two light chains connected via several disulfide bonds. Functionally, the IgG molecule can be separated into fragment antigen binding (F(ab)-fragment) and fragment crystallizable (Fc-fragment), which contains a single N-linked sugar domain attached to each heavy chain at the residue 297 (N297) (Nimmerjahn et al. 2023a). Upon recognizing target structures on infected cells or upon deposition in tissues during autoimmune diseases, such as rheumatoid arthritis (RA) or systemic lupus erythematosus (SLE), IgG antibodies gain the ability to activate the complement system and bind to cellular Fcγ-receptors (FcγRs) expressed ubiquitously on innate immune effector cells (Nimmerjahn and Ravetch 2008; Nimmerjahn et al. 2023b; Takai 2005). Crosslinking FcγRs on mast cells, basophils, eosinophils, neutrophils, monocytes, or macrophages leads to a large variety of downstream effector functions, including the release of pro-inflammatory mediators, the phagocytosis of opsonized particles, complement-dependent cytotoxicity (CDC), and antibody-dependent cellular cytotoxicity (Nimmerjahn et al. 2023b; Pincetic et al. 2014). Moreover, pooled serum IgG from thousands of donors (so-called intravenous immunoglobulins (IVIg)) interferes with inflammatory processes if infused into patients at doses of 1 g/kg (Bohm et al. 2012, 2014). Here, different C-type lectin receptors, such as SIGNR1, DC-SIGN, CD23, or DCIR were suggested to be required for these anti-inflammatory activities in various mouse models of inflammation (Schwab and Nimmerjahn 2013; Seeling et al. 2017; Pincetic et al. 2014). Of note, for both, the pro- and anti-inflammatory activities of IgG glycosylation are crucial. More recently, an impact of glycosylation on antibody function has also been noted for other antibody isotypes, such as IgE (Shade et al. 2015, 2019, 2020), which is, however, beyond the scope of this review. Before discussing the role of IgG glycosylation for IgG function in more depth, a brief description of the molecular and cellular pathways underlying IgG function with a focus on the interaction of IgG with cellular FcγRs will be given.

2 Fcγ-Receptors as Mediators of IgG Activity In Vivo

FcγRs bind the IgG Fc fragment near the hinge-proximal region in a 1:1 complex and include several activating FcγRs (FcγRI, FcγRIII, FcγRIV in mice, and FcγRIA, FcγRII, FcγRIIc, FcγRIIIa, FcγRIIIb in humans) and one inhibitory FcγR (FcγRIIb in mice and humans) (Sondermann et al. 2000). With the exception of human FcγRIIa, all activating FcγRs associate with a dimer of the common FcRγ chain, containing an immunoreceptor tyrosine-based activation motif (ITAM), while the inhibitory FcγRIIb transmits signals via an immunoreceptor tyrosine-based inhibitory motif (ITIM) present in its cytoplasmic region (Fig. 1) (Brandsma et al. 2016; Nimmerjahn et al. 2023a). As both activating and the inhibitory FcγRIIb (more recently also referred to as type I FcγRs) are co-expressed on most innate immune cells, a balanced signaling is initiated preventing an overshooting immune response. Besides FcγR expression levels, the different affinities of individual IgG subclasses for specific FcγRs modulate the magnitude of downstream effector pathways. For instance, the high-affinity receptor FcγRI in both mouse and human binds IgG2a (in mice) or IgG1 and IgG3 (in humans) with an affinity of 10^8–10^9 M^{-1}, while all other FcγRs exhibit a 100- to 1,000-fold lower affinity for IgG subclasses. Additionally, mouse IgG1 and IgG2a/c or IgG2b bind to activating or inhibitory FcγRs with different affinities, which leads to an IgG subclass-specific negative regulation via the inhibitory FcγRIIb (Nimmerjahn and Ravetch 2005). For instance, IgG1 binds with an approximately tenfold higher affinity to the inhibitory FcγRIIb compared to its activating receptor, FcγRIII. Thus, IgG1 activity in mice is strongly suppressed by the inhibitory FcγRIIb. Conversely, IgG2a/c or IgG2b shows a higher affinity for activating FcγRs, FcγRI, and FcγRIV, resulting in a lower threshold for cell activation (Nimmerjahn and Ravetch 2005; Bruhns and Jonsson 2015; Takai 2005). To use IgG affinities for FcγRs as a predictive tool for IgG subclass activity in vivo, the concept of the A/I ratio, where the affinity of an individual subclass for the activating receptor is divided by the affinity for the inhibitory FcγRIIb, has been developed and validated by mathematical modeling (Nimmerjahn and Ravetch 2005; Robinett et al. 2018; Tan et al. 2023). Based on this concept, cytotoxic antibody affinities have been optimized for binding to activating FcγRs. As we will discuss later, one potent way of increasing the affinity for specific activating FcγRs is by Fc-glycoengineering (Nimmerjahn et al. 2023a).

With respect to effector cells underlying IgG activity in vivo, a major change in paradigm has occurred more recently. Thus, myeloid cells including monocytes and macrophages and to a much lesser extent natural killer cells (NK cells) seem to play a dominant role for depleting target cells in vivo (Biburger et al. 2011, 2014; Gordan et al. 2015; Wohner et al. 2024). During complex autoimmune diseases such as rheumatoid arthritis or systemic lupus erythematosus, a whole spectrum of tissue-specific innate immune cells can be triggered, including mast cells, tissue resident macrophages, neutrophils and eosinophils (Seeling et al. 2017). Moreover, IgG immune complexes can feed-back on the adaptive immune response via binding to the inhibitory FcγRIIb on B cells (Nimmerjahn and Ravetch 2008). The interested reader may be directed to the many excellent review articles cited in this manuscript,

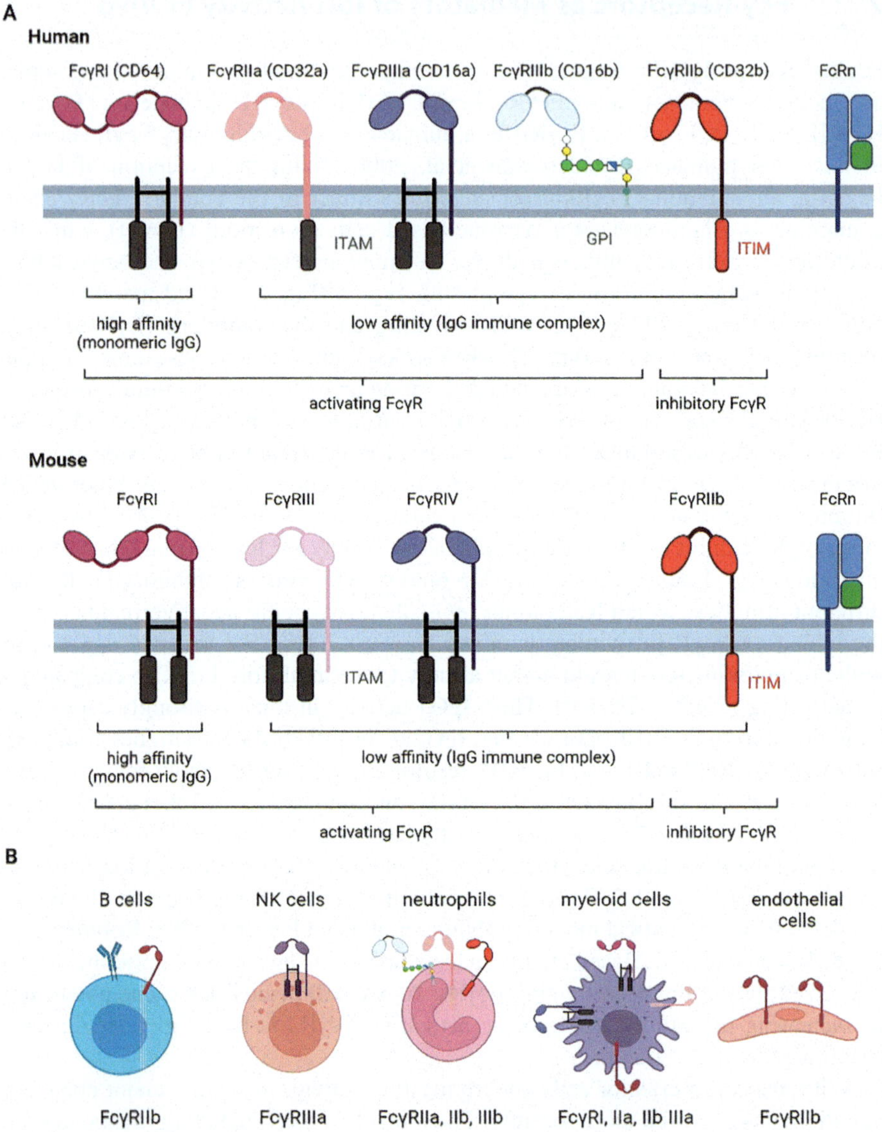

Fig. 1 Overview of the mouse and human Fcγ-receptor system. (**a**) Shown are the members of the human (top) and mouse (bottom) Fcγ-receptor family. In addition, the neonatal Fc-receptor (FcRn) is depicted, which is critical for IgG half-life. Note that in mice and humans, one high-affinity activating FcγR (FcγRI), which has the ability to interact with monomeric IgG, and one inhibitory FcγR (FcγRIIb) exist. Most FcγRs in mice and humans are of low affinity and require cross-linking by multimeric IgG complexes (immune complexes) for productive signaling. (**b**) Depicted is the expression of FcγRs on various immune and non-immune cells. Whereas neutrophils and myeloid cells express both activating and the inhibitory FcγRIIb, B cells and endothelial cells selectively express the inhibitory FcγRIIb. NK cells exclusively express one activating FcγR in mice and humans. See text for further details

which have covered this complex topic in more depth. In summary, monocytes, macrophages, and neutrophils have been identified to be central for both the activity of autoreactive and for the function of therapeutic antibodies.

3 Cell Biology of Antibody Glycosylation

In general, glycan biosynthesis occurs in the endoplasmic reticulum (ER) and the Golgi apparatus of cells including B cells as well as antibody producing plasma cells and plasma blasts (van Kooyk et al. 2013). Initially, high mannose N-linked glycans are attached to recipient proteins in the ER and the glycosylated proteins are transferred to the Golgi, where various enzymes catalyze trimming or addition of further sugar moieties. These trimming glycosidases and glycosyltransferases, expressed differentially within the Golgi, can be influenced by factors such as transcription factors, sugar donor availability, or cytokines like tumor necrosis factor-alpha (TNFα) (Garcia-Vallejo et al. 2005; Gringhuis et al. 2005; Klaric et al. 2020; Rajput et al. 1996). Consequently, the ultimate glycan structure of glycoproteins, such as IgG, is largely determined by the selective activity of glycosidases, glycosyltransferases, and individual metabolic conditions within the cell, allowing IgG glycosylation to adapt to altered environmental conditions (Wang et al. 2011). In addition, protein-specific structural constraints such as those present in the IgG-Fc domain may impact the composition of the IgG-linked sugar domain, as we will discuss later. Recently, it was proposed that IgG sialylation may occur independently of the secretory pathway, allowing for B cell-independent modification of IgG glycosylation (Dougher et al. 2017; Jones et al. 2016; Lee et al. 2014; Manhardt et al. 2017). However, more recent studies suggest that the efficiency of this B cell-extrinsic IgG sialylation pathway (compared to B cell-intrinsic IgG sialylation) may be very low. Thus, the injection of desialylated IgG into B cell-deficient mice revealed that B cell-extrinsic IgG glycosylation occurs at very low levels and exclusively in the more accessible F(ab) fragment of the IgG molecule if acceptor sites for glycosylation have been generated during somatic hypermutation and affinity maturation of the IgG molecule (Ohmi et al. 2016; Schaffert et al. 2019). Consistent with this notion, mice deficient for ST6Gal1, the enzyme that adds terminal sialic acid residues, specifically in B cells entirely lacked sialylation of serum IgG on the IgG Fc-domain, suggesting that B cell-extrinsic IgG sialylation is not contributing to sialylation of the IgG Fc-domain.

4 Variability of Serum IgG Glycosylation

Similar to other immunoglobulin isotypes, the four different IgG subclasses are glycosylated (Fig. 2a–c). In contrast to IgM, IgA, and IgE, however, IgG molecules only carry one N-linked sugar moiety attached to the asparagin 297 (N297) residue in each of the two IgG Fc-domains (Fig. 2b). Upon affinity maturation, additional N-linked sugar moieties may be added to the IgG variable domain in the heavy or

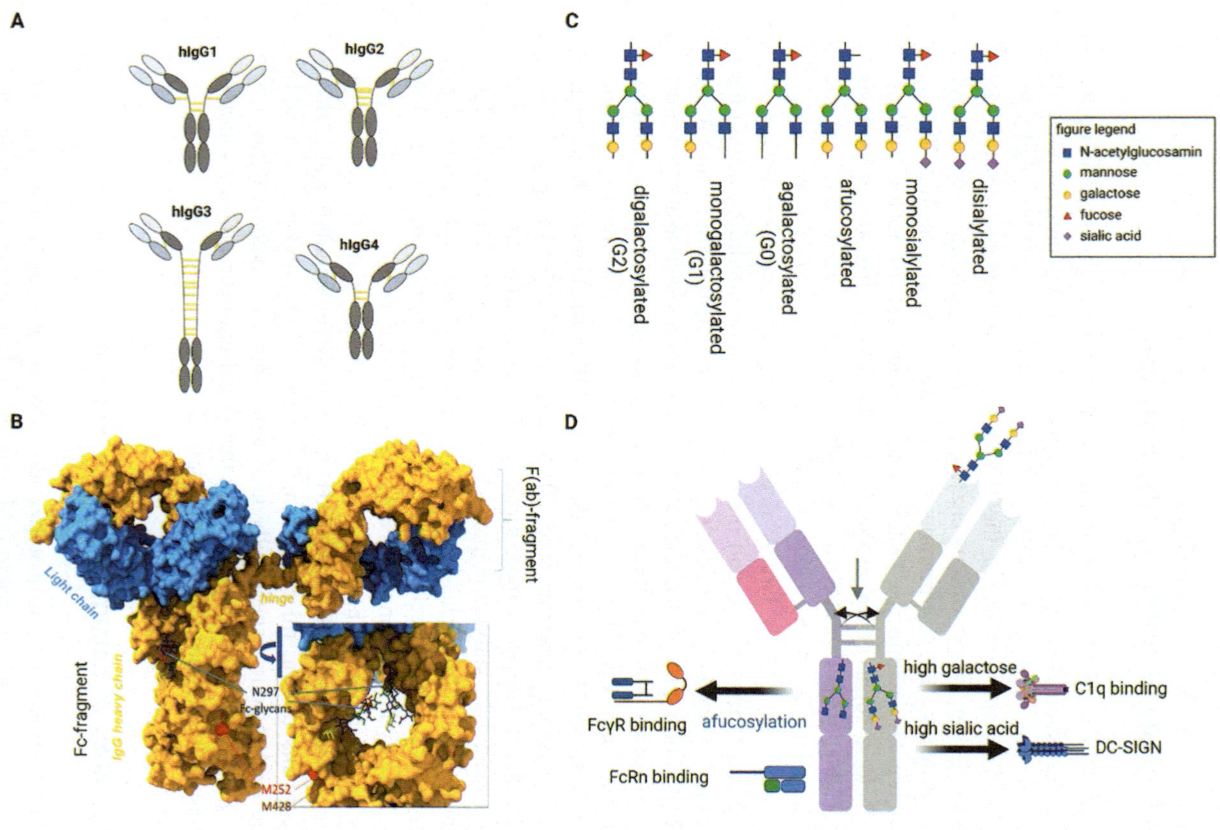

Fig. 2 IgG structure and glycosylation forms. (**a**) Depicted are the four human IgG sublasses. Note the different length of the IgG subclass Fc-domains and the varying amounts of intra- and intermolecule disulphide bonds (yellow lines) connecting different parts of the IgG molecule. (**b**) Three-dimensional structural model of the IgG molecule including the sugar domain attached to each of the IgG heavy chains at the asparagin 297 (N297) residue. The different parts of the IgG molecule are labeled. (**c**) Shown are the different sugar moieties frequently found attached to IgG Fc-domains. Depending on the level of galactose, G0, G1, and G2 glycoforms can be distinguished. Note that each individual IgG Fc domain may contain different sugar moieties. (**d**) Depicted is the impact of different sugar moieties on activation of different effector pathways such as enhanced binding of afucosylated IgG to FcγRIIIa in humans or FcγRIV in mice, enhanced binding of G2 glycoforms to C1q, or the interaction of highly sialylated glycoforms with DC-SIGN. Note that IgG binding to FcRn is independent of IgG glycosylation

light chain (Fig. 2d) (van de Bovenkamp et al. 2018). Similar to other serum proteins, the sugar moiety of IgG is highly processed and can contain terminal galactose and sialic acid residues (Arnold et al. 2007; Nimmerjahn et al. 2023a; Seeling et al. 2017). Of note, more than thirty different IgG glycosylation variants can be identified for any of the four different human IgG subclasses generating a great heterogeneity. Creating even more complexity, there is evidence that the sugar moieties of the two different Fc-fragments may differ, theoretically enabling in theory more than 500 IgG glycoforms (Dekkers et al. 2017; Nimmerjahn et al. 2023a; Kobata 2008; Raju 2008). Of note, the IgG Fc-linked sugar moiety is not exposed on the surface of the molecule but rather buried in the hydrophobic pocket formed by the two IgG heavy chain constant domains two and three (CH2 and CH3) (Fig. 2b). Based on the presence or absence of terminal galactose residues, three major IgG glycoforms have been defined. Whereas the IgG-G0 form (~20–25% of IgG pool) has no terminal galactose residues, the IgG-G1 form (35–45% of IgG pool) has one and the IgG-G2 form (~16–27% of IgG pool) has two galactose residues (Fig. 2c). This distribution can change during states of active inflammation and lead to an overrepresentation of the IgG-G0 glycoform (>55% of the glycan pool). With respect to terminal sialic acids, about 10–12% of IgG glycans contain one terminal sialic acid residue, whereas disialylated structures can only be identified in trace amounts. There seems to be a predominant sialylation on the 1,3-arm of the biantennary structure, and monosialylation on both sugar moieties occurs in about 1% of the glycan pool (Wormald et al. 1997; Stadlmann et al. 2010). In addition to the sugar moiety of the Fc-domain, up to 15–20% of serum IgG molecules may become glycosylated in the heavy or light chain of the F(ab)-fragment due to the de novo generation of glycosylation sites during somatic hypermutation of the IgG repertoire (Arnold et al. 2007; Jefferis and Lund 1997; Stadlmann et al. 2010). In contrast to the N297 attached sugar moiety, more than 46% of the F(ab)-associated sugar structures may contain terminal sialic acid (predominantly on the 1,6-arm) and galactose residues and have higher levels of branching GlcNAc residues (van de Bovenkamp et al. 2018; Arnold et al. 2007). In contrast to the high flexibility of the F (ab)-associated sugar moiety, which can hardly be resolved in crystal structure analysis, the Fc-linked sugar moiety is more rigid due to the confined space within the hydrophobic pocket formed by the two Fc-fragments. Whereas the 1,6-arm comes in close contact with the amino acid backbone of the CH2 domain, the 1,3-arms of the two sugar moieties are orientated toward the center of the hydrophobic pocket (Deisenhofer 1981; Sondermann et al. 2000; Radaev et al. 2001). Crystal structural analysis revealed that glycosylated IgG Fc fragments can occur in open or closed conformations, while aglycosylated human Fc fragments predominantly exist in the closed IgG-Fc conformation, hindering functional Fc receptor binding (Ahmed et al. 2014; Borrok et al. 2012; Jefferis and Lund 1997; Krapp et al. 2003). Interestingly, individual IgG subclasses seem to have a differential requirement for the length of the N297-linked sugar moiety. While a single GlcNAc residue is sufficient to maintain the activity of mouse IgG2a and human IgG1 subclasses, trimming down the Fc-linked sugar moiety to one residue largely abolishes the activity of mouse IgG1 and IgG2b subclasses (Kao et al. 2015). Moreover, the four murine IgG subclasses (IgG1, IgG2b, IgG2a/c, and IgG3) are differentially glycosylated during the steady state, with 50% of IgG1 being present in the G0 form, while IgG2b and IgG2a/c molecules show an increased proportion of galactosylated and sialylated glycan species, resulting in a complex scenario when attempting to decipher the contribution of individual sugar moieties to IgG function (Kao et al. 2017). Another factor rarely considered when discussing the impact of IgG glycosylation on IgG function is the significant impact of higher-order immune complexes, the natural ligands of low-affinity FcγRs. It was shown, for example, that large immune complexes still can bind to the high-affinity and several allelic variants of low affinity FcγRs even in the absence of glycosylation (Lux et al. 2013). To what extent this binding triggers effector functions remains to be established, however.

Of note, the impact of IgG glycosylation on antibody half-life is minimal. This relates to the interaction of IgG with the neonatal Fc receptor (FcRn), which is responsible for determining IgG half-life. FcRn binds IgG molecules at the CH2-CH3 interface of the IgG Fc-fragment, which is far from the N297-linked sugar domain (Fig. 2d). However, certain IgG glycosylation variants have been shown to have an altered half-life. For example, antibodies with terminal sialic acid residues in F(ab)-linked sugar moieties may have a slightly longer half-life compared to non-sialylated antibodies(van de Bovenkamp et al. 2016). Conversely, glycoproteins containing high mannose structures may exhibit a reduced serum half-life due to their ability to bind to the mannose receptor on macrophages (van de Bovenkamp et al. 2016; Alessandri et al. 2012). Thus, while IgG glycosylation is not critical for FcRn binding, certain sugar moieties may exert slight influences on IgG half-life in vivo via binding to other receptor systems, such as the hepatic asialoreceptor or mannose receptors on macrophages.

5 Impact of Vaccination and T Helper Cell Responses on Serum IgG Glycosylation

In mice, vaccination studies have provided insights into how different pathways of B cell activation can impact IgG glycan composition. Immunization with T cell-independent (TI) antigens or T cell-dependent (TD) antigens without adjuvants mostly leads to the formation of sialylated IgG molecules, whereas stimulation with TD antigens under inflammatory conditions induces IgG antibodies with low levels of sialic acid residues (Hess et al. 2013; Oefner et al. 2012). Moreover, IgG subclass-specific changes have been observed after stimulation with TI and TD antigens (Kao et al. 2017). Similarly, human IgG1 has higher levels of galactose-containing sugar structures than IgG2 and IgG4, suggesting a role for IgG subclass structure or genetic factors linking IgG class switch to glycosylation (Plomp et al. 2017; Sonneveld et al. 2018). Furthermore, there is evidence suggesting that specific types of T helper cell responses (Th1, Th2, or Th17) can impose a distinct IgG glycosylation pattern. For example, treatment of B cells with Interleukin-21 (IL-21) and CpG significantly increases the generation of galactosylated and sialylated

N-glycan species while decreasing bisecting GlcNAc (Wang et al. 2011). Conversely, IL-21 seems to have the opposite effect on IgG glycosylation (Hess et al. 2013; Pfeifle et al. 2017). In a mouse model of rheumatoid arthritis (RA), IL-21 triggered the downregulation of ST6Gal1 expression in newly developing plasma blasts and plasma cells via IL-23-activated Th17 cells, thus shifting IgG glycosylation to pro-inflammatory sugar structures and triggering the clinical onset of autoimmune arthritis (Lubberts 2015; Pfeifle et al. 2017). Moreover, IL-21 may act in an autocrine loop specifically supporting the proliferation and differentiation of Th17 cells and induce IL-23 production by B cells, further enhancing the autoimmune process (Korn et al. 2007; Lee et al. 2015). The pivotal role of T follicular helper 17 (Tfh17) cells in modulating IgG sialylation was confirmed recently by showing that Tfh17 induction by IL-6 and IL-23 leads to a downregulation of ST6Gal1, allowing the production of desialylated IgG molecules (Bartsch et al. 2020). Similarly, water-in-oil adjuvants, which trigger IFNγ-producing Tfh1 cells, also inhibit ST6Gal1 expression in germinal center B cells, affecting IgG glycosylation (Riteau et al. 2016). In summary, these findings highlight the complex and still incompletely understood regulatory networks involved in regulation of IgG glycosylation upon immune system activation.

6 Changes in IgG Glycosylation During Diseases

The notion that antibody glycosylation is not stable but can change upon deviation from the steady state, that is inflammatory diseases, is almost 40 years old (Arnold et al. 2007; Parekh et al. 1985). Since then, it has become established that during microbial infection, vaccination, autoimmune inflammation, aging, and pregnancy, characteristic changes in serum IgG glycosylation, predominantly affecting the terminal sugar residues galactose and sialic acid, can occur (Bartsch et al. 2020; de Haan et al. 2016, 2020; Jansen et al. 2016; Kaneko et al. 2006b; Kao et al. 2017; Kristic et al. 2014; Mahan et al. 2016; Mehta et al. 2008; Moore et al. 2005; Parekh et al. 1988; Pezer et al. 2016; Pucic et al. 2012; Selman et al. 2012; Twisselmann et al. 2018; Vadrevu et al. 2018). Furthermore, socioeconomic factors may impact antibody glycosylation. For instance, individuals living in developing countries exhibit serum IgG glycome traits characterized by high levels of G0 glycan forms, while industrial countries with well-developed health systems and sufficient access to water and food display IgG glycan traits with higher levels of galactosylated and sialylated glycan species (Mahan et al. 2016; Stambuk et al. 2020). Additionally, alterations in IgG core fucosylation, usually highly abundant in IgG sugar domains, have been observed in platelet-specific alloantibodies in fetal or neonatal alloimmune thrombocytopenia (FNAIT) (Kapur et al. 2014; Sonneveld et al. 2016). Moreover, low levels of IgG fucosylation were detected on dengue virus-specific antibodies and correlated with more severe infections (Wang et al. 2017). The impact of the presence or absence of these variable sugar moieties will be discussed in more detail later. Of note, the likelihood of developing an autoimmune disease such as rheumatoid arthritis increases with age. Interestingly, age-related differences in serum IgG glycosylation have been observed in several studies (Parekh et al. 1988; Shikata et al. 1998; Yamada et al. 1997). It has been demonstrated that galactosylation increases from early childhood to 25 years, reaching a peak of 80% serum IgG in either the IgG-G1 or IgG-G2 glycoform, followed by a decline in these glycoforms later in life (Parekh et al. 1988). The age-related increase in IgG lacking terminal sialic acid and galactose residues appears to be strongly influenced by sex, as only women showed a significant increase in IgG-G0 variants. This aspect must be considered when interpreting results from studies that include both male and female individuals (Shikata et al. 1998). Thus, the IgG-G0 glycoform was present in 20% of the serum IgG pool of women in their twenties, increasing to about 40% in women in their sixties, while the IgG-G0 level remained relatively stable at 25% in men. Importantly, these changes occurred independently of the development of overt autoimmune disease. However, the age-related increase in IgG-G0 never reached the levels observed during acute inflammation in patients with rheumatoid arthritis, where IgG-G0 can exceed 50%. A recent study investigating age-related changes in IgG1 and IgG2 subclass glycosylation confirmed these findings and further suggested that decreased levels of bisecting GlcNAc residues on IgG-G0 glycoforms below the age of 60 may be associated with human longevity (Ruhaak et al. 2010). One possible interpretation of these results is that age-related physiological changes and sex hormones may impact IgG glycosylation. Whether this provides the basis for or lowers the threshold for the development of autoimmune diseases remains unknown.

7 The Impact of Galactose on IgG Activity

As the lack of terminal galactose (and sialic acid) residues coincided with high autoimmune disease activity in rheumatoid arthritis, it was suggested early on that the absence of galactose might increase the pro-inflammatory activity of IgG (or reversely that terminal galactose residues may be negative regulatory of IgG activity). Indeed, a serum transfer study with arthritogenic serum either left untreated or digested with galactosidase to remove terminal galactose residues, the galactosidase treated serum induced a more severe arthritis in recipient mice (Rademacher et al. 1994). A subsequent study demonstrated that the IgG-G0 glycoform, due to its exposure of the mannose-rich core sugar structure, acquires the capacity to bind to mannose-binding lectin (MBL), providing a plausible explanation for this enhanced pro-inflammatory activity in vivo (Malhotra et al. 1995). Similar to C1q, MBL is a multimeric protein consisting of two subunits (MBL-A and MBL-B) that can bind to mannose-rich structures on the surface of bacteria, resulting in the activation of the complement pathway. However, some studies reported the opposite phenotype, showing that IgG glycovariants containing high levels of galactose residues may have higher pro-inflammatory or lytic activity, although others could not confirm this enhanced activity in independent model systems (Hadley et al. 1995; Kumpel et al. 1994, 1995; Boyd et al. 1995). At the time of these experiments, MBL knockout mice were not available to test this prediction in vivo. With the generation of these mice, it was possible to revisit the role of the lectin pathway for the pro-inflammatory activity of IgG-G0. Interestingly, MBL-deficient mice did not show impaired activity of IgG-G0 or IgG-G1/2 variants in several different model systems in vivo (Nimmerjahn et al. 2007; Ji et al. 2002). Similar results were obtained in complement C3-deficient mice, showing no impaired IgG-G0 activity (Nimmerjahn et al. 2007). In contrast, IgG-G0 activity was abrogated in the absence of activating FcγRs, consistent with results obtained with IgG-G1/2 glycoforms. Moreover, the absence of galactose residues resulted in a slight decrease in affinity for the inhibitory FcγRIIB, which may account for increased activity in some model systems (Nimmerjahn et al. 2007). A recent study investigated the presence of different MBL alleles resulting in either normal, intermediate, or low/absent MBL serum levels in women with arthritis during and right after pregnancy (van de Geijn et al. 2011). In accordance with the in vivo studies in MBL knockout mice, no correlation between MBL levels and rheumatoid arthritis disease activity could be observed. Data from cross-sectional studies with rheumatoid arthritis patients even suggested that the absence of MBL may accelerate the development of disease and joint destruction, indicating that MBL might have a protective role (Garred et al. 2000; Graudal et al. 2000). Apart from lectin complement pathway activation, IgG antibodies containing high levels of terminal galactose residues have been suggested to allow a better activation of the classical complement pathway via enabling a better IgG multimerization and C1q binding (Nimmerjahn et al. 2023a; Peschke et al. 2017; Dekkers et al. 2017; Van Osch et al. 2022). Thus, at least for certain conditions, including virus-specific immune responses (SARS-CoV-2), high levels of terminal galactose residues may increase the pro-inflammatory activity of IgG via classical complement pathway activation. With respect to IgG-G0 glycoforms, no clear picture emerges for an impact on IgG activity in vivo. There is no doubt, however, that this glycoform is a good biomarker correlating with disease activity.

8 The Role of Fucose in the Pro-inflammatory Activity of IgG

The impact of fucose on IgG-dependent effector functions is well documented, particularly in enhancing antibody-dependent cellular cytotoxicity (ADCC), which has led to the introduction of cytotoxic antibodies (e.g., the CD20-specific antibody obinutuzumab) with low fucose content into clinical application (Nimmerjahn et al. 2023a). Initial studies suggested that IgG1 molecules bearing bisecting N-acetylglucosamine residues exhibit higher ADCC activity. However, subsequent research revealed that this effect was primarily due to the concomitant absence of fucose residues in these IgG preparations (Shinkawa et al. 2003). Non-fucosylated antibodies of all IgG subclasses bind with higher affinity to the activating FcγRIIIA, significantly increasing ADCC activity regardless of IgG subclass. Interestingly, antigen binding and complement-dependent cytotoxicity (CDC) appear unaffected by the removal of core fucose (Niwa et al. 2005a, b; Shields et al. 2002; Nimmerjahn and Ravetch 2005). Consequently, generating afucosylated IgG has emerged as a general strategy to selectively enhance therapeutic antibody binding to FcγRIIIA.

Mechanistically, crystal structural analysis has shown that the penultimate fucose residue of IgG interacts closely with an N-linked sugar moiety present in FcγRIIIA, hindering a tight protein-protein interaction and resulting in a lower-affinity interaction (Ferrara et al. 2011). It is worth noting that this IgG glycosylation variant selectively modulates binding to FcγRIIIA but does not affect IgG binding to other activating or the inhibitory FcγRIIB. Of note, fucosylation on serum IgG antibodies in mice and humans is present in more than 90% of serum IgG and it has been very difficult to generate vaccination conditions under which antigen-specific IgG antibodies would be low in fucose (Dekkers et al. 2017; Kao et al. 2017). Of note, however, recent studies have identified that during certain virus infections or autoimmune diseases, virus- or self-reactive antibodies with a low level of fucose are generated (Thulin et al. 2020; Wang et al. 2017; Kapur et al. 2014). Understanding which factors trigger a downregulation of fucosyltransferases during B cell differentiation toward plasma cells or plasma blasts will be key to allow generating IgG antibodies with low levels of fucose via vaccination in the future (Oosterhoff et al. 2022; Nimmerjahn et al. 2023a).

9 The Role of Sialic Acid for the Anti-inflammatory Activity of IgG

The final sugar residue, which has been associated with impacting IgG function, is terminal sialic acid. It is well known from human clinical application that IgG infused intravenously at high doses (also referred to as IVIg therapy) of 1-3 g/kg can have an active anti-inflammatory activity (Schwab and Nimmerjahn 2013). Thus, an alternative explanation for the association of inflammation with an increased level of IgG-G0 glycoforms might be a loss of anti-inflammatory terminal sialic acid containing IgG glycoforms which become replenished by IVIg infusion (Durandy et al. 2009; Schwab and Nimmerjahn 2013; Seeling et al. 2017). The first evidence that the N297 attached sugar moiety might be an important component of the anti-inflammatory activity of IgG was provided by a study demonstrating that aglycosylated IVIg lost its ability to suppress joint inflammation in a murine model of rheumatoid arthritis (Kaneko et al. 2006b). A more detailed analysis demonstrated that terminal sialic acid residues are the major factor determining the anti-inflammatory activity of IgG as IVIg treated with neuraminidase also lost its capacity to suppress rheumatoid arthritis or inflammatory lung disease (Kaneko et al. 2006b; Massoud et al. 2014). Conversely, generating IVIg preparations rich in terminal sialic acid residues enhanced the anti-inflammatory activity (Kaneko et al. 2006b; Washburn et al. 2015). Consistent with previous studies in mice and humans, showing that the IgG Fc- but not the Fab-fragment is essential for the anti-inflammatory activity, it was demonstrated that the Fc- and not the Fab-linked sugar moiety is crucial for the anti-inflammatory activity (Anthony et al. 2008a, b; Kaneko et al. 2006b; Schwab and Nimmerjahn 2013). Importantly, the results obtained in inbred mouse models were confirmed in humanized mice, in which all human immune cell subsets are present (Schwab et al. 2015). Of note, other serum proteins such as transferrin and fetuin, containing a sugar moiety with a similar structure and high levels of terminal sialic acid residues, were not able to recapitulate IVIg activity, indicating that not simply the presence of sialic acid but more likely structural changes imposed by high levels of sialic acid on the Fc-fragment might be responsible for the anti-inflammatory activity (Fiebiger et al. 2015; Kaneko et al. 2006b; Sondermann et al. 2013). Moreover, only the addition of 2,6-linked but not 2,3-linked sialic acid residues was able to recapitulate IVIg activity and suppress rheumatoid arthritis (Anthony et al. 2008a, b). Although the identification of the essential role of sialic acid-rich IgG glycovariants for the therapeutic activity of IVIg was considered a major breakthrough, the underlying molecular and cellular pathways triggered by this molecule remain not fully resolved. Several studies indicated that the inhibitory FcγRIIB is an important component of the anti-inflammatory pathway triggered by IVIg (Anthony et al. 2011; Kaneko et al. 2006b; Schwab and Nimmerjahn 2013). Thus, activity of normal as well as siailic acid-rich IVIg was abrogated in FcγRIIB-deficient mice in models of ITP, nephrotoxic nephritis, and serum transfer arthritis (Anthony et al. 2011; Kaneko et al. 2006b). As discussed before, the inhibitory Fcγ-receptor is co-expressed with activating FcγRs on the majority of innate immune effector cells and establishes a threshold for cell activation by immune complexes (Nimmerjahn and Ravetch 2008). Further studies in mice revealed that FcγRIIB either became upregulated on effector macrophages or that more FcγRIIB-positive myeloid cells were present at sites of inflammation after IVIg treatment (Bruhns et al. 2003; Kaneko et al. 2006a, b; Samuelsson et al. 2001). Additional results obtained in a model of nephrotoxic nephritis demonstrated that, in addition to upregulation of FcγRIIB, a decreased expression of activating FcγRs could be observed resulting in a strong change of the A/I ratio toward immune cell inhibition (Kaneko et al. 2006a). More recently, the relevance of this observation for the human system was demonstrated in patients with chronic demyelinating inflammatory polyneuropathy (CIDP) demonstrating an upregulation of FcγRIIB upon IVIg infusion (Tackenberg et al. 2009, 2010). Taken together, a convincing set of data supports the concept that IVIg and the sialic acid-rich IgG fraction within the IVIg preparation may increase the threshold for innate immune effector cell activation. A critical question is how IVIg infusion triggers this upregulation of FcγRIIB on innate immune effector cells. In principle, IVIg could act directly on the effector cell population or indirectly via other immunomodulatory pathways. At present, the latter possibility seems to be the more likely pathway. Thus, IVIg activity was abrogated in osteopetrotic op/op mice lacking colony stimulating factor 1 (CSF1)-dependent monocytes and macrophages including the splenic marginal zone macrophage population expressing the cell surface receptors MARCO and SIGNR1 (Bruhns et al. 2003). An important role of the spleen and an intact splenic structure was suggested by the inability of IVIg to protect splenectomized mice and RAG mice lacking B- and T cells and therefore a normal splenic architecture (Anthony et al. 2008b). By using a variety of blocking antibodies and knockout mice, the authors could show that the SIGNR1 receptor is crucial for IVIg activity in mice (Anthony et al. 2008b; Schwab et al. 2014). Moreover, sialic acid-rich IgG but not asialylated IgG could bind to SIGNR1 and its human orthologue DC-SIGN (Anthony et al. 2008b; Choi et al. 2023). In contrast to the restricted expression pattern of mouse SIGNR1, human DC-SIGN can be detected on most myeloid-derived cells including certain monocyte subpopulations, dendritic cells, and macrophages. To study the role of human DC-SIGN for IVIg activity in vivo, a transgenic mouse expressing human DC-SIGN on a mouse SIGN-R1 background was generated recently (Anthony et al. 2011). Indeed, the anti-inflammatory activity of sialic acid-rich Fc-fragments (sFC) which were absent in SIGN-R1 knockout mice was restored in human DC-SIGN transgenic animals. Moreover, IVIg injection resulted in the production of IL33, which can initiate the production of Th2 cytokines including IL4 production which is known to upregulate FcγRIIB and to downmodulate activating FcγRs on innate immune effector cells. Although IL4 production could not be shown directly in vivo, knockout mice for IL4, the IL4Ra chain, and Stat6 which is essential for IL4 dependent signal transduction were no longer protected by IVIg from development of rheumatoid arthritis (Wermeling et al. 2013; Anthony et al. 2011). Of note, basophils were identified as a potential cell type responsible for IL33-dependent production of IL4. An important point to consider, however, is that this pathway is specifically triggered under preventive treatment conditions. IVIg-mediated resolution of established inflammation, while dependent on terminal sialic acid residues, did not require SIGNR1 (Schwab et al. 2014, 2015). Apart from the effect of highly sialylated IgG on the innate immune system, more recent studies have identified that sialylated IgG species may bind to CD23 on B cells, providing a feedback loop to modulate and fine-tune virus-specific immune responses (Wang et al. 2015).

10 Outlook

In summary, it has become established that the sugar moiety attached to each of the two IgG heavy chains is way more than just a component of IgG structural integrity. Instead, individual components of the sugar moiety represent molecular switches, which are able to modulate IgG binding to cellular receptors and signaling pathways, allowing to turn IgG molecules from pro-inflammatory mediators during infection and autoimmunity to active anti-inflammatory molecules that maintain immune homeostasis during the steady state. Importantly, we are only at the beginning of understanding many of the underlying cellular and molecular pathways in more detail.

References

Ahmed AA, Giddens J, Pincetic A, Lomino JV, Ravetch JV, Wang LX, Bjorkman PJ (2014) Structural characterization of anti-inflammatory immunoglobulin G Fc proteins. J Mol Biol 426: 3166–3179

Alessandri L, Ouellette D, Acquah A, Rieser M, Leblond D, Saltarelli M, Radziejewski C, Fujimori T, Correia I (2012) Increased serum clearance of oligomannose species present on a human IgG1 molecule. MAbs 4:509–520

Anthony RM, Nimmerjahn F, Ashline DJ, Reinhold VN, Paulson JC, Ravetch JV (2008a) Recapitulation of IVIG anti-inflammatory activity with a recombinant IgG Fc. Science 320:373–376

Anthony RM, Wermeling F, Karlsson MC, Ravetch JV (2008b) Identification of a receptor required for the anti-inflammatory activity of IVIG. Proc Natl Acad Sci USA 105:19571–19578

Anthony RM, Kobayashi T, Wermeling F, Ravetch JV (2011) Intravenous gammaglobulin suppresses inflammation through a novel T(H)2 pathway. Nature 475:110–113

Arnold JN, Wormald MR, Sim RB, Rudd PM, Dwek RA (2007) The impact of glycosylation on the biological function and structure of human immunoglobulins. Annu Rev Immunol 25:21–50

Bartsch YC, Eschweiler S, Leliavski A, Lunding HB, Wagt S, Petry J, Lilienthal GM, Rahmoller J, De Haan N, Holscher A, Erapaneedi R, Giannou AD, Aly L, Sato R, De Neef LA, Winkler A, Braumann D, Hobusch J, Kuhnigk K, Kremer V, Steinhaus M, Blanchard V, Gemoll T, Habermann JK, Collin M, Salinas G, Manz RA, Fukuyama H, Korn T, Waisman A, Yogev N, Huber S, Rabe B, Rose-John S, Busch H, Berberich-Siebelt F, Holscher C, Wuhrer M, Ehlers M (2020) IgG Fc sialylation is regulated during the germinal center reaction following immunization with different adjuvants. J Allergy Clin Immunol 146:652–666 e11

Biburger M, Aschermann S, Schwab I, Lux A, Albert H, Danzer H, Woigk M, Dudziak D, Nimmerjahn F (2011) Monocyte subsets responsible for immunoglobulin G-dependent effector functions in vivo. Immunity 35:932–944

Biburger M, Lux A, Nimmerjahn F (2014) How immunoglobulin G antibodies kill target cells: revisiting an old paradigm. Adv Immunol 124:67–94

Bohm S, Schwab I, Lux A, Nimmerjahn F (2012) The role of sialic acid as a modulator of the anti-inflammatory activity of IgG. Semin Immunopathol 34:443–453

Bohm S, Kao D, Nimmerjahn F (2014) Sweet and sour: the role of glycosylation for the anti-inflammatory activity of immunoglobulin G. Curr Top Microbiol Immunol 382:393–417

Borrok MJ, Jung ST, Kang TH, Monzingo AF, Georgiou G (2012) Revisiting the role of glycosylation in the structure of human IgG fc. ACS Chem Biol 7:1596–1602

Boyd PN, Lines AC, Patel AK (1995) The effect of the removal of sialic acid, galactose and total carbohydrate on the functional activity of Campath-1H. Mol Immunol 32:1311–1318

Brandsma AM, Hogarth PM, Nimmerjahn F, Leusen JH (2016) Clarifying the confusion between cytokine and Fc receptor "common gamma chain". Immunity 45:225–226

Bruhns P, Jonsson F (2015) Mouse and human FcR effector functions. Immunol Rev 268:25–51

Bruhns P, Samuelsson A, Pollard JW, Ravetch JV (2003) Colony-stimulating factor-1-dependent macrophages are responsible for IVIG protection in antibody-induced autoimmune disease. Immunity 18:573–581

Choi H, Yang SW, Joo JS, Park M, Jin Y, Kim JW, Lee SY, Lee SV, Yun TJ, Cho ML, Hwang HS, Kang YS (2023) Sialylated IVIg binding to DC-SIGN(+) Hofbauer cells induces immune tolerance through the caveolin-1/NF-kB pathway and IL-10 secretion. Clin Immunol 246: 109215

De Haan N, Reiding KR, Driessen G, Van Der Burg M, Wuhrer M (2016) Changes in healthy human IgG Fc-glycosylation after birth and during early childhood. J Proteome Res 15:1853–1861

De Haan N, Falck D, Wuhrer M (2020) Monitoring of immunoglobulin N- and O-glycosylation in health and disease. Glycobiology 30:226–240

Deisenhofer J (1981) Crystallographic refinement and atomic models of a human Fc fragment and its complex with fragment B of protein A from Staphylococcus aureus at 2.9- and 2.8-A resolution. Biochemistry 20:2361–2370

Dekkers G, Treffers L, Plomp R, Bentlage AEH, De Boer M, Koeleman CAM, Lissenberg-Thunnissen SN, Visser R, Brouwer M, Mok JY, Matlung H, van den Berg TK, van Esch WJE, Kuijpers TW, Wouters D, Rispens T, Wuhrer M, Vidarsson G (2017) Decoding the human immunoglobulin G-glycan repertoire reveals a Spectrum of Fc-receptor- and complement-mediated-effector activities. Front Immunol 8:877

Dougher CWL, Buffone A, Nemeth MJ, Nasirikenari M, Irons EE, Bogner PN, Lau JTY (2017) The blood-borne sialyltransferase ST6Gal-1 is a negative systemic regulator of granulopoiesis. J Leukoc Biol 102:507–516

Durandy A, Kaveri SV, Kuijpers TW, Basta M, Miescher S, Ravetch JV, Rieben R (2009) Intravenous immunoglobulins – understanding properties and mechanisms. Clin Exp Immunol 158(Suppl 1):2–13

Ferrara C, Grau S, Jager C, Sondermann P, Brunker P, Waldhauer I, Hennig M, Ruf A, Rufer AC, Stihle M, Umana P, Benz J (2011) Unique carbohydrate-carbohydrate interactions are required for high affinity binding between FcgammaRIII and antibodies lacking core fucose. Proc Natl Acad Sci USA 108:12669–12674

Fiebiger BM, Maamary J, Pincetic A, Ravetch JV (2015) Protection in antibody- and T cell-mediated autoimmune diseases by antiinflammatory IgG Fcs requires type II FcRs. Proc Natl Acad Sci USA 112:E2385–E2394

Garcia-Vallejo JJ, van Dijk W, van Die I, Gringhuis SI (2005) Tumor necrosis factor-alpha up-regulates the expression of beta1,4-galactosyltransferase I in primary human endothelial cells by mRNA stabilization. J Biol Chem 280:12676–12682

Garred P, Madsen HO, Marquart H, Hansen TM, Sorensen SF, Petersen J, Volck B, Svejgaard A, Graudal NA, Rudd PM, Dwek RA, Sim RB, Andersen V (2000) Two edged role of mannose binding lectin in rheumatoid arthritis: a cross sectional study. J Rheumatol 27:26–34

Gordan S, Biburger M, Nimmerjahn F (2015) bIgG time for large eaters: monocytes and macrophages as effector and target cells of antibody-mediated immune activation and repression. Immunol Rev 268:52–65

Graudal NA, Madsen HO, Tarp U, Svejgaard A, Jurik G, Graudal HK, Garred P (2000) The association of variant mannose-binding lectin genotypes with radiographic outcome in rheumatoid arthritis. Arthritis Rheum 43:515–521

Gringhuis SI, Garcia-Vallejo JJ, Van Het Hof B, van Dijk W (2005) Convergent actions of I kappa B kinase beta and protein kinase C delta modulate mRNA stability through phosphorylation of 14-3-3 beta complexed with tristetraprolin. Mol Cell Biol 25:6454–6463

Hadley AG, Zupanska B, Kumpel BM, Pilkington C, Griffiths HL, Leader KA, Jones J, Booker DJ, Stamps R, Sokol RJ (1995) The glycosylation of red cell autoantibodies affects their functional activity in vitro. Br J Haematol 91:587–594

Hess C, Winkler A, Lorenz AK, Holecska V, Blanchard V, Eiglmeier S, Schoen AL, Bitterling J, Stoehr AD, Petzold D, Schommartz T, Mertes MM, Schoen CT, Tiburzy B, Herrmann A, Kohl J, Manz RA, Madaio MP, Berger M, Wardemann H, Ehlers M (2013) T cell-independent B cell activation induces immunosuppressive sialylated IgG antibodies. J Clin Invest 123:3788–3796

Jansen BC, Bondt A, Reiding KR, Lonardi E, De Jong CJ, Falck D, Kammeijer GS, Dolhain RJ, Rombouts Y, Wuhrer M (2016) Pregnancy-associated serum N-glycome changes studied by high-throughput MALDI-TOF-MS. Sci Rep 6:23296

Jefferis R, Lund J (1997) Glycosylation of antibody molecules: structural and functional significance. Chem Immunol 65:111–128

Ji H, Ohmura K, Mahmood U, Lee DM, Hofhuis FM, Boackle SA, Takahashi K, Holers VM, Walport M, Gerard C, Ezekowitz A, Carroll MC, Brenner M, Weissleder R, Verbeek JS, Duchatelle V, Degott C, Benoist C, Mathis D (2002) Arthritis critically dependent on innate immune system players. Immunity 16:157–168

Jones MB, Oswald DM, Joshi S, Whiteheart SW, Orlando R, Cobb BA (2016) B-cell-independent sialylation of IgG. Proc Natl Acad Sci USA 113:7207–7212

Kaneko Y, Nimmerjahn F, Madaio MP, Ravetch JV (2006a) Pathology and protection in nephrotoxic nephritis is determined by selective engagement of specific Fc receptors. J Exp Med 203: 789–797

Kaneko Y, Nimmerjahn F, Ravetch JV (2006b) Anti-inflammatory activity of immunoglobulin G resulting from Fc sialylation. Science 313:670–673

Kao D, Danzer H, Collin M, Gross A, Eichler J, Stambuk J, Lauc G, Lux A, Nimmerjahn F (2015) A monosaccharide residue is sufficient to maintain mouse and human IgG subclass activity and directs IgG effector functions to cellular Fc receptors. Cell Rep 13:2376–2385

Kao D, Lux A, Schaffert A, Lang R, Altmann F, Nimmerjahn F (2017) IgG subclass and vaccination stimulus determine changes in antigen specific antibody glycosylation in mice. Eur J Immunol 47:2070–2079

Kapur R, Kustiawan I, Vestrheim A, Koeleman CA, Visser R, Einarsdottir HK, Porcelijn L, Jackson D, Kumpel B, Deelder AM, Blank D, Skogen B, Killie MK, Michaelsen TE, De Haas M, Rispens T, van der Schoot CE, Wuhrer M, Vidarsson G (2014) A prominent lack of IgG1-Fc fucosylation of platelet alloantibodies in pregnancy. Blood 123:471–480

Klaric L, Tsepilov YA, Stanton CM, Mangino M, Sikka TT, Esko T, Pakhomov E, Salo P, Deelen J, McGurnaghan SJ, Keser T, Vuckovic F, Ugrina I, Kristic J, Gudelj I, Stambuk J, Plomp R, Pucic-Bakovic M, Pavic T, Vilaj M, Trbojevic-Akmacic I, Drake C, Dobrinic P, Mlinarec J, Jelusic B, Richmond A, Timofeeva M, Grishchenko AK, Dmitrieva J, Bermingham ML, Sharapov SZ, Farrington SM, Theodoratou E, Uh HW, Beekman M, Slagboom EP, Louis E, Georges M, Wuhrer M, Colhoun HM, Dunlop MG, Perola M, Fischer K, Polasek O, Campbell H, Rudan I, Wilson JF, Zoldos V, Vitart V, Spector T, Aulchenko YS, Lauc G, Hayward C (2020) Glycosylation of immunoglobulin G is regulated by a large network of genes pleiotropic with inflammatory diseases. Sci Adv 6:eaax0301

Kobata A (2008) The N-linked sugar chains of human immunoglobulin G: their unique pattern, and their functional roles. Biochim Biophys Acta 1780:472–478

Korn T, Bettelli E, Gao W, Awasthi A, Jager A, Strom TB, Oukka M, Kuchroo VK (2007) IL-21 initiates an alternative pathway to induce proinflammatory T(H)17 cells. Nature 448:484–487

Krapp S, Mimura Y, Jefferis R, Huber R, Sondermann P (2003) Structural analysis of human IgG-Fc glycoforms reveals a correlation between glycosylation and structural integrity. J Mol Biol 325:979–989

Kristic J, Vuckovic F, Menni C, Klaric L, Keser T, Beceheli I, Pucic-Bakovic M, Novokmet M, Mangino M, Thaqi K, Rudan P, Novokmet N, Sarac J, Missoni S, Kolcic I, Polasek O, Rudan I, Campbell H, Hayward C, Aulchenko Y, Valdes A, Wilson JF, Gornik O, Primorac D, Zoldos V, Spector T, Lauc G (2014) Glycans are a novel biomarker of chronological and biological ages. J Gerontol A Biol Sci Med Sci 69:779–789

Kumpel BM, Rademacher TW, Rook GA, Williams PJ, Wilson IB (1994) Galactosylation of human IgG monoclonal anti-D produced by EBV-transformed B-lymphoblastoid cell lines is dependent on culture method and affects Fc receptor-mediated functional activity. Hum Antibodies Hybridomas 5:143–151

Kumpel BM, Wang Y, Griffiths HL, Hadley AG, Rook GA (1995) The biological activity of human monoclonal IgG anti-D is reduced by beta-galactosidase treatment. Hum Antibodies Hybridomas 6:82–88

Lee MM, Nasirikenari M, Manhardt CT, Ashline DJ, Hanneman AJ, Reinhold VN, Lau JT (2014) Platelets support extracellular sialylation by supplying the sugar donor substrate. J Biol Chem 289:8742–8748

Lee Y, Mitsdoerffer M, Xiao S, Gu G, Sobel RA, Kuchroo VK (2015) IL-21R signaling is critical for induction of spontaneous experimental autoimmune encephalomyelitis. J Clin Invest 125: 4011–4020

Lubberts E (2015) The IL-23-IL-17 axis in inflammatory arthritis. Nat Rev Rheumatol 11:415–429

Lux A, Yu X, Scanlan CN, Nimmerjahn F (2013) Impact of immune complex size and glycosylation on IgG binding to human FcgammaRs. J Immunol 190:4315–4323

Mahan AE, Jennewein MF, Suscovich T, Dionne K, Tedesco J, Chung AW, Streeck H, Pau M, Schuitemaker H, Francis D, Fast P, Laufer D, Walker BD, Baden L, Barouch DH, Alter G (2016) Antigen-specific antibody glycosylation is regulated via vaccination. PLoS Pathog 12: e1005456

Malhotra R, Wormald MR, Rudd PM, Fischer PB, Dwek RA, Sim RB (1995) Glycosylation changes of IgG associated with rheumatoid arthritis can activate complement via the mannose-binding protein. Nat Med 1:237–243

Manhardt CT, Punch PR, Dougher CWL, Lau JTY (2017) Extrinsic sialylation is dynamically regulated by systemic triggers in vivo. J Biol Chem 292:13514–13520

Massoud AH, Yona M, Xue D, Chouiali F, Alturaihi H, Ablona A, Mourad W, Piccirillo CA, Mazer BD (2014) Dendritic cell immunoreceptor: A novel receptor for intravenous immunoglobulin mediates induction of regulatory T cells. J Allergy Clin Immunol 133:853–863 e5

Mehta AS, Long RE, Comunale MA, Wang M, Rodemich L, Krakover J, Philip R, Marrero JA, Dwek RA, Block TM (2008) Increased levels of galactose-deficient anti-Gal immunoglobulin G in the sera of hepatitis C virus-infected individuals with fibrosis and cirrhosis. J Virol 82:1259–1270

Moore JS, Wu X, Kulhavy R, Tomana M, Novak J, Moldoveanu Z, Brown R, Goepfert PA, Mestecky J (2005) Increased levels of galactose-deficient IgG in sera of HIV-1-infected individuals. AIDS 19:381–389

Nimmerjahn F, Ravetch JV (2005) Divergent immunoglobulin g subclass activity through selective Fc receptor binding. Science 310:1510–1512

Nimmerjahn F, Ravetch JV (2008) Fcgamma receptors as regulators of immune responses. Nat Rev Immunol 8:34–47

Nimmerjahn F, Anthony RM, Ravetch JV (2007) Agalactosylated IgG antibodies depend on cellular Fc receptors for in vivo activity. Proc Natl Acad Sci USA 104:8433–8437

Nimmerjahn F, Vidarsson G, Cragg MS (2023a) Effect of posttranslational modifications and subclass on IgG activity: from immunity to immunotherapy. Nat Immunol 24:1244–1255

Nimmerjahn F, Vidarsson G, Cragg MS (2023b) Publisher correction: effect of posttranslational modifications and subclass on IgG activity: from immunity to immunotherapy. Nat Immunol 24: 1961

Niwa R, Natsume A, Uehara A, Wakitani M, Iida S, Uchida K, Satoh M, Shitara K (2005a) IgG subclass-independent improvement of antibody-dependent cellular cytotoxicity by fucose removal from Asn297-linked oligosaccharides. J Immunol Methods 306:151–160

Niwa R, Sakurada M, Kobayashi Y, Uehara A, Matsushima K, Ueda R, Nakamura K, Shitara K (2005b) Enhanced natural killer cell binding and activation by low-fucose IgG1 antibody results in potent antibody-dependent cellular cytotoxicity induction at lower antigen density. Clin Cancer Res 11:2327–2336

Oefner CM, Winkler A, Hess C, Lorenz AK, Holecska V, Huxdorf M, Schommartz T, Petzold D, Bitterling J, Schoen AL, Stoehr AD, Vu Van D, Darcan-Nikolaisen Y, Blanchard V, Schmudde I, Laumonnier Y, Strover HA, Hegazy AN, Eiglmeier S, Schoen CT, Mertes MM, Loddenkemper C, Lohning M, Konig P, Petersen A, Luger EO, Collin M, Kohl J, Hutloff A, Hamelmann E, Berger M, Wardemann H, Ehlers M (2012) Tolerance induction with T cell-dependent protein antigens induces regulatory sialylated IgGs. J Allergy Clin Immunol 129: 1647–55 e13

Ohmi Y, Ise W, Harazono A, Takakura D, Fukuyama H, Baba Y, Narazaki M, Shoda H, Takahashi N, Ohkawa Y, Ji S, Sugiyama F, Fujio K, Kumanogoh A, Yamamoto K, Kawasaki N, Kurosaki T, Takahashi Y, Furukawa K (2016) Sialylation converts arthritogenic IgG into inhibitors of collagen-induced arthritis. Nat Commun 7:11205

Oosterhoff JJ, Larsen MD, van der Schoot CE, Vidarsson G (2022) Afucosylated IgG responses in humans – structural clues to the regulation of humoral immunity. Trends Immunol 43:800–814

Parekh RB, Dwek RA, Sutton BJ, Fernandes DL, Leung A, Stanworth D, Rademacher TW, Mizuochi T, Taniguchi T, Matsuta K et al (1985) Association of rheumatoid arthritis and primary osteoarthritis with changes in the glycosylation pattern of total serum IgG. Nature 316:452–457

Parekh R, Roitt I, Isenberg D, Dwek R, Rademacher T (1988) Age-related galactosylation of the N-linked oligosaccharides of human serum IgG. J Exp Med 167:1731–1736

Peschke B, Keller CW, Weber P, Quast I, Lunemann JD (2017) Fc-galactosylation of human immunoglobulin gamma isotypes improves C1q binding and enhances complement-dependent cytotoxicity. Front Immunol 8:646

Pezer M, Stambuk J, Perica M, Razdorov G, Banic I, Vuckovic F, Gospic AM, Ugrina I, Vecenaj A, Bakovic MP, Lokas SB, Zivkovic J, Plavec D, Devereux G, Turkalj M, Lauc G (2016) Effects of allergic diseases and age on the composition of serum IgG glycome in children. Sci Rep 6:33198

Pfeifle R, Rothe T, Ipseiz N, Scherer HU, Culemann S, Harre U, Ackermann JA, Seefried M, Kleyer A, Uderhardt S, Haugg B, Hueber AJ, Daum P, Heidkamp GF, Ge C, Bohm S, Lux A, Schuh W, Magorivska I, Nandakumar KS, Lonnblom E, Becker C, Dudziak D, Wuhrer M, Rombouts Y, Koeleman CA, Toes R, Winkler TH, Holmdahl R, Herrmann M, Bluml S, Nimmerjahn F, Schett G, Kronke G (2017) Regulation of autoantibody activity by the IL-23-T(H)17 axis determines the onset of autoimmune disease. Nat Immunol 18:104–113

Pincetic A, Bournazos S, Dilillo DJ, Maamary J, Wang TT, Dahan R, Fiebiger BM, Ravetch JV (2014) Type I and type II Fc receptors regulate innate and adaptive immunity. Nat Immunol 15: 707–716

Plomp R, Ruhaak LR, Uh HW, Reiding KR, Selman M, Houwing-Duistermaat JJ, Slagboom PE, Beekman M, Wuhrer M (2017) Subclass-specific IgG glycosylation is associated with markers of inflammation and metabolic health. Sci Rep 7:12325

Pucic M, Muzinic A, Novokmet M, Skledar M, Pivac N, Lauc G, Gornik O (2012) Changes in plasma and IgG N-glycome during childhood and adolescence. Glycobiology 22:975–982

Radaev S, Motyka S, Fridman WH, Sautes-Fridman C, Sun PD (2001) The structure of a human type III Fcgamma receptor in complex with fc. J Biol Chem 276:16469–16477

Rademacher TW, Williams P, Dwek RA (1994) Agalactosyl glycoforms of IgG autoantibodies are pathogenic. Proc Natl Acad Sci USA 91:6123–6127

Rajput B, Shaper NL, Shaper JH (1996) Transcriptional regulation of murine beta1,4-galactosyltransferase in somatic cells. Analysis of a gene that serves both a housekeeping and a mammary gland-specific function. J Biol Chem 271:5131–5142

Raju TS (2008) Terminal sugars of Fc glycans influence antibody effector functions of IgGs. Curr Opin Immunol 20:471–478

Riteau N, Radtke AJ, Shenderov K, Mittereder L, Oland SD, Hieny S, Jankovic D, Sher A (2016) Water-in-oil-only adjuvants selectively promote T follicular helper cell polarization through a type I IFN and IL-6-dependent pathway. J Immunol 197:3884–3893

Robinett RA, Guan N, Lux A, Biburger M, Nimmerjahn F, Meyer AS (2018) Dissecting FcgammaR regulation through a multivalent binding model. Cell Syst 7:41–48 e5

Ruhaak LR, Uh HW, Beekman M, Koeleman CA, Hokke CH, Westendorp RG, Wuhrer M, Houwing-Duistermaat JJ, Slagboom PE, Deelder AM (2010) Decreased levels of bisecting GlcNAc glycoforms of IgG are associated with human longevity. PLoS One 5:e12566

Samuelsson A, Towers TL, Ravetch JV (2001) Anti-inflammatory activity of IVIG mediated through the inhibitory Fc receptor. Science 291:484–486

Schaffert A, Hanic M, Novokmet M, Zaytseva O, Kristic J, Lux A, Nitschke L, Peipp M, Pezer M, Hennig R, Rapp E, Lauc G, Nimmerjahn F (2019) Minimal B cell extrinsic IgG glycan modifications of pro- and anti-inflammatory IgG preparations in vivo. Front Immunol 10:3024

Schwab I, Nimmerjahn F (2013) Intravenous immunoglobulin therapy: how does IgG modulate the immune system? Nat Rev Immunol 13:176–189

Schwab I, Mihai S, Seeling M, Kasperkiewicz M, Ludwig R, Nimmerjahn F (2014) Broad requirement for terminal sialic acid residues and FcgRIIB for the preventive and therapeutic activity of intravenous immunoglobulins in vivo. Eur J Immunol 44:1444

Schwab I, Lux A, Nimmerjahn F (2015) Pathways responsible for human autoantibody and therapeutic intravenous IgG activity in humanized mice. Cell Rep 13:610–620

Seeling M, Bruckner C, Nimmerjahn F (2017) Differential antibody glycosylation in autoimmunity: sweet biomarker or modulator of disease activity? Nat Rev Rheumatol 13:621–630

Selman MH, De Jong SE, Soonawala D, Kroon FP, Adegnika AA, Deelder AM, Hokke CH, Yazdanbakhsh M, Wuhrer M (2012) Changes in antigen-specific IgG1 Fc N-glycosylation upon influenza and tetanus vaccination. Mol Cell Proteomics 11:M111 014563

Shade KT, Platzer B, Washburn N, Mani V, Bartsch YC, Conroy M, Pagan JD, Bosques C, Mempel TR, Fiebiger E, Anthony RM (2015) A single glycan on IgE is indispensable for initiation of anaphylaxis. J Exp Med 212:457–467

Shade KT, Conroy ME, Anthony RM (2019) IgE glycosylation in health and disease. Curr Top Microbiol Immunol 423:77–93

Shade KC, Conroy ME, Washburn N, Kitaoka M, Huynh DJ, Laprise E, Patil SU, Shreffler WG, Anthony RM (2020) Sialylation of immunoglobulin E is a determinant of allergic pathogenicity. Nature 582:265–270

Shields RL, Lai J, Keck R, O'Connell LY, Hong K, Meng YG, Weikert SH, Presta LG (2002) Lack of fucose on human IgG1 N-linked oligosaccharide improves binding to human Fcgamma RIII and antibody-dependent cellular toxicity. J Biol Chem 277:26733–26740

Shikata K, Yasuda T, Takeuchi F, Konishi T, Nakata M, Mizuochi T (1998) Structural changes in the oligosaccharide moiety of human IgG with aging. Glycoconj J 15:683–689

Shinkawa T, Nakamura K, Yamane N, Shoji-Hosaka E, Kanda Y, Sakurada M, Uchida K, Anazawa H, Satoh M, Yamasaki M, Hanai N, Shitara K (2003) The absence of fucose but not the presence of galactose or bisecting N-acetylglucosamine of human IgG1 complex-type oligosaccharides shows the critical role of enhancing antibody-dependent cellular cytotoxicity. J Biol Chem 278:3466–3473

Sondermann P, Huber R, Oosthuizen V, Jacob U (2000) The 3.2-A crystal structure of the human IgG1 Fc fragment-Fc gammaRIII complex. Nature 406:267–273

Sondermann P, Pincetic A, Maamary J, Lammens K, Ravetch JV (2013) General mechanism for modulating immunoglobulin effector function. Proc Natl Acad Sci USA 110:9868–9872

Sonneveld ME, Natunen S, Sainio S, Koeleman CA, Holst S, Dekkers G, Koelewijn J, Partanen J, van der Schoot CE, Wuhrer M, Vidarsson G (2016) Glycosylation pattern of anti-platelet IgG is stable during pregnancy and predicts clinical outcome in alloimmune thrombocytopenia. Br J Haematol 174:310–320

Sonneveld ME, Koeleman CAM, Plomp HR, Wuhrer M, van der Schoot CE, Vidarsson G (2018) Fc-glycosylation in human IgG1 and IgG3 is similar for both Total and anti-red-blood cell anti-K antibodies. Front Immunol 9:129

Stadlmann J, Pabst M, Altmann F (2010) Analytical and functional aspects of antibody sialylation. J Clin Immunol 30(Suppl 1):S15–S19

Stambuk J, Nakic N, Vuckovic F, Pucic-Bakovic M, Razdorov G, Trbojevic-Akmacic I, Novokmet M, Keser T, Vilaj M, Stambuk T, Gudelj I, Simurina M, Song M, Wang H, Salihovic MP, Campbell H, Rudan I, Kolcic I, Eller LA, Mckeigue P, Robb ML, Halfvarson J, Kurtoglu M, Annese V, Skaric-Juric T, Molokhia M, Polasek O, Hayward C, Kibuuka H, Thaqi K, Primorac D, Gieger C, Nitayaphan S, Spector T, Wang Y, Tillin T, Chaturvedi N, Wilson JF, Schanfield M, Filipenko M, Wang W, Lauc G (2020) Global variability of the human IgG glycome. Aging (Albany NY) 12:15222–15259

Tackenberg B, Jelcic I, Baerenwaldt A, Oertel WH, Sommer N, Nimmerjahn F, Lunemann JD (2009) Impaired inhibitory Fcgamma receptor IIB expression on B cells in chronic inflammatory demyelinating polyneuropathy. Proc Natl Acad Sci USA 106:4788–4792

Tackenberg B, Nimmerjahn F, Lunemann JD (2010) Mechanisms of IVIG efficacy in chronic inflammatory demyelinating polyneuropathy. J Clin Immunol 30(Suppl 1):S65–S69

Takai T (2005) Fc receptors and their role in immune regulation and autoimmunity. J Clin Immunol 25:1–18

Tan ZC, Lux A, Biburger M, Varghese P, Lees S, Nimmerjahn F, Meyer AS (2023) Mixed IgG Fc immune complexes exhibit blended binding profiles and refine FcR affinity estimates. Cell Rep 42:112734

Thulin NK, Brewer RC, Sherwood R, Bournazos S, Edwards KG, Ramadoss NS, Taubenberger JK, Memoli M, Gentles AJ, Jagannathan P, Zhang S, Libraty DH, Wang TT (2020) Maternal anti-dengue IgG fucosylation predicts susceptibility to dengue disease in infants. Cell Rep 31: 107642

Twisselmann N, Bartsch YC, Pagel J, Wieg C, Hartz A, Ehlers M, Hartel C (2018) IgG Fc glycosylation patterns of preterm infants differ with gestational age. Front Immunol 9:3166

Vadrevu SK, Trbojevic-Akmacic I, Kossenkov AV, Colomb F, Giron LB, Anzurez A, Lynn K, Mounzer K, Landay AL, Kaplan RC, Papasavvas E, Montaner LJ, Lauc G, Abdel-Mohsen M (2018) Frontline science: plasma and immunoglobulin G galactosylation associate with HIV persistence during antiretroviral therapy. J Leukoc Biol 104:461–471

Van De Bovenkamp FS, Hafkenscheid L, Rispens T, Rombouts Y (2016) The emerging importance of IgG Fab glycosylation in immunity. J Immunol 196:1435–1441

Van de Bovenkamp FS, Derksen NIL, Ooijevaar-De Heer P, van Schie KA, Kruithof S, Berkowska MA, van der Schoot CE, IJspeert H, van der Burg M, Gils A, Hafkenscheid L, Toes REM, Rombouts Y, Plomp R, Wuhrer M, van Ham SM, Vidarsson G, Rispens T (2018) Adaptive antibody diversification through N-linked glycosylation of the immunoglobulin variable region. Proc Natl Acad Sci USA 115:1901–1906

Van De Geijn FE, De Man YA, Wuhrer M, Willemsen SP, Deelder AM, Hazes JM, Dolhain RJ (2011) Mannose-binding lectin does not explain the course and outcome of pregnancy in rheumatoid arthritis. Arthritis Res Ther 13:R10

Van Kooyk Y, Kalay H, Garcia-Vallejo JJ (2013) Analytical tools for the study of cellular glycosylation in the immune system. Front Immunol 4:451

Van Osch TLJ, Oosterhoff JJ, Bentlage AEH, Nouta J, Koeleman CAM, Geerdes DM, Mok JY, Heidt S, Mulder A, van Esch WJE, Kapur R, Porcelijn L, van der Schoot CE, De Haas M, Wuhrer M, Voorberg J, Vidarsson G (2022) Fc galactosylation of anti-platelet human IgG1 alloantibodies enhances complement activation on platelets. Haematologica 107:2432–2444

Wang J, Balog CI, Stavenhagen K, Koeleman CA, Scherer HU, Selman MH, Deelder AM, Huizinga TW, Toes RE, Wuhrer M (2011) Fc-glycosylation of IgG1 is modulated by B-cell stimuli. Mol Cell Proteomics 10:M110 004655

Wang TT, Maamary J, Tan GS, Bournazos S, Davis CW, Krammer F, Schlesinger SJ, Palese P, Ahmed R, Ravetch JV (2015) Anti-HA glycoforms drive B cell affinity selection and determine influenza vaccine efficacy. Cell 162:160–169

Wang TT, Sewatanon J, Memoli MJ, Wrammert J, Bournazos S, Bhaumik SK, Pinsky BA, Chokephaibulkit K, Onlamoon N, Pattanapanyasat K, Taubenberger JK, Ahmed R, Ravetch JV (2017) IgG antibodies to dengue enhanced for FcgammaRIIIA binding determine disease severity. Science 355:395–398

Washburn N, Schwab I, Ortiz D, Bhatnagar N, Lansing JC, Medeiros A, Tyler S, Mekala D, Cochran E, Sarvaiya H, Garofalo K, Meccariello R, Meador JW 3rd, Rutitzky L, Schultes BC, Ling L, Avery W, Nimmerjahn F, Manning AM, Kaundinya GV, Bosques CJ (2015) Controlled tetra-Fc sialylation of IVIg results in a drug candidate with consistent enhanced anti-inflammatory activity. Proc Natl Acad Sci USA 112:E1297–E1306

Wermeling F, Anthony RM, Brombacher F, Ravetch JV (2013) Acute inflammation primes myeloid effector cells for anti-inflammatory STAT6 signaling. Proc Natl Acad Sci USA 110: 13487–13491

Wohner M, Brechtelsbauer S, Friedrich N, Vorsatz C, Bulang J, Liang C, Schorr L, Beschin A, Guilliams M, Ravetch J, Nimmerjahn F, Biburger M (2024) Tissue niche occupancy determines the contribution of fetal- versus bone-marrow-derived macrophages to IgG effector functions. Cell Rep 43:113757

Wormald MR, Rudd PM, Harvey DJ, Chang SC, Scragg IG, Dwek RA (1997) Variations in oligosaccharide-protein interactions in immunoglobulin G determine the site-specific glycosylation profiles and modulate the dynamic motion of the Fc oligosaccharides. Biochemistry 36: 1370–1380

Yamada E, Tsukamoto Y, Sasaki R, Yagyu K, Takahashi N (1997) Structural changes of immunoglobulin G oligosaccharides with age in healthy human serum. Glycoconj J 14:401–405

Diagnostic and Therapeutic Approaches in Congenital Disorders of Glycosylation

Alexandre Raynor, Élodie Lebredonchel, François Foulquier, François Fenaille, and Arnaud Bruneel

Contents

A. Raynor · É. Lebredonchel
AP-HP, Biochimie Métabolique et Cellulaire, Hôpital Bichat, Paris, France

F. Foulquier
Université de Lille, CNRS, UMR 8576 – UGSF – Unité de Glycobiologie Structurale et Fonctionnelle, Lille, France

F. Fenaille
Université Paris-Saclay, CEA, INRAE, Département Médicaments et Technologies pour la Santé (DMTS), MetaboHUB, Gif sur Yvette, France

A. Bruneel (✉)
AP-HP, Biochimie Métabolique et Cellulaire, Hôpital Bichat, Paris, France

INSERM UMR1193, Faculté de Pharmacie, Université Paris-Saclay, Orsay, France
e-mail: arnaud.bruneel@aphp.fr

F. Pfrengle et al. (eds.), *Complex Carbohydrates in Health and Disease*, Handbook of Experimental Pharmacology 288, https://doi.org/10.1007/164_2025_745

Abstract

Congenital disorders of glycosylation (CDG) constitute an increasing group of inborn metabolic disorders, with more than 170 described diseases to date. A disturbed glycosylation process characterizes them, with molecular defects localized in distinct cell compartments. In CDG, N-glycosylation, O-glycosylation, glycosylation of lipids (including phosphatidylinositol) as well as the glycosaminoglycan synthesis can be affected. Owing to the importance of glycosylation for the function of concerned proteins and lipids, glycosylation defects have diverse clinical consequences. CDG affected individuals often present with a non-specific multivisceral syndrome including neurological involvement, intellectual disability, dysmorphia, and hepatopathy. As CDG are rare diseases frequently lacking distinctive symptoms, biochemical and genetic testing bear important and complementary diagnostic roles.

After an introduction on glycosylation and CDG, we review current biomarkers and analytical techniques in the field. Furthermore, we illustrate their interests in the follow-up of proven therapeutic approaches including D-mannose in MPI-CDG, D-galactose in PGM1-CDG, and manganese ($MnSO_4$) in TMEM165-CDG.

Keywords

Apolipoprotein C-III · Bikunin · CDG · Galactose · Glycosylation · Manganese · Mannose · MPI-CDG · PGM1-CDG · TMEM165-CDG · Transferrin

Abbreviations

2-DE	Two-dimensional electrophoresis
ALG	Asparagine-linked glycosyltransferase
apoC-III	Apolipoprotein C-III
Asn	Asparagine
Bkn	Bikunin
CDG	Congenital disorder(s) of glycosylation
CDT	Carbohydrate deficient transferrin
CE	Capillary electrophoresis
COG	Conserved oligomeric Golgi
CS	Chondroitin sulfate

DS	Dermatan sulfate
EndoH	Endo-β-*N*-acetylglucosaminidase H
ER	Endoplasmic reticulum
ESI	Electrospray ionization
GAG	Glycosaminoglycan(s)
Gal	Galactose
GalNAc	*N*-Acetylgalactosamine
GDP	Guanosine diphosphate
Glc	Glucose
GlcA	Glucuronic acid
GlcNAc	*N*-Acetylglucosamine
GPI	Glycophosphatidylinositol
GSD1	Glycogen storage disease 1
HPLC	High-pressure liquid chromatography
ICAM-1	Intercellular adhesion molecule-1
IEF	Isoelectric focusing
ITI	Inter-α-trypsin inhibitor
KS	Keratan sulfate
LAMP2	Lysosomal-associated membrane protein 2
MALDI-TOF	Matrix-assisted laser desorption/ionization – time of flight
Man	Mannose
MPI-CDG	Mannose-phosphate isomerase-CDG
MRI	Magnetic resonance imaging
MS	Mass spectrometry
MW	Molecular weight
PαI	Pro-α-trypsin inhibitor
PGM1-CDG	Phosphoglucomutase 1-CDG
PIG-CDG	Phosphatidylinositol glycan-CDG
PMM2-CDG	Phosphomannomutase 2-CDG
PNGase F	Peptide-N-glycosidase F
Q-TOF	Quadrupole – time of flight
RFMS	Rapifluor^R mass spectrometry
SA	Sialic acid
Ser	Serine
Tf	Transferrin
TGN46	Trans-Golgi network protein 46
Thr	Threonine
UDP	Uridine-diphosphate
VUS	Variant of unknown significance
Xyl	Xylose

1 Introduction

Glycosylation is a co-/post-translational modification corresponding to the covalent linkage of a glycan chain to a protein or a lipid (Varki 2017). Protein-bound glycans are synthesized through sequential additions and removals of monosaccharides catalyzed by specific enzymes. In humans, glycans are mainly bound to proteins via the nitrogen of an asparagine (Asn) (N-linked glycans) or the oxygen of a serine (Ser) or threonine (Thr) (O-linked glycans). N-glycosylation begins in the endoplasmic reticulum (ER), followed by trimming and maturation in the Golgi apparatus, fueled by nucleotide-sugars synthesized in the cytoplasm (Colley et al. 2022). With regard to O-glycans, it can be distinguished as follows: mucin-type O-glycosylation, where the glycan chain is linked via an *N*-acetylgalactosamine (GalNAc); O-xylosylation, where it is linked via a xylose (Xyl) residue (leading to glycosaminoglycans chains); O-fucosylation; and O-mannosylation. Similar to the terminal steps of N-glycosylation, O-glycosylation mainly occurs in the Golgi apparatus (Brockhausen et al. 2022), and a fine regulation of the homeostasis of this organite, notably in terms of pH, trafficking, and ions transport, is mandatory (Rivinoja et al. 2009).

Over 170 inborn errors affecting glycosylation have been described and are designated as congenital disorders of glycosylation (CDG). Besides CDG affecting protein glycosylation, CDG affecting phosphatidyl inositol-linked glycans (PIG-CDG) and lipid glycosylation has also been described (Ng and Freeze 2018). The clinical phenotype of CDG greatly varies with multisystemic diseases including neurological disorders, intellectual disability, bone/cartilage defects, hepatopathy, and coagulopathy (Ondruskova et al. 2021). In view of these unspecific and broad presentations, biomarkers have been developed to study glycosylation and screen for CDG. While robust blood biomarkers are available to highlight defects of N-glycosylation (transferrin, N-glycome) and mucin core1 O-glycosylation (apolipoprotein C-III), the screening/diagnosis of other types of glycosylation defects is more difficult.

Inspired by a recently published (2024) synthesis work from our team (Raynor et al. 2024a), we provide here a simplified review regarding blood and cellular CDG biomarkers applicable in routine by specialized laboratories. We notably emphasized the diagnostic value of serum transferrin (Tf) and serum N-glycome for N-glycosylation defects, serum apolipoprotein C-III (apoC-III) for mucin core1 O-glycosylation defects, and serum bikunin for glycosaminoglycans (GAG) synthesis defects. We also consider more recently developed glycoproteomic tools for the study of protein-specific N-glycosylation defects. In addition, we illustrate the interests of these biomarkers in the currently available proven CDG treatments including oral mannose in PMI-CDG (CDG-1b), oral galactose in PGM1-CDG, and oral manganese in TMEM165-CDG.

2 Glycosylation Types and Congenital Disorders of Glycosylation

2.1 N-Glycosylation

N-glycosylation is the most frequent glycosylation type in humans (Brockhausen et al. 2022). Here, the glycan moiety is bound to the protein via the amide function of an asparagine (Asn), classically located in a consensus sequence Asn-X-Ser/Thr (X: any amino acid except proline). N-glycosylation is a two-step process that begins in the ER and ends in the Golgi apparatus. First, the glycan moiety is synthesized and attached to the nascent proteins in the ER and, second, the N-linked glycan chain undergoes maturation in the ER and Golgi apparatus.

As shown in Fig. 1, the first step begins at the cytosolic face of the ER membrane: sugars are attached sequentially to an anchored lipid named dolichol-phosphate (dol-P) independently of the protein translation process. These monosaccharides are available as nucleotide-sugars synthesized in the cytosol and are linked, thanks to

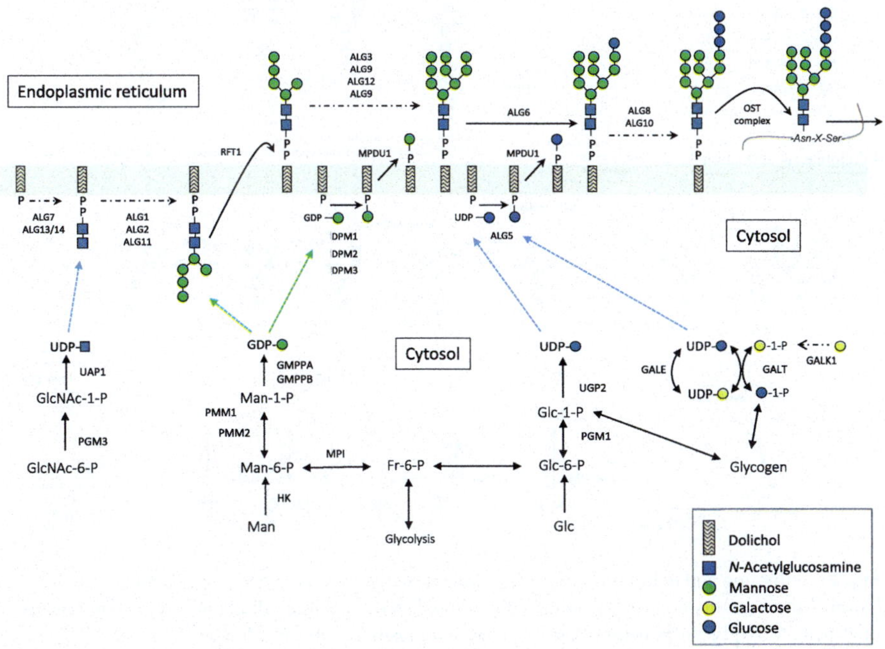

Fig. 1 First steps of N-glycosylation in the cytosol and endoplasmic reticulum. Nucleotide-sugars (UDP-GlcNAc, GDP-Man, UDP-Glc) synthesized in the cytosol through specific enzymatic machineries are the substrates for cytosolic and ER enzymes which participate in the step-by-step elongation of the precursor glycan linked to the lipidic anchor dolichol. Once synthesized, this precursor is transferred to the acceptor Asn of a nascent protein, thanks to the oligosaccharyl transferase (OST) complex. Enzymes abbreviations correspond to official gene names

the catalysis of glycosyltransferases of the ALG (Asn-linked glycosyltransferase) family. First, a *N*-acetylglucosamine-phosphate (GlcNAc-P) is attached to dol-P. Then, a GlcNAc and five mannoses (Man) are attached, yielding a Man_3-(Man)-Man-$GlcNAc_2$-PP-dolichol, which is translocated into the ER lumen. Then, four Man followed by three glucoses (Glc) are linked, yielding the biantennary N-glycan Glc_3-Man_3-(Man_2-(Man_2)-Man)-Man-$GlcNAc_2$-PP-dolichol. Here, Man and Glc originate from the cytosol, where they are attached to dol-P before translocation into the ER. Then, the biantennary polysaccharidic chain is transferred from the dol-P to the protein being translated, through the catalysis of the oligosaccharyltransferase (OST) complex (Fig. 1) (Breitling and Aebi 2013).

The second step is the maturation of the N-linked glycan (Fig. 2) that begins in the ER, where the glycoprotein undergoes folding and quality control. First, glucosidases I and II remove two terminal Glc, yielding a Glc_1-Man_3-(Man_5)-Man-$GlcNAc_2$ motif, which is recognized by calnexin and calreticulin, two lectins

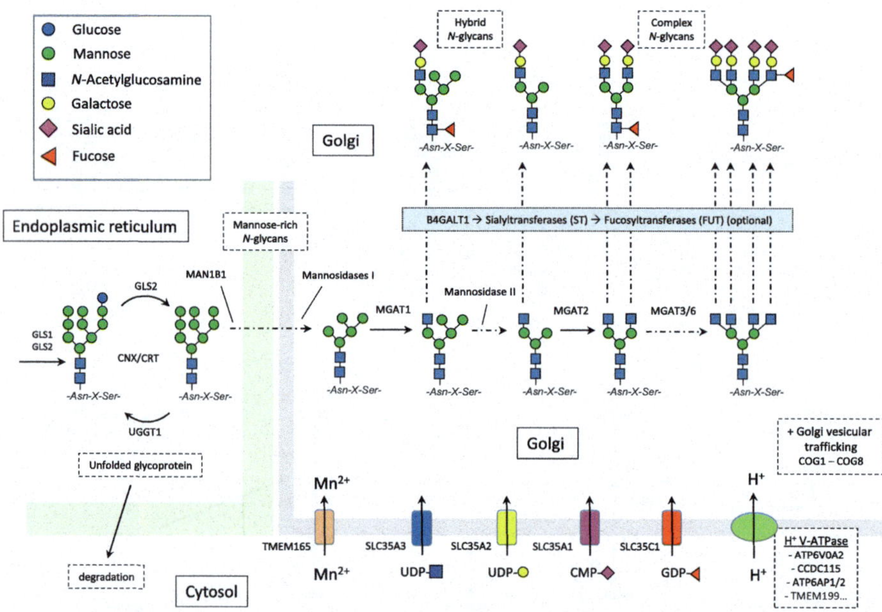

Fig. 2 Trimming and maturation of protein-linked N-glycans in the late endoplasmic reticulum and Golgi. After a quality control involving calnexin (CNX) and calreticulin (CRT), the "mannose-rich" intermediate N-glycan is trimmed by various mannosidases in the ER and the Golgi. Thereafter, GlcNAc, Gal, SA, and fucoses are successively added in this order (thanks to several glycosyltransferases), yielding a N-linked oligosaccharide of "complex-type" or "hybrid-type." Nucleotide-sugars (UDP-GlcNAc, UDP-Gal, CMP-SA, and GDP-Fuc) are synthesized in the cytosol and enter the Golgi through specific transporters of the SLC35 family. Golgi homeostasis (notably pH, Mn^{2+} concentration, and vesicular trafficking) plays a crucial role in the regulation of most molecular steps related to N-glycosylation. Enzymes abbreviations correspond to official gene names

which facilitate the glycoprotein folding (Ellgaard and Frickel 2003). At the end of the folding process, glucosidase II removes the remaining terminal Glc (Fig. 2) (Ellgaard and Helenius 2003). If the glycoprotein is incorrectly folded, the UDP-glucose glycoprotein glycosyltransferase (UGGT) adds a new Glc, and the glycoprotein can be recognized again by the lectins and undergoes a new folding process (Ellgaard and Helenius 2003). If correct folding is not achieved, three to four Man are removed and the unfolded glycoprotein is then exported to the cytosol for degradation (Hosokawa et al. 2003). If correct folding is achieved, the glycoprotein is transferred to the Golgi apparatus for further maturation (Fig. 2). Golgi mannosidases remove three Man; then, a GlcNAc transferase catalyzes the addition of a GlcNAc on one of the two antennae of the chain and mannosidase II removes the two terminal Man from the other antenna, resulting in a GlcNAc-Man_2-(Man)-$GlcNAc_2$ chain. Other sugars can then be sequentially attached, including (in this order) GlcNAc, galactose (Gal), terminal sialic acid (SA), and fucose (Fuc) (Stanley 2011). These sugars are available as nucleotide-sugars, which are transported from the cytosol into the Golgi apparatus, thanks to membrane transporters of the SLC35 family (Fig. 2).

In healthy individuals, N-glycans are mostly matured in “complex” N-glycans (Schachter 2014) characterized by the presence of terminal SA. Other types include “mannose-rich” and “hybrid-type” N-glycans (Fig. 2) (Huhn et al. 2009).

2.2 Mucin-Type O-Glycosylation

Mucin-type O-glycosylation takes place in the Golgi apparatus and involves the sequential building of the glycan directly on the protein. In mucin-type O-glycosylation, polypeptide-*N*-acetylgalactosamine transferases (ppGalNAcTs) first link a GalNAc on the oxygen of a Ser or a Thr (Ten Hagen et al. 2003) (no consensus sequence is yet clearly identified). Thereafter, other glycosyltransferases can attach other monosaccharides to build more complex O-glycan chains (core1 to core8 mucin-type structures) (Gill et al. 2011). In O-glycosylation, nucleotide-sugars also originate from the cytosol and are translocated to the Golgi trough SLC35 family transporters.

A mucin-type O-glycosylated protein of interest is serum apolipoprotein C-III (apoC-III) (Vaith et al. 1978). After the linkage of the GalNAc on Thr^{74}, a Gal is added by the core1 synthase (C1GalT1) (Fig. 3a). The Gal-GalNAc disaccharide constitutes the core1 motif upon which a GlcNAc can be added generating minor core2 glycoforms. For the majority core1 glycoforms of apoC-III, the addition of a first SA is catalyzed by β-galactoside α-2,3-sialyltransferase 1 (ST3GAL1), giving rise to the so-called “apoC-III_1” glycoform. Then, the optional attachment of a second SA onto the GalNAc is catalyzed by α-*N*-acetylgalactosaminide α-2,6-sialyltransferase 1 (ST6GalNAc I) leading to the “apoC-III_2” glycoform (Fig. 3a) (Wopereis et al. 2003).

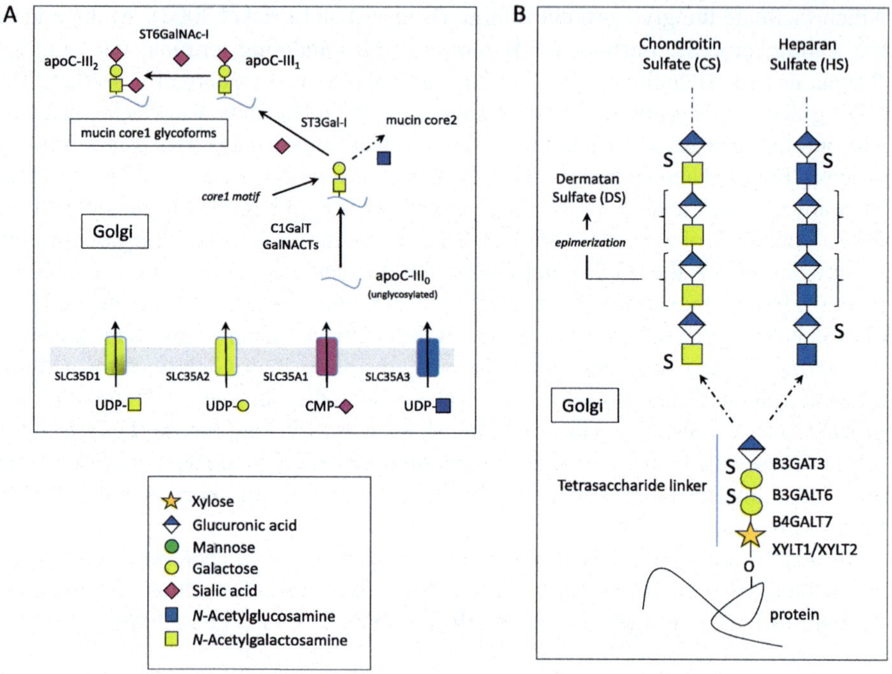

Fig. 3 Mucin core1 O-glycosylation and glycosaminoglycans biosynthesis. (**a**) For the prototype mucin core1 O-glycosylated apolipoprotein C-III (apoC-III), a GalNAc is first attached to the protein. Then, the linkage of a Gal leads to the core1 structure upon which one or two SA can be added, generating the apoC-III$_1$ and apoC-III$_2$ glycoforms. These are found in blood together with the minor unglycosylated apoC-III$_0$ glycoform. The core1 motif can also be substituted by a GlcNAc generating several minor mucin core2 glycoforms. Nucleotide-sugars (UDP-GalNAc, UDP-Gal, CMP-SA, and UDP-GlcNAc) are synthesized in the cytosol and enter the Golgi through specific transporters of the SLC35 family. Golgi homeostasis (notably pH, Mn^{2+} level, and membrane trafficking) plays a crucial role in the regulation of most molecular steps related to mucin core O-glycosylation. Enzymes abbreviations correspond to official gene names. (**b**) Upon the core protein-linked tetrasaccharide linkage region, GAG chain polymerization starts with the transfer of a first amino sugar (GalNAc for chondroitin sulfate [CS] and GlcNAc for heparan sulfate [HS]), followed by the addition of repeat and specific disaccharide moieties. Epimerization from CS to dermatan sulfate (DS) can occur. The GAG chains are typically heavily sulfated. Involved nucleotide-sugars are synthesized in the cytosol and enter the Golgi through specific transporters of the SLC35 family. Heparin and keratan sulfate (KS) biosynthesis pathways are not shown here. Enzymes abbreviations (here limited to the tetrasaccharide linker elongation) correspond to official gene names

2.3 Glycosaminoglycans (GAG) Synthesis

Glycosaminoglycans (GAG) chains are characterized by repetitive disaccharide motifs comprising an amino-sugar (e.g., GlcNAc or GalNAc) and a uronic acid (e.g., glucuronic acid: GlcA). In most of GAG, the chain is O-linked to the protein by a common GlcA-Gal-Gal-Xyl tetrasaccharide linker. This is the case with heparan sulfate, chondroitin sulfate (CS), and dermatan sulfate (DS) (Fig. 3b). A notable exception is keratan sulfate (KS), which consists of a (GlcNAc-Gal)$_x$ chain linked through alternate O- or N-linkages (Casale and Crane 2024). GAG synthesis starts in the ER, at the interface with the Golgi apparatus. Once the tetrasaccharide linker synthetized, various glycosyltransferases add a variable number of repeated disaccharide motifs. Similar to other Golgi-localized glycosylation processes, sugar moieties used in GAG synthesis originate from the cytosol, where they are "activated" through their linkage to a UDP. Furthermore, GAGs are abundantly sulfated, making them very negatively charged and hydrophilic (Haouari et al. 2021) (Fig. 3b).

2.4 Congenital Disorders of Glycosylation (CDG)

The high number of enzymes involved in glycosylation, their specificity, their sensitivity to the biochemical microenvironment (pH, ions, etc.), in addition to several transporters and molecular actors, make glycosylation a complex and fragile process. CDG comprises over 170 inborn diseases linked to the inheritance of pathogenic gene variants. They are typically autosomal recessive; however, rare cases of X-linked and autosomal dominant pathogenic variants have been described (Lipiński and Tylki-Szymańska 2021). CDG are rare diseases, with one to less than a hundred described affected individuals. Causative variants can affect genes coding for enzymes (e.g., mutases, glycosyltransferases, mannosidases), Golgi trafficking proteins (e.g., COG complex subunits), nucleotide-sugar transporters (SLC35 family), or ionic transporters (e.g., V-ATPase proton pumps) (Lipiński and Tylki-Szymańska 2021). The first cases of CDG (PMM2-CDG, MGAT2-CDG, MPI-CDG, etc.) were identified on the basis of defective protein glycosylation; subsequently, cases with faulty glycosylation of lipids or glycosylphosphatidylinositol (GPI) anchors (PIG-CDG) have also been reported (Jaeken et al. 1984, 1994; Francisco et al. 2019; Pedersen and Tygstrup 1980).

Clinically, CDG are heterogeneous, typically comprising non-specific symptoms of variable severity. Most commonly, affected individuals present with neurological signs and intellectual disability, failure to thrive, hepatopathy, coagulopathy, and dysmorphia. It is believed that part of these clinical signs results from disturbed glycosylation before birth. Indeed, numerous studies have pointed to the role of glycoproteins in embryogenesis and fetal development, and more particularly in neurological and bone/cartilage development (Yanagishita 1993; Mulkey et al. 2019; Flanagan-Steet and Steet 2013; Schwartz and Domowicz 2018). Glycosylation defects could be responsible for abnormal protein folding, transport, and ligand-receptor binding, thereby altering signaling pathways essential to embryonic development (e.g., TGF-β, sonic hedgehog) and/or directly impacting protein function (e.g., proteoglycans in bone/cartilage growth).

Historically, first CDG cases were discovered based on an abnormal Tf glycosylation pattern, underlying the first classification in type I: CDG-I and type II: CDG-II (Aebi et al. 1999; Marklová and Albahri 2007). CDG-I involves the first part of the

N-glycosylation process (synthesis of the precursor N-glycan attached to dol-P, or its transfer to the nascent protein), where pathogenic variants concern genes coding for cytosolic or ER actors. In CDG-I, Tf loses one or two complete N-glycan chains. The most common CDG, PMM2-CDG (CDG-Ia), belongs to this category (Jaeken et al. 1984). CDG-IIs involve the second part of the N-glycosylation process (after the linkage to the protein), where pathogenic variants are found in genes coding for ER or Golgi actors. In CDG-II, N-glycan chains are present on Tf but lack variable number of sugars, including terminal SA.

Historical CDG nomenclature was built upon this CDG-I/CDG-II classification; nowadays, each disorder is named after the mutated gene (e.g., PMM2-CDG instead of CDG-Ia) (Jaeken et al. 2009).

3 CDG Biomarkers and Related Laboratory Tools

As clinical presentations of CDG are usually unspecific, their screening/diagnosis relies on biochemical and genetic testing. Biochemical testing can highlight abnormal glycosylation, thanks to the study of specific glycoproteins, which may change in charge and/or molecular weight. Besides, characteristic patterns may orientate the diagnosis toward a specific etiology (Marklová and Albahri 2007). Genetic testing can be performed concurrently to biochemical testing, confirming the presence of causative gene variants. Both approaches are complementary, as genetic testing often results in the discovery of variants of unknown significance (VUS), which necessitate functional biochemical confirmations to assert their pathogenicity. As previously mentioned, the historical biomarker of CDG is serum transferrin (Tf).

3.1 Serum Tf Analysis by Using Isoelectric Focusing, Capillary Electrophoresis or Mass Spectrometry (N-Glycosylation Defects)

Tf is a blood-circulating N-glycoprotein synthesized by the liver and plays a key role in iron metabolism. It contains two N-glycan acceptor sites on Asn^{413} and Asn^{611} (MacGillivray et al. 1983); in healthy individuals, the two sites are occupied by complex bi-antennary N-glycans and Tf is mainly 4-sialylated (4-sialo Tf) while other glycoforms (8- to 5-, 3- and 2-sialo Tf) are present in minor proportions (Fig. 4a) (Coddeville et al. 1998).

The first technique developed to study Tf glycoforms was gel isoelectric focusing (IEF) (van Eijk et al. 1982). Tf glycoforms are gel-separated based on their charge, which depends on their terminal SA content. Elevated proportions of hyposialylated Tf glycoforms highlighted using IEF were first described in individuals with chronic alcohol consumption leading to the establishment of "carbohydrate-deficient Tf" (CDT) as a biomarker for chronic ethylism (van Eijk et al. 1983; Kapur et al. 1989). The discovery of CDG resulted from the observation of elevated % of hyposialylated Tf glycoforms in children suffering from unexplained neurological disorders (Jaeken

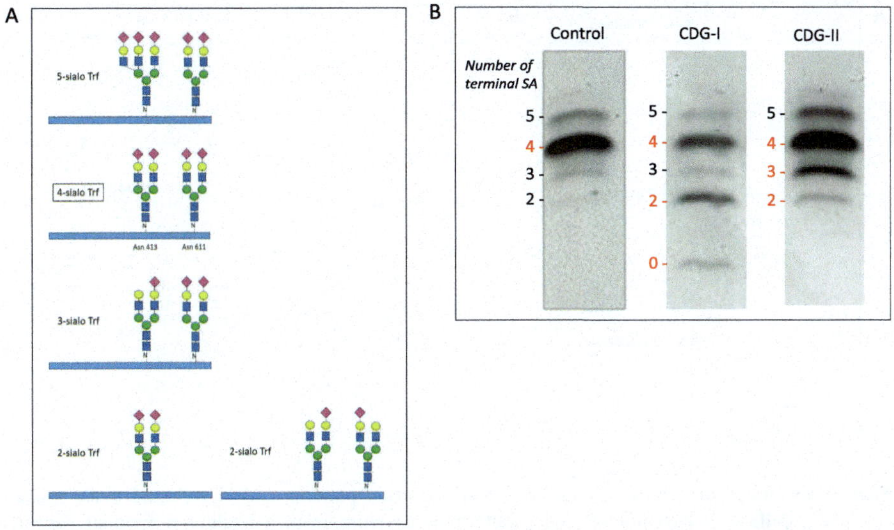

Fig. 4 Normal serum transferrin glycoforms and typical IEF profiles. (**a**) In the serum of healthy controls, the most common serum Tf glycoform is 4-sialoTf, bearing two mature N-glycan chains. 3-sialoTf and 2-sialoTf are minor glycoforms respectively missing one or two SA. The 2-sialoTf glycoform can result from the absence of a complete N-glycan chain or the loss of two terminal SA. Polyantennarity can occur, generating, e.g., the 5-sialo Tf glycoform. Although not represented, serum Tf glycoforms can also be fucosylated. (**b**) In CDG-I, the loss of one or two entire N-glycan chain(s) can be visualized after IEF with the increases of protein bands corresponding to the 2-sialo and the 0-sialo Tf glycoforms. In CDG-II, the maturation defects of the protein-linked N-glycans are associated to an increase of bands corresponding to various hyposialylated glycoforms (3-sialo and 2-sialo Tf glycoforms for the presented CDG-II profile)

et al. 1984). As previously mentioned, two types of Tf IEF patterns were described: a pattern with increased 2-sialo Tf and 0-sialo Tf (CDG-I profile), and a pattern also comprising elevated 3- and 1-sialo Tf, in variable proportions (CDG-II profile) (Fig. 4b). While most often only pointing to CDG-I or CDG-II, some peculiar Tf IEF patterns may orientate toward the causative gene, as it has been shown in MGAT2-CDG (major cathodical shift), PGM1-CDG (mixed CDG-I/CDG-II Tf pattern), or MAN1B1-CDG (major increase of the 3-sialoTf) (Jaeken et al. 1994; Rymen et al. 2013; Altassan et al. 2021).

Capillary electrophoresis (CE) is currently an interesting alternative technique for CDG screening (Parente et al. 2010; Oda et al. 1997). Here, Tf glycoforms are separated according to their electrophoretic mobility, which depends on their charge and size. The number of terminal SA affects charge, while the number and length of N-glycan chains affect size. Different commercial kits, developed for CDT measurement, were shown to provide results similar to IEF in CDG screening (Fig. 5a), with easier separation and better quantification of Tf glycoforms (Parente et al. 2010).

Both IEF and CE are sensitive to certain Tf protein polymorphisms, which more often than not modify its charge (Kamboh and Ferrell 1987) and could lead to an erroneous CDG screening. The majority of Tf polymorphisms may be detected by

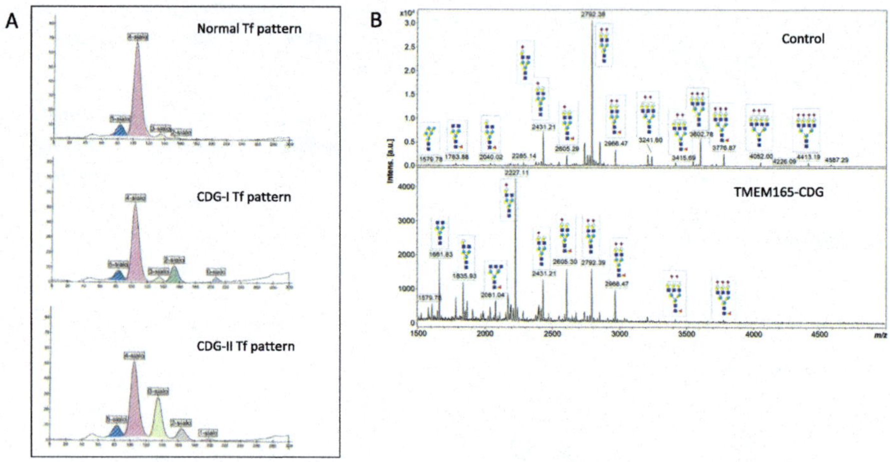

Fig. 5 Examples of serum Tf capillary electrophoresis patterns and N-glycome mass spectra. (**a**) The capillary electrophoresis (CE) pattern of a serum from a healthy individual shows a major peak corresponding to 4-sialoTf, with 5-sialo, 3-sialo, and 2-sialoTf also being present in minor proportions. The CE pattern of a CDG-I affected individual is characterized by increased proportions of 2- and 0-sialoTf at the expense of 4-sialoTf. In CDG-II, elevated proportions of various hyposialylated serum glycoforms can be visualized and accurately quantified. (**b**) MALDI-TOF mass spectra of permethylated N-glycans released through the action of PNGase F from serum samples of a healthy control and of an individual affected with TMEM165-CDG (CDG-II). Several additional peaks corresponding to major hypogalactosylation can be observed. Measurements were performed in positive-ion mode and all ions are present in sodiated form

pre-treatment of samples with sialidase, which removes all terminal SA residues from Tf N-glycan chains (Parker and Bearn 1962). In homozygous individuals, the sialidase-treated Tf pattern will display a single 0-sialo Tf band/peak, while two 0-sialo Tf moieties will be displayed in heterozygous individuals.

In some liver diseases, abnormal Tf N-glycosylation may also affect glycoforms separation. This has notably been observed (increased fucosylation and polyantennarity) in metabolic dysfunction-associated steatosis liver disease or cirrhosis. Therefore, in the context of CDG with hepatopathy, reanalysis by high pressure liquid chromatography (HPLC) could be recommended since it has been shown to permit the recovery of a significant part of uninterpretable CE profiles, thanks to better peak resolution (Stewart et al. 2010). Furthermore, a drawback in CDG screening with Tf is that neither IEF nor CE (and HPLC) is sensitive to fucosylation defects, as fucose is uncharged. Tf also lacks diagnostic value in CDG without N-glycosylation defects. Finally, we must be aware that some proven CDG linked to N-glycosylation defects could be falsely or transiently normal when analyzing Tf. Notably, this has been well described, but not clearly understood, in cases of PMM2-CDG, ATP6V0A2-CDG, COG-CDG as well as in GNE myopathy (an inherited SA biosynthesis defect) (Willems et al. 1860; Lefeber et al. 2011; Raynor et al. 2021a). Therefore, this should be kept in mind when facing unexpected normal CDG screening tests or when considering Tf as a potential biomarker for therapeutic monitoring in CDG clinical trials (Witters et al. 2021; Taday et al. 2021).

Mass spectrometry (MS)-based methods have been developed for highlighting abnormal glycan structures attached to Tf and to more precisely orientate toward causative genes. Analytical workflows involving liquid chromatography (LC) coupled to electrospray ionization (ESI) and quadruple time-of-flight (Q-TOF) mass spectrometer have been implemented in specialized laboratories for analyzing immunopurified intact Tf (Jansen et al. 2020; Bakar et al. 2018). Of note, while the MS study of intact Tf can theoretically not distinguish between CDG sharing a partial absence of entire N-glycan chains (CDG-I), it could present a direct diagnostic value for some CDG-II with specific abnormal N-glycan structures such as MGAT2-CDG or MAN1B-CDG (Van Scherpenzeel et al. 2014; Poskanzer et al. 2021).

Nevertheless, MS analysis of intact Tf has some analytical limitations. Notably, ESI mass spectra of Tf show a distribution of multiply charged ions that needs to be deconvoluted into "zero-charge" spectra of Tf glycoforms and potential variants. This step facilitates data interpretation but should be a source of inaccuracy. Thus, high spectrum complexity due to insufficient resolution or mass measurement accuracy, low sensitivity for minor Tf glycoforms, as well as improper deconvolution parameters should be considered for good data quality and proper interpretation (Wada et al. 2022). Jansen et al. recently performed ESI-QTOF-MS analysis of intact Tf on samples from 247 patients and 40 healthy controls and obtained interpretable spectral data only for approximately half of the cases, which could be regarded as rather limiting for CDG screening (Jansen et al. 2020). Furthermore, as mentioned for other techniques, falsely normal MS Tf patterns have also been reported in some cases of MOGS-CDG (Shimada et al. 2022) as well as in CDG with fucose deficiency (e.g., SLC35C1-CDG), probably due to the low basal level of fucosylation of Tf (Hipgrave Ederveen et al. 2020).

3.2 N-Glycome Analysis

In the field of CDG screening/diagnosis, the analysis of total serum/plasma N-glycans, the so-called N-glycome, has been shown to be a useful complement to Tf glycoprofiling (Bruneel et al. 2018a; Xia et al. 2013). Serum/plasma N-glycome analysis usually starts by the release of N-glycans from circulating glycoproteins by using peptide-N-glycosidase F (PNGase F) which cleaves the bond between the first GlcNAc of N-linked oligosaccharides and the Asn residue of the carrier protein. Released N-glycans are then derivatized by using chemical (e.g., permethylation) or fluorescent tags (e.g., 2-aminobenzamide) prior to their analysis. Most of the methods used for CDG diagnosis are based on MS, matrix-assisted laser desorption/ionization (MALDI), and/or LC coupled to high-resolution ESI MS (Bruneel et al. 2018a; Beimdiek et al. 2022; Park et al. 2020). When high throughput and ease of operation and data interpretation are required, MALDI-TOF-based approaches are often preferred for N-glycan profiling (Fig. 5b). Of interest, recent approaches involving labeling of N-glycans with a *N*-hydroxysuccinimide carbamate tag (RapiFluor-MS® or RFMS) were reported to bring high detection sensitivity and clinical specificity for CDG diagnosis when coupled to high-resolution ESI Q-TOF instruments (Messina et al. 2020; Chen et al. 2019). Lastly, N-glycomics can also be used for highlighting specific N-glycosylation defects of isolated plasma-derived glycoproteins such as Tf, IgG, or α_1-antitrypsin (Hipgrave Ederveen et al. 2020; Beimdiek et al. 2022; Sturiale et al. 2021).

Many examples thus demonstrate that total N-glycan profiling can be very informative for CDG-II diagnosis (Fig. 5b) (Hipgrave Ederveen et al. 2020; Bruneel et al. 2020). Recent reports also proved the usefulness of sensitive RFMS-based strategies for CDG-I diagnosis, for instance, through the detection of a low-intensity specific N-linked tetrasaccharide, and its fucosylated counterpart, in samples from individuals affected by different CDG-I such as ALG1-CDG (CDG-Ik), PMM2-CDG (CDG-Ia), and MPI-CDG (CDG-Ib) (Raynor et al. 2024b; Zhang et al. 2016; Abu Bakar et al. 2022).

Lastly, the use of endo-β-*N*-acetylglucosaminidase H (EndoH), which preferentially cleaves between the two GlcNAc of high-mannose and hybrid-type *N*-linked oligosaccharides, has been recently reported as facilitating the N-glycome-based diagnosis of MAN1B1-CDG or SLC37A4-CDG (Raynor et al. 2021b; Duvet et al. 2018).

3.3 Serum N-Glycoproteins Analysis by Using Western Blot or Two-Dimensional Electrophoresis

Besides Tf, the analysis of haptoglobin, alpha-1-antitrypsin, and alpha-1-acid glycoprotein by classical Western blot and/or two-dimensional electrophoresis (2-DE) can help orientate toward CDG-I or CDG-II (Bruneel et al. 2017; Seta et al. 1996). In classical Western blot, the absence of complete N-glycan chains in CDG-I will produce additional bands of lower molecular weight compared to controls (Fig. 6a), while the partial loss of monosaccharides in CDG-II will, in most cases, not significantly alter protein migration (Seta et al. 1996). Using 2-DE, additional protein spots of less acidic isoelectric point (cathodical shift) can be observed in CDG-I and CDG-II, while combined losses in molecular weight will typically be displayed in CDG-I only (Fig. 6c) (Bruneel et al. 2017). Of note, Western blot and 2-DE can be applied to very practical dried blood spot samples, by contrast with most of other described CDG screening methods.

3.4 Serum Apolipoprotein C-III Analysis (Mucin Core 1 O-Glycosylation)

Apolipoprotein C-III (apoC-III) is an O-linked glycoprotein (~10 kDa) primarily synthesized by hepatocytes and participating in the metabolism of lipoproteins (Norata et al. 2015). As mentioned above, it contains a single O-glycosylation

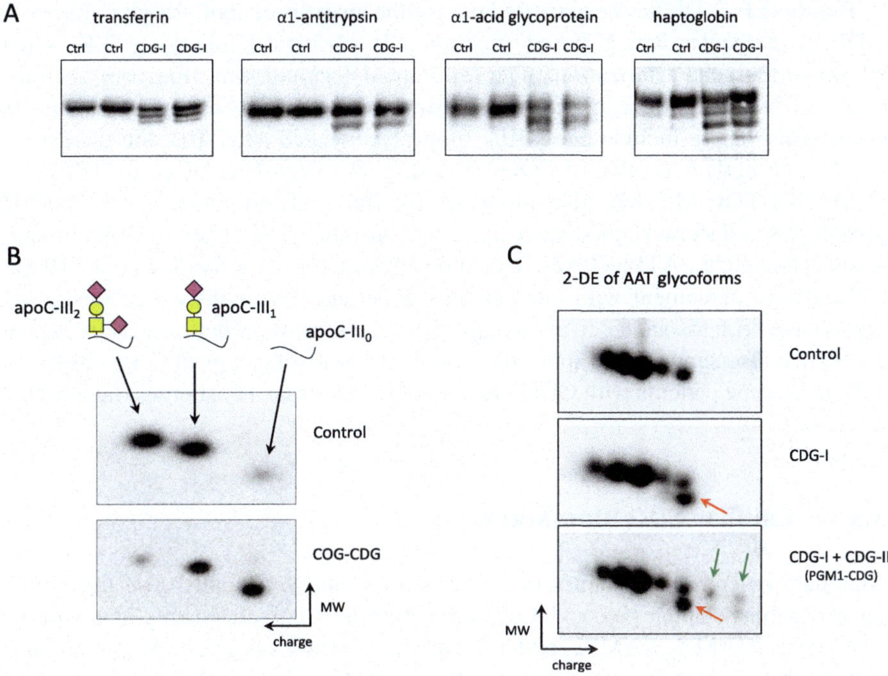

Fig. 6 Western blot and 2-DE patterns of various serum N-glycoproteins and mucin core1 O-glycosylated apolipoprotein C-III. (**a**) By comparison with controls (Ctrl), Western blot profiles of serum Tf, α1-antitrypsin, α1-acid glycoprotein, and haptoglobin in CDG-I affected individuals (CDG-I) typically display additional bands of decreased molecular weights. (**b**) In control, the 2-DE pattern of serum O-glycosylated apolipoprotein C-III (apoC-III) typically displays the major apoC-III_1 and apoC-III_2 glycoforms with trace amounts of unglycosylated apoC-III_0. In most of samples from COG-CDG-affected individuals (CDG-II with Golgi trafficking defects), a major and quasi-specific increase of the apoC-III_0 glycoform is observed. (**c**) Using 2-DE, serum α1-antitrypsin (AAT) can be separated into five major protein spots in control. In CDG-I, an additional spot (red arrow) corresponding to the loss of an entire N-glycan chain is classically observed. In PGM1-CDG (a mixed CDG-I + CDG-II), 2-DE of serum AAT displays the typical additional spot of CDG-I (red arrow) together with other additional spots typical of CDG-II (green arrows)

acceptor site on Thr^{74} (Vaith et al. 1978). In healthy individuals, this site is occupied by a mono- or bi-sialylated core1 O-glycan chain with trace amounts of asialylated apoC-III glycoforms (Wopereis et al. 2003) (Fig. 3a). The first technique developed to study the glycosylation of apoC-III for CDG screening was IEF (Wopereis et al. 2003). Three types of patterns were described: "apoC-III_0," "apoC-III_1," and "apoC-III_2," characterized respectively by increased % of 0-, 1-, and 2-sialo-apoC-III. Notably, apoC-III_0 and apoC-III_1 patterns were reported in CDG related to the vesicular Golgi trafficking (COG-CDG) or to the V-ATPase proton pump (ATP6V0A2-CDG) (Wopereis et al. 2003).

Beside IEF, 2-DE has been carried out for the analysis of apoC-III glycoforms. In 2-DE, a molecular weight-based separation is performed following IEF, which allows further characterization of core1 apoC-III glycoforms (Fig. 6b) (Bruneel et al. 2007). In addition, this technique can separate various asialylated apoC-III_0 glycoforms. These include apoC-III_{0a} (non-glycosylated apoC-III), the increase of which is strongly evocative of COG-CDG (Fig. 6b) (Yen-Nicolaÿ et al. 2015).

MALDI-TOF MS was also proposed for the study of intact core1 apoC-III glycoforms and showed good capacity to screen some CDG linked to Golgi homeostasis defects (e.g., ATP6V0A2-CDG and COG-CDG) (Yen-Nicolaÿ et al. 2015).

Lastly, in agreement with our personal experience, liver diseases, obesity and sepsis have been associated to altered apoC-III glycoforms proportions (Harvey et al. 2009). We also showed that apoC-III patterns could display typical O-glycosylation defects in some patients with CDG presenting with normal Tf patterns (Raynor et al. 2021a).

3.5 Cellular CDG Biomarkers

Although essential to characterize deficiencies and eventually test therapeutic options, robust cellular glycosylation biomarkers are relatively lacking in the routine CDG field. ICAM-1, LAMP2, and TGN46 are currently studied by Western blot to detect and characterize glycosylation deficiencies. Classically, glycosylation defects lead to abnormalities in their electrophoretic profiles and/or a decrease in their abundance. The intercellular cell adhesion molecule-1 (ICAM-1) has been identified as a useful cellular CDG-I biomarker as the lack of N-glycans leads to its degradation (He et al. 2012; Ferrer et al. 2020). Concerning N- and/or O-glycan defects, LAMP2 (heavily N-glycosylated) and TGN46 (heavily O-glycosylated) are interesting cellular biomarkers that have been widely used to identify glycosylation defects mainly in skin-derived fibroblasts of patients with Golgi homeostasis deficiencies (e.g., COG-CDG, TMEM165-CDG) (Houdou et al. 2019; Blackburn et al. 2018; Lebredonchel et al. 2019). Lastly, some of these cellular biomarkers such as LAMP2 and ICAM-1 have shown interests in the research and development of novel pharmacological agents (Ligezka et al. 2021; Durin et al. 2024).

3.6 Serum Bikunin Analysis (GAG Biosynthesis)

Bikunin is synthesized and excreted by hepatocytes under various isoforms in the blood. These include (i) the free core protein (Bkn), (ii) the proteoglycan consisting of Bkn bearing a CS chain (Bkn-CS) attached via the common tetrasaccharide linker, and (iii) Bkn-CS esterified by one or two "heavy chain" (HC) proteins leading to pro-α-trypsin inhibitor (PαI) and inter-α-trypsin inhibitor (ITI) (Fig. 7a).

We implemented serum Bkn-CS analysis as a biomarker of GAG biosynthesis defects such as linkeropathies, i.e., genetic diseases affecting the enzymes involved in the elongation of the GlcA-Gal-Gal-Xyl-O tetrasaccharide (Haouari et al. 2020;

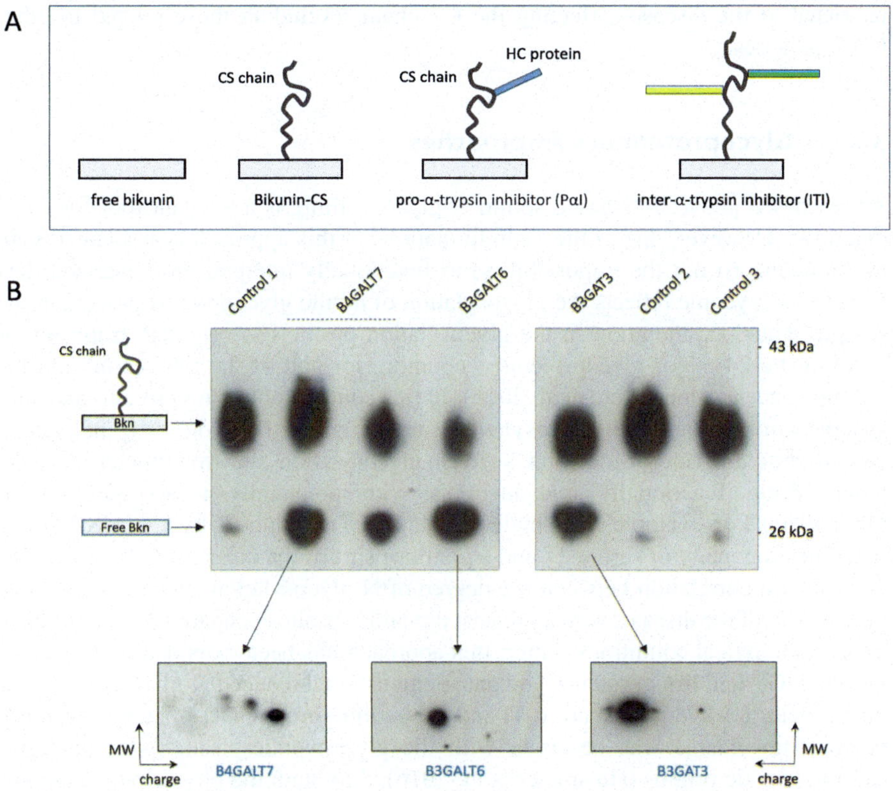

Fig. 7 Serum bikunin isoforms and Western blot patterns (1D and 2D) of linkeropathy-affected patients. (**a**) The major serum bikunin isoforms include (i) the free bikunin protein, (ii) the proteoglycan consisting of bikunin linked to a CS chain (bikunin-CS), and (iii) bikunin-CS linked to one or two heavy chain (HC) proteins leading to pro-α-trypsin inhibitor (PαI) and inter-α-trypsin inhibitor (ITI). (**b**) Compared to controls (control 1, 2, and 3), the Western blot of serum bikunin light forms (free Bkn and Bkn-CS) from linkeropathy-affected individuals (B4GALT7, B3GALT6, B3GAT3) show a large ~27 kDa band traducing a major defect in the initial biosynthesis of the CS chain. After 2-DE-based separation of the protein content of these abnormal bands, apparently gene-specific patterns could be observed

Bruneel et al. 2018b). We showed that Western blot of serum bikunin in individuals affected with linkeropathies (e.g., B4GALT7-CDG, B3GALT6-CDG) showed increased levels of abnormal light forms pointing to a defect in the tetrasaccharide linker biosynthesis (Fig. 7b). Furthermore, 2-DE of these abnormal forms displayed apparently specific patterns hinting at the causative mutated gene (Haouari et al. 2020) (Fig. 7b). While an interesting biomarker for the diagnosis of linkeropathies, Bkn-CS analysis has some limitations. Since mainly synthesized by hepatocytes, Bkn-CS analysis can only detect deficiencies in enzymes expressed in the liver. Furthermore, beside linkeropathies, the diagnostic interest of bikunin is theoretically restricted to the diseases affecting the CS chain, excluding those related to other GAG moieties.

3.7 Glycoproteomics Approaches

As discussed above, MS-based serum N-glycoprofiling is a relevant tool for CDG diagnosis. However, one of the main limitations of this approach is associated with the difficulty (if not the impossibility) to functionally interpret the generated data because N-glycome reflects the glycosylation of all the glycoproteins present in the sample. Thus, an alteration in the glycosylation profile can originate from one or more of the 24 major glycoproteins accounting for half of the total serum/plasma protein concentration (Clerc et al. 2016). In this context, obtaining protein- and site-specific information about glycosylation site occupancy (macroheterogeneity) and glycan structure heterogeneity at a given glycosylation site (microheterogeneity) would bring functionally relevant data with potentially meaningful related biomarkers. For example, Hülsmeier et al. (2007) evaluated by LC-MS/MS the macroheterogeneity of serum Tf and α_1-antitrypsin from a dozen of CDG-I patients and found a correlation between the degree of N-glycosylation site occupancy and the severity of the disease, with a greatest dynamics in site occupancy of Asn^{611} from Tf in pathological conditions. Later, this approach has been extended to Tf, haptoglobin, IgG_2, and IgA_1 proteins, and data demonstrated reduced N-glycosylation site occupancies for liver-synthesized Tf and haptoglobin from CDG-I patients samples, whereas no change was observed with the glycopeptides from IgA_1 and IgG_2 (of non-hepatic origin) (Hülsmeier et al. 2016). Although the physiological significance of such site occupancy-related data remains to be demonstrated, these approaches hold great potentials for bringing meaningful protein- and site-specific information.

These studies also confirmed that individual glycoproteins could respond differently to CDG and that glycopeptide-based diagnostic could present greater specificity and sensitivity. This can be illustrated by a very recent work of Garapati et al. (2024) which, using a multistep MS-based N-glycoproteomics approach, highlighted that a complement-C4 derived $Man_5GlcNAc_2$ glycopeptide can be regarded as a potentially sensitive and specific biomarker of PMM2-CDG.

4 Therapeutic Approaches in CDG and Interests of Biomarkers

Most of CDG are yet incurable. However, in a few disorders, therapeutic options exist that can correct or improve glycosylation status and/or symptoms. Notably oral D-mannose (D-Man), oral D-galactose (D-Gal), and oral manganese (Mn^{2+}) therapies showed beneficial clinical and/or biological effects in MPI-CDG,

PGM1-CDG, and TMEM165-CDG, respectively (Durin et al. 2024; Verheijen et al. 2020).

4.1 Oral D-Mannose in MPI-CDG (CDG-1b)

Mannose phosphate isomerase (MPI)-congenital disorder of glycosylation or MPI-CDG (formerly named CDG-1b; OMIM #602579) is an autosomal recessive inherited defect of the N-linked glycosylation of proteins (Niehues et al. 1998; Vuillaumier-Barrot et al. 2002). The enzymatic deficiency is caused by pathogenic variants affecting the *MPI* gene on chromosome 15. Mannose phosphate isomerase is a cytosolic enzyme that catalyzes the conversion of fructose-6-phosphate into mannose-6-phosphate, one of the first steps of the GDP-Man biosynthesis (Fig. 1). The clinical phenotype significantly differs from other CDG since affected individuals mostly present hypoglycemia (of variable severity), gastrointestinal symptoms (diarrhea, vomiting, protein losing enteropathy), liver disease (elevated transaminases, progressive fibrosis), and coagulopathy (mostly thrombosis), but in the absence of neurological defects or intellectual disability (Girard et al. 2020). Furthermore, MPI-CDG is one of the few treatable CDG subtypes with a proven efficacy of oral D-mannose therapy that can bypass the enzymatic deficiency (Fig. 8a) (Niehues et al. 1998; Girard et al. 2020).

Biochemical screening of MPI-CDG can be performed by previously detailed analyses of Tf, classically showing a marked CDG-I pattern. Although simple, fast, and very sensitive, these Tf-based methods lack specificity since MPI-CDG shares the same abnormal pattern with all other CDG-I subtypes. Diagnostic suspicion can be reinforced by determining MPI enzymatic activity from skin fibroblasts or blood leukocytes (Van Schaftingen and Jaeken 1995), which typically shows a significant decrease compared to controls. The diagnosis must be imperatively confirmed by sequencing of the *MPI* gene in the affected individual and both parents.

The recommended dose of oral D-mannose varies between 150 and 170 mg/kg bodyweight, three to five times a day (0.5 to 0.75 g/kg bodyweight/day). D-mannose is well absorbed orally, and its half-life is of approximately 4 h. The normal serum concentration of mannose varies between 46 μM and 65 μM. The concentration peak is obtained 1–2 h after ingestion of 0.07–0.2 g/kg. In adults, an intake of 12–20 g/day (to be modulated according to the digestive tolerability) should be sufficient (Čechová et al. 2020).

Mannose treatment improves the general condition and the digestive symptoms of the MPI-CDG affected individual. It normalizes glycemia and coagulation factors but its efficacy on the liver fibrosis regression is not demonstrated. Diarrhea, vomiting episodes, and hypoalbuminemia have been corrected in all patients presenting a protein losing enteropathy. In most cases, mannose treatment also normalizes (at least significantly improves) the glycosylation profile of Tf. The clinical response is classically observed within the first week of treatment while biological abnormalities often persisted several weeks (at least 15 days for Tf glycosylation) or months (Girard et al. 2020). In the French experience, thrombosis

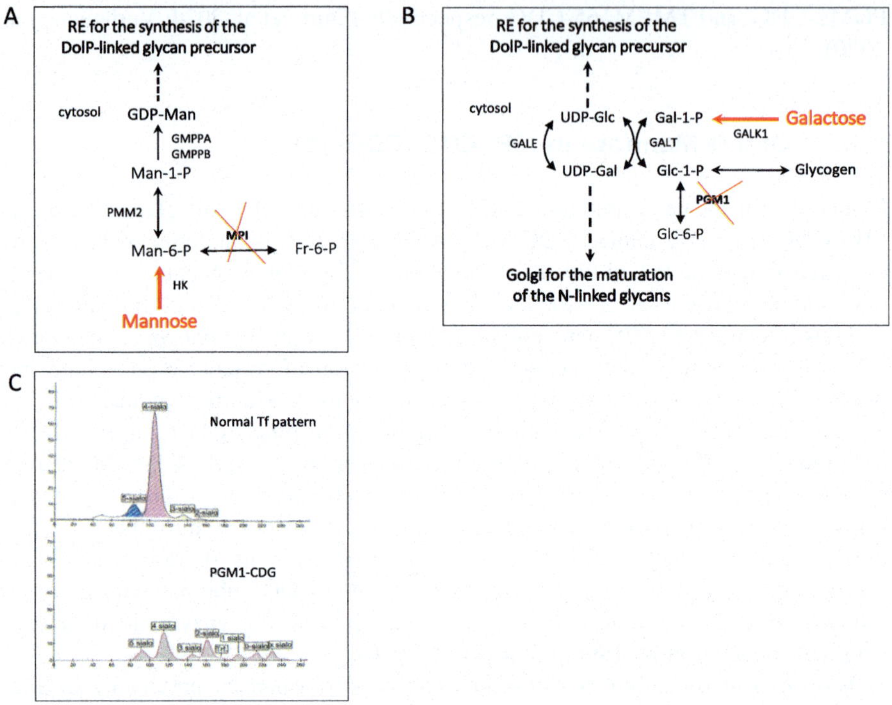

Fig. 8 Altered metabolic pathways in MPI-CDG and PGM1-CDG. Typical CE profile of untreated PGM1-CDG. (**a**) In MPI-CDG (CDG-Ib), mannose phosphate isomerase enzymatic deficiency can be bypassed by oral D-mannose supplementation that restores cytosolic biosynthesis of GDP-Man necessary to the N-glycosylation process in the endoplasmic reticulum (ER). (**b**) In PGM1-CDG (CDG-It), phosphoglucomutase 1 (PGM1) enzymatic deficiency can be bypassed by oral D-galactose supplementation, fueling in Gal-1P the biosynthesis of UDP-Gal and its conversion in UDP-Glc, both necessary to the N-glycosylation process (in the ER and Golgi). (**c**) Typical mixed CDG-I/CDG-II serum transferrin CE profile of a PGM1-CDG-affected individual compared to a control

episodes could occur at the discontinuation of the treatment (non-compliant patient), strongly suggesting that mannose must be administered for life.

The major adverse effects of oral mannose are abdominal pain and diarrhea, which could also reflect underdosage or bad compliance. Concerning this last point, we have shown a correlation between poor compliance for the oral mannose treatment and the worsening of the Tf glycosylation profile (Girard et al. 2020). The development of more convenient galenic formulations than the powder form (slow-release forms) would be a significant improvement for this very beneficial long-life treatment.

4.2 Oral D-Galactose in PGM1-CDG

PGM1-CDG (formerly CDG-It; OMIM #614921) corresponds to the inherited deficiency of the phosphoglucomutase 1 enzyme (Pérez et al. 2013) which catalyzes the cytosolic conversion of Glc-6P to Glc-1P, the latter being importantly involved in glycogen metabolism and UDP-Glc and UDP-Gal biosynthesis (Fig. 8b) (Radenkovic et al. 2019). It is an autosomal recessive CDG with biallelic mutations of the *PGM1* gene associated to defects in the N-glycosylation of proteins. Less than one hundred affected individuals have been described to date. Clinically, two distinct phenotypes can be differentiated (Altassan et al. 2021; Tegtmeyer et al. 2014) – one with a predominant muscular symptomatology, typically with an onset during adulthood, and presenting similarities to other glycogen storage diseases (acute rhabdomyolysis, muscular weakness, etc.) and the other phenotype is multivisceral with early symptoms consisting of liver disease (with elevated aminotransferases), rhabdomyolysis, hypoglycemia, congenital malformations (cleft palate, bifid uvula), coagulopathy, myopathy, and possibly severe cardiac involvement. Overall, neurological involvement and intellectual disability are less frequently observed than in other CDG. Furthermore, PGM1-CDG belongs to the rare subtypes for which an effective treatment is available. Indeed, oral galactose supplementation can bypass PGM1 deficiency by fueling in Gal-1P the biosynthesis of UDP-Gal and its conversion in UDP-Glc, both necessary to the N-glycosylation process (Fig. 8b) (Wong et al. 2017).

Besides possible specialized measurement of the PGM1 enzyme activity in skin fibroblasts (Tegtmeyer et al. 2014), biochemical screening of PGM1-CDG can be performed by previously detailed analyses of Tf, classically showing a very peculiar CDG-I/CDG-II mixed pattern (Tegtmeyer et al. 2014). More precisely, PGM1-CDG Tf patterns commonly associate elevated 2-sialo and 0-sialo Tf, reflecting the absence of entire N-glycan chains (type I; ER-localized origin), to elevated 3-sialo and 1-sialo Tf, linked to altered maturation of N-glycans in the Golgi (type II). It is tempting to hypothesize that, in this CDG, deficiencies in UDP-Glc (involved in ER steps of N-glycosylation) and UDP-Gal (involved in Golgi maturation steps) result in CDG-I and CDG-II abnormalities, respectively. Using CE, related mixed profiles appear quite puzzling with unusual abnormalities (in term of relative proportions of each Tf fractions) and frequent additional abnormal peaks (Fig. 8c). Using 2-DE of serum glycoproteins, superimposed CDG-I and CDG-II patterns can be more easily differentiated (Bruneel et al. 2017) (Fig. 6c). In our experience, such Tf patterns are quite specific since only found, in very attenuated form, in rare cases of fructosemia and chronic alcoholism. Nevertheless, isolated CDG-I or CDG-II Tf patterns have been described in a few number of PGM1-CDG cases (Altassan et al. 2021). In this context, more sensitive MS methods could perhaps bring additional clues. Furthermore, it has been suggested, using MS analysis of purified Tf, that the intensity of observed alterations (notably the aglycosylation and hypogalactosylation) could be related to the initial severity of the disease (Bakar et al. 2018). Lastly, as with other CDG, the diagnosis of PGM1-CDG must be imperatively confirmed by sequencing of the *PGM1* gene in the affected individual and both parents.

Recommended oral D-galactose doses range between 0.5 and 3 g/kg/day (to be gradually increased; single dose or divided in 4 doses daily) with a maximum of 50 g/day to avoid excessive levels of toxic Gal-1P and galactitol. Under this dosage, no significant adverse event has been noted with the exception of one patient who did not tolerate higher doses. In most of the patients, this therapy has been shown to improve several clinical and biological signs such as exercise intolerance, rhabdomyolysis, frequency of hypoglycemic episodes, as well as CK, aminotransferases, and coagulation factors levels (Altassan et al. 2021). Nevertheless, if initially present, the cardiac phenotype of the disease might be resistant to the treatment (Radenkovic et al. 2023). Concerning Tf glycosylation, the intakes of UDP-Glc and UDP-Gal related to galactose supplementation led to significant improvements (albeit without complete normalization) in the majority of cases. Lastly, as with mannose in MPI-CDG and although data about compliance to this treatment are quite scarce, galenic improvements of this oral powder D-galactose therapy would be also highly appreciated by concerned affected individuals.

4.3 Oral Manganese in TMEM165-CDG

We recently reported the ongoing manganese (Mn^{2+}) + D-galactose-based treatment in a young TMEM165-CDG-affected individual (Durin et al. 2024). In this case, TMEM165-CDG is related to a homozygous gene variant affecting the expression of the membrane transporter TMEM165 involved in the import of Mn^{2+} from the cytosol into the Golgi lumen (Fig. 2). As Mn^{2+} is crucial as a cofactor of numerous glycosylation enzymes (notably galactosyltransferases) (Durin et al. 2023; Foulquier and Legrand 2020), its TMEM165-CDG-related deficiency in the Golgi is not only accompanied by defects in N-glycosylation (maturation steps) but also in mucin-type O-glycosylation and GAG biosynthesis (Haouari et al. 2020; Foulquier et al. 2012; Khan et al. 2021).

The patient presented at 2 months old with a multivisceral syndrome combining recurrent hypovolemic shocks, diarrhea, failure to thrive, skeletal abnormalities, hepatomegaly with elevated transaminases, psychomotor delay, and abnormal coagulation. The few other described patients were diagnosed at older age and presented essentially with major skeletal abnormalities (scoliosis, osteoporosis), joint laxity, dwarfism, and psychomotor retardation (Foulquier et al. 2012; Zeevaert et al. 2013). Based essentially on the research of the team of François Foulquier on TMEM165 KO cell lines and on the skin fibroblasts of the patient, a $MnSO_4$ + D-galactose treatment was early initiated at 1 year old. Notably, it had been shown that the $MnCl_2$ treatment of TMEM165 KO cells fully rescued all abnormalities affecting N-glycosylation (LAMP2 profile), O-glycosylation (TGN46 profile), and GAG biosynthesis (Houdou et al. 2019; Durin et al. 2024; Foulquier et al. 2012; Khan et al. 2021). Furthermore, the Mn^{2+} treatment of the cultivated patient's fibroblasts showed beneficial effects on the altered migration profile of TGN46.

In first line, the biochemical screening of this TMEM165-CDG was performed by CE of serum Tf, showing a very abnormal CDG-II pattern with increased levels of all hyposialylated Tf glycoforms, at the expense of 4-sialoTf. In second line, MS-based serum N-glycome revealed major galactosylation defects of serum N-glycoproteins (Fig. 5b), as previously described for other TMEM165-CDG cases. In combination with these N-glycosylation defects, the 2-DE pattern of apoC-III showed major defects in mucin core1 O-glycosylation, highly suggestive for a Golgi homeostasis defect affecting all types of glycosylation. Furthermore, the Western blot of serum bikunin also showed marked abnormalities suggestive for a linkeropathy-like deficiency of the GAG biosynthesis (Durin et al. 2024). Altogether, these results prompted us to urgently sequence the *TMEM165* gene, with the discovery of a new homozygous pathogenic variant.

Oral $MnSO_4$ treatment was initiated at 11 months-old in combination with D-galactose. Although previously described as only improving the N-glycosylation defects observed in TMEM165 KO cells, D-galactose was added in view of its absence of toxicity and the high degree of hypogalactosylation in patient's samples. Oral galactose therapy was started at 0.5 g/kg bodyweight/day, progressively increasing (by steps of 6 weeks) to 1.5 g/kg/day. Combined Mn^{2+} treatment (MnSO4; H_2O) was started at 2 mg/day for 4 months, then 12 mg/day for 3 months, 30 mg/day for 1.5 months, to finally reach 10 mg/kg bodyweight/day.

After 1 year of treatment, the patient's digestive symptoms completely disappeared. Skeletal abnormalities and hepatomegaly did not improve but stabilized. Discrete psychomotor delay is still present, but the patient progressed at her own pace. Body weight increased under enteral nutrition. By contrast with the mostly stable clinical phenotype, we observed major biological improvements of aminotransferases and coagulation factors (both quasi-normal after 1 year). Regarding glycosylation, results were also spectacular with a quasi-normalization of all glycosylation types as evaluated by CE of serum Tf and MS-based serum N-glycome (N-glycosylation), 2-DE of apoC-III (mucin core1 O-glycosylation), and Western blot of serum bikunin (GAG biosynthesis). Furthermore, the kinetics of improvement of these glycosylation biomarkers varied, with belated improvements of apoC-III and bikunin patterns, suggesting a time and/or dose-dependent effects (Durin et al. 2024).

Regarding the potential toxicity of manganese, overexposure to this metal element can cause "manganism," which is a Parkinson-like syndrome related to irreversible neurological damages (Zheng et al. 2011). No efficient biological biomarker of manganese toxicity has been yet identified, notably due to large discrepancies between its blood and tissular half-life (Zheng et al. 2000, 2011). In these conditions, it has been shown that magnetic resonance imaging (MRI) can reflect manganese accumulation in the brain with good sensitivity and specificity. In the case of our patient, after 1 year of treatment, brain MRI did not show any manganese accumulation. This possible adverse effect will however require cautious evaluation of the patient during follow-up consultations.

Therefore, metal (i.e., cofactor) supplementation might offer novel opportunities as pharmacological agents to improve symptomatology in other CDG yet without therapy.

5 Conclusions

The glycosylation process is a very complex modification of proteins and lipids involving a multitude of enzymes and other molecular actors, in distinct subcellular compartments. Related congenital disorders of glycosylation (CDG) constitute an ever-growing group of inborn errors of metabolism. Their biochemical screening/diagnosis still mainly relies on the analysis of serum N-glycosylated transferrin using various gel-based and/or MS techniques. Beside, additional MS-based N-glycoprofiling and other biomarkers, such as apolipoprotein C-III (O-glycosylation) and bikunin (GAG biosynthesis), are precious tools to orientate toward the causative gene defect or to assert its pathogenicity. Furthermore, new N-glycomics and glycoproteomics approaches offer exiting perspectives in term of sensitivity and specificity. Nevertheless, several important glycosylation pathways remain poorly understood and currently lack biomarkers easily usable in clinical laboratories. Established and efficient pharmacological therapies for CDG are still very rare and quite limited to oral monosaccharide supplementation, with potential for improvement in galenic formulations (e.g., late-release forms). In this context, we believe that ongoing research on new metal-based CDG therapies look promising and offer exciting prospects regarding future clinical developments.

References

Abu Bakar N, Ashikov A, Brum JM, Smeets R, Kersten M, Huijben K, Keng WT, Speck-Martins CE, de Carvalho DR, de Rizzo IMPO, de Mello WD, Heiner-Fokkema R, Gorman K, Grunewald S, Michelakakis H, Moraitou M, Martinelli D, van Scherpenzeel M, Janssen M, de Boer L, van den Heuvel LP, Thiel C, Lefeber DJ (2022) Synergistic use of glycomics and single-molecule molecular inversion probes for identification of congenital disorders of glycosylation type-1. J Inherit Metab Dis 45:769–781. https://doi.org/10.1002/jimd.12496

Aebi M, Helenius A, Schenk B, Barone R, Fiumara A, Berger EG, Hennet T, Imbach T, Stutz A, Bjursell C, Uller A, Wahlström JG, Briones P, Cardo E, Clayton P, Winchester B, Cormier-Dalre V, de Lonlay P, Cuer M, Dupré T, Seta N, de Koning T, Dorland L, de Loos F, Kupers L (1999) Carbohydrate-deficient glycoprotein syndromes become congenital disorders of glycosylation: an updated nomenclature for CDG. First international workshop on CDGS. Glycoconj J 16:669–671. https://doi.org/10.1023/a:1017249723165

Altassan R, Radenkovic S, Edmondson AC, Barone R, Brasil S, Cechova A, Coman D, Donoghue S, Falkenstein K, Ferreira V, Ferreira C, Fiumara A, Francisco R, Freeze H, Grunewald S, Honzik T, Jaeken J, Krasnewich D, Lam C, Lee J, Lefeber D, Marques-da-Silva D, Pascoal C, Quelhas D, Raymond KM, Rymen D, Seroczynska M, Serrano M, Sykut-Cegielska J, Thiel C, Tort F, Vals M-A, Videira P, Voermans N, Witters P, Morava E (2021) International consensus guidelines for phosphoglucomutase 1 deficiency (PGM1-CDG): diagnosis, follow-up, and management. J Inherit Metab Dis 44:148–163. https://doi.org/10.1002/jimd.12286

Bakar NA, Lefeber DJ, van Scherpenzeel M (2018) Clinical glycomics for the diagnosis of congenital disorders of glycosylation. J Inherit Metab Dis 41:499–513. https://doi.org/10.1007/s10545-018-0144-9

Beimdiek J, Hennig R, Burock R, Puk O, Biskup S, Rapp E, Lesinski-Schiedat A, Buettner FFR, Das AM (2022) Serum N-glycomics of a novel CDG-IIb patient reveals aberrant IgG glycosylation. Glycobiology 32:380–390. https://doi.org/10.1093/glycob/cwac003

Blackburn JB, Kudlyk T, Pokrovskaya I, Lupashin VV (2018) More than just sugars: conserved oligomeric Golgi complex deficiency causes glycosylation-independent cellular defects. Traffic 19:463–480. https://doi.org/10.1111/tra.12564

Breitling J, Aebi M (2013) N-linked protein glycosylation in the endoplasmic reticulum. Cold Spring Harb Perspect Biol 5:a013359. https://doi.org/10.1101/cshperspect.a013359

Brockhausen I, Wandall HH, Hagen KGT, Stanley P (2022) O-GalNAc Glycans. In: Varki A, Cummings RD, Esko JD, Stanley P, Hart GW, Aebi M, Mohnen D, Kinoshita T, Packer NH, Prestegard JH, Schnaar RL, Seeberger PH (eds) Essentials of Glycobiology, 4th edn. Cold Spring Harbor Laboratory Press, Cold Spring Harbor (NY). http://www.ncbi.nlm.nih.gov/books/NBK579921/ (accessed April 22, 2024)

Bruneel A, Robert T, Lefeber DJ, Benard G, Loncle E, Djedour A, Durand G, Seta N (2007) Two-dimensional gel electrophoresis of apolipoprotein C-III and other serum glycoproteins for the combined screening of human congenital disorders of O-and N-glycosylation. PROTEOMICS – Clin Appl 1:321–324. https://doi.org/10.1002/prca.200600777

Bruneel A, Habarou F, Stojkovic T, Plouviez G, Bougas L, Guillemet F, Brient N, Henry D, Dupré T, Vuillaumier-Barrot S, Seta N (2017) Two-dimensional electrophoresis highlights haptoglobin beta chain as an additional biomarker of congenital disorders of glycosylation. Clin Chim Acta 470:70–74. https://doi.org/10.1016/j.cca.2017.04.022

Bruneel A, Cholet S, Drouin-Garraud V, Jacquemont M-L, Cano A, Mégarbané A, Ruel C, Cheillan D, Dupré T, Vuillaumier-Barrot S, Seta N, Fenaille F (2018a) Complementarity of electrophoretic, mass spectrometric, and gene sequencing techniques for the diagnosis and characterization of congenital disorders of glycosylation. Electrophoresis 39:3123–3132. https://doi.org/10.1002/elps.201800021

Bruneel A, Dubail J, Roseau C, Prada P, Haouari W, Huber C, Dupré T, Poüs C, Cormier-Daire V, Seta N (2018b) Serum bikunin is a biomarker of linkeropathies. Clin Chim Acta 485:178–180. https://doi.org/10.1016/j.cca.2018.06.044

Bruneel A, Cholet S, Tran NT, Mai TD, Fenaille F (2020) CDG biochemical screening: where do we stand? Biochim Biophys Acta Gen Subj 1864:129652. https://doi.org/10.1016/j.bbagen.2020.129652

Casale J, Crane JS (2024) Biochemistry, Glycosaminoglycans. In: StatPearls. StatPearls Publishing, Treasure Island (FL). http://www.ncbi.nlm.nih.gov/books/NBK544295/ (accessed April 22, 2024)

Čechová A, Altassan R, Borgel D, Bruneel A, Correia J, Girard M, Harroche A, Kiec-Wilk B, Mohnike K, Pascreau T, Pawliński Ł, Radenkovic S, Vuillaumier-Barrot S, Aldamiz-Echevarria L, Couce ML, Martins EG, Quelhas D, Morava E, de Lonlay P, Witters P, Honzík T (2020) Consensus guideline for the diagnosis and management of mannose phosphate isomerase-congenital disorder of glycosylation. J Inherit Metab Dis 43:671–693. https://doi.org/10.1002/jimd.12241

Chen J, Li X, Edmondson A, Meyers GD, Izumi K, Ackermann AM, Morava E, Ficicioglu C, Bennett MJ, He M (2019) Increased clinical sensitivity and specificity of plasma protein N-glycan profiling for diagnosing congenital disorders of glycosylation by use of flow injection–electrospray ionization–quadrupole time-of-flight mass spectrometry. Clin Chem 65:653–663. https://doi.org/10.1373/clinchem.2018.296780

Clerc F, Reiding KR, Jansen BC, Kammeijer GSM, Bondt A, Wuhrer M (2016) Human plasma protein N-glycosylation. Glycoconj J 33:309–343. https://doi.org/10.1007/s10719-015-9626-2

Coddeville B, Carchon H, Jaeken J, Briand G, Spik G (1998) Determination of glycan structures and molecular masses of the glycovariants of serum transferrin from a patient with carbohydrate deficient syndrome type II. Glycoconj J 15:265–273. https://doi.org/10.1023/a:1006997012617

Colley KJ, Varki A, Haltiwanger RS, Kinoshita T (2022) Cellular Organization of Glycosylation. In: Varki A, Cummings RD, Esko JD, Stanley P, Hart GW, Aebi M, Mohnen D, Kinoshita T, Packer NH, Prestegard JH, Schnaar RL, Seeberger PH (eds) Essentials of Glycobiology, 4th edn. Cold Spring Harbor Laboratory Press, Cold Spring Harbor (NY). http://www.ncbi.nlm.nih.gov/books/NBK579926/ (accessed April 22, 2024)

Durin Z, Houdou M, Legrand D, Foulquier F (2023) Metalloglycobiology: the power of metals in regulating glycosylation. Biochim Biophys Acta Gen Subj 1867:130412. https://doi.org/10.1016/j.bbagen.2023.130412

Durin Z, Raynor A, Fenaille F, Cholet S, Vuillaumier-Barrot S, Alili J-M, Poupon J, Oussedik ND, Tuchmann-Durand C, Attali J, Touzé R, Dupré T, Lebredonchel E, Akaffou MA, Legrand D, de Lonlay P, Bruneel A, Foulquier F (2024) Efficacy of oral manganese and D-galactose therapy in a patient bearing a novel TMEM165 variant. Transl Res 266:57–67. https://doi.org/10.1016/j.trsl.2023.11.005

Duvet S, Mouajjah D, Péanne R, Matthijs G, Raymond K, Jaeken J, Morava E, Foulquier F (2018) Use of endoglycosidase H as a diagnostic tool for MAN1B1-CDG patients. Electrophoresis 39: 3133–3141. https://doi.org/10.1002/elps.201800020

Ellgaard L, Frickel E-M (2003) Calnexin, calreticulin, and ERp57: teammates in glycoprotein folding. Cell Biochem Biophys 39:223–247. https://doi.org/10.1385/CBB:39:3:223

Ellgaard L, Helenius A (2003) Quality control in the endoplasmic reticulum. Nat Rev Mol Cell Biol 4:181–191. https://doi.org/10.1038/nrm1052

Ferrer A, Starosta RT, Ranatunga W, Ungar D, Kozicz T, Klee E, Rust LM, Wick M, Morava E (2020) Fetal glycosylation defect due to ALG3 and COG5 variants detected via amniocentesis: complex glycosylation defect with embryonic lethal phenotype. Mol Genet Metab 131:424–429. https://doi.org/10.1016/j.ymgme.2020.11.003

Flanagan-Steet HR, Steet R (2013) "Casting" light on the role of glycosylation during embryonic development: insights from zebrafish. Glycoconj J 30:33–40. https://doi.org/10.1007/s10719-012-9390-5

Foulquier F, Legrand D (2020) Biometals and glycosylation in humans: congenital disorders of glycosylation shed lights into the crucial role of Golgi manganese homeostasis. Biochim Biophys Acta Gen Subj 1864:129674. https://doi.org/10.1016/j.bbagen.2020.129674

Foulquier F, Amyere M, Jaeken J, Zeevaert R, Schollen E, Race V, Bammens R, Morelle W, Rosnoblet C, Legrand D, Demaegd D, Buist N, Cheillan D, Guffon N, Morsomme P, Annaert W, Freeze HH, Van Schaftingen E, Vikkula M, Matthijs G (2012) TMEM165 deficiency causes a congenital disorder of glycosylation. Am J Hum Genet 91:15–26. https://doi.org/10.1016/j.ajhg.2012.05.002

Francisco R, Marques-da-Silva D, Brasil S, Pascoal C, Dos Reis Ferreira V, Morava E, Jaeken J (2019) The challenge of CDG diagnosis. Mol Genet Metab 126:1–5. https://doi.org/10.1016/j.ymgme.2018.11.003

Garapati K, Budhraja R, Saraswat M, Kim J, Joshi N, Sachdeva GS, Jain A, Ligezka AN, Radenkovic S, Ramarajan MG, Udainiya S, Raymond K, He M, Lam C, Larson A, Edmondson AC, Sarafoglou K, Larson NB, Freeze HH, Schultz MJ, Kozicz T, Morava E, Pandey A (2024) A complement C4-derived glycopeptide is a biomarker for PMM2-CDG. JCI Insight 9: e172509. https://doi.org/10.1172/jci.insight.172509

Gill DJ, Clausen H, Bard F (2011) Location, location, location: new insights into O-GalNAc protein glycosylation. Trends Cell Biol 21:149–158. https://doi.org/10.1016/j.tcb.2010.11.004

Girard M, Douillard C, Debray D, Lacaille F, Schiff M, Vuillaumier-Barrot S, Dupré T, Fabre M, Damaj L, Kuster A, Torre S, Mention K, McLin V, Dobbelaere D, Borgel D, Bauchard E, Seta N, Bruneel A, De Lonlay P (2020) Long term outcome of MPI-CDG patients on D-mannose therapy. J Inherit Metab Dis 43:1360–1369. https://doi.org/10.1002/jimd.12289

Haouari W, Dubail J, Lounis-Ouaras S, Prada P, Bennani R, Roseau C, Huber C, Afenjar A, Colin E, Vuillaumier-Barrot S, Seta N, Foulquier F, Poüs C, Cormier-Daire V, Bruneel A (2020) Serum bikunin isoforms in congenital disorders of glycosylation and linkeropathies. J Inherit Metab Dis 43:1349–1359. https://doi.org/10.1002/jimd.12291

Haouari W, Dubail J, Poüs C, Cormier-Daire V, Bruneel A (2021) Inherited proteoglycan biosynthesis defects-current laboratory tools and bikunin as a promising blood biomarker. Genes 12: 1654. https://doi.org/10.3390/genes12111654

Harvey SB, Zhang Y, Wilson-Grady J, Monkkonen T, Nelsestuen GL, Kasthuri RS, Verneris MR, Lund TC, Ely EW, Bernard GR, Zeisler H, Homoncik M, Jilma B, Swan T, Kellogg TA (2009)

O-glycoside biomarker of apolipoprotein C3: responsiveness to obesity, bariatric surgery, and therapy with metformin, to chronic or severe liver disease and to mortality in severe sepsis and graft vs host disease. J Proteome Res 8:603–612. https://doi.org/10.1021/pr800751x

He P, Ng BG, Losfeld M-E, Zhu W, Freeze HH (2012) Identification of intercellular cell adhesion molecule 1 (ICAM-1) as a hypoglycosylation marker in congenital disorders of glycosylation cells. J Biol Chem 287:18210–18217. https://doi.org/10.1074/jbc.M112.355677

Hipgrave Ederveen AL, de Haan N, Baerenfaenger M, Lefeber DJ, Wuhrer M (2020) Dissecting total plasma and protein-specific glycosylation profiles in congenital disorders of glycosylation. Int J Mol Sci 21:7635. https://doi.org/10.3390/ijms21207635

Hosokawa N, Tremblay LO, You Z, Herscovics A, Wada I, Nagata K (2003) Enhancement of endoplasmic reticulum (ER) degradation of misfolded null Hong Kong alpha1-antitrypsin by human ER mannosidase I. J Biol Chem 278:26287–26294. https://doi.org/10.1074/jbc.M303395200

Houdou M, Lebredonchel E, Garat A, Duvet S, Legrand D, Decool V, Klein A, Ouzzine M, Gasnier B, Potelle S, Foulquier F (2019) Involvement of thapsigargin- and cyclopiazonic acid-sensitive pumps in the rescue of TMEM165-associated glycosylation defects by Mn2. FASEB J 33:2669–2679. https://doi.org/10.1096/fj.201800387R

Huhn C, Selman MHJ, Ruhaak LR, Deelder AM, Wuhrer M (2009) IgG glycosylation analysis. Proteomics 9:882–913. https://doi.org/10.1002/pmic.200800715

Hülsmeier AJ, Paesold-Burda P, Hennet T (2007) N-glycosylation site occupancy in serum glycoproteins using multiple reaction monitoring liquid chromatography-mass spectrometry. Mol Cell Proteomics MCP 6:2132–2138. https://doi.org/10.1074/mcp.M700361-MCP200

Hülsmeier AJ, Tobler M, Burda P, Hennet T (2016) Glycosylation site occupancy in health, congenital disorder of glycosylation and fatty liver disease. Sci Rep 6:33927. https://doi.org/10.1038/srep33927

Jaeken J, van Eijk HG, van der Heul C, Corbeel L, Eeckels R, Eggermont E (1984) Sialic acid-deficient serum and cerebrospinal fluid transferrin in a newly recognized genetic syndrome. Clin Chim Acta 144:245–247. https://doi.org/10.1016/0009-8981(84)90059-7

Jaeken J, Schachter H, Carchon H, De Cock P, Coddeville B, Spik G (1994) Carbohydrate deficient glycoprotein syndrome type II: a deficiency in Golgi localised N-acetyl-glucosaminyltransferase II. Arch Dis Child 71:123–127. https://doi.org/10.1136/adc.71.2.123

Jaeken J, Hennet T, Matthijs G, Freeze HH (2009) CDG nomenclature: time for a change! Biochim Biophys Acta 1792:825–826. https://doi.org/10.1016/j.bbadis.2009.08.005

Jansen JC, van Hoek B, Metselaar HJ, van den Berg AP, Zijlstra F, Huijben K, van Scherpenzeel M, Drenth JPH, Lefeber DJ (2020) Screening for abnormal glycosylation in a cohort of adult liver disease patients. J Inherit Metab Dis 43:1310–1320. https://doi.org/10.1002/jimd.12273

Kamboh MI, Ferrell RE (1987) Human transferrin polymorphism. Hum Hered 37:65–81. https://doi.org/10.1159/000153680

Kapur A, Wild G, Milford-Ward A, Triger DR (1989) Carbohydrate deficient transferrin: a marker for alcohol abuse. BMJ 299:427–431. https://doi.org/10.1136/bmj.299.6696.427

Khan S, Sbeity M, Foulquier F, Barré L, Ouzzine M (2021) TMEM165 a new player in proteoglycan synthesis: loss of TMEM165 impairs elongation of chondroitin- and heparan-sulfate glycosaminoglycan chains of proteoglycans and triggers early chondrocyte differentiation and hypertrophy. Cell Death Dis 13:11. https://doi.org/10.1038/s41419-021-04458-1

Lebredonchel E, Houdou M, Potelle S, de Bettignies G, Schulz C, Krzewinski Recchi M-A, Lupashin V, Legrand D, Klein A, Foulquier F (2019) Dissection of TMEM165 function in Golgi glycosylation and its Mn2+ sensitivity. Biochimie 165:123–130. https://doi.org/10.1016/j.biochi.2019.07.016

Lefeber DJ, Morava E, Jaeken J (2011) How to find and diagnose a CDG due to defective N-glycosylation. J Inherit Metab Dis 34:849–852. https://doi.org/10.1007/s10545-011-9370-0

Ligezka AN, Radenkovic S, Saraswat M, Garapati K, Ranatunga W, Krzysciak W, Yanaihara H, Preston G, Brucker W, McGovern RM, Reid JM, Cassiman D, Muthusamy K, Johnsen C, Mercimek-Andrews S, Larson A, Lam C, Edmondson AC, Ghesquière B, Witters P,

Raymond K, Oglesbee D, Pandey A, Perlstein EO, Kozicz T, Morava E (2021) Sorbitol is a severity biomarker for PMM2-CDG with therapeutic implications. Ann Neurol 90:887–900. https://doi.org/10.1002/ana.26245

Lipiński P, Tylki-Szymańska A (2021) Congenital disorders of glycosylation: what clinicians need to know? Front Pediatr 9:715151. https://doi.org/10.3389/fped.2021.715151

MacGillivray RT, Mendez E, Shewale JG, Sinha SK, Lineback-Zins J, Brew K (1983) The primary structure of human serum transferrin. The structures of seven cyanogen bromide fragments and the assembly of the complete structure. J Biol Chem 258:3543–3553

Marklová E, Albahri Z (2007) Screening and diagnosis of congenital disorders of glycosylation. Clin Chim Acta 385:6–20. https://doi.org/10.1016/j.cca.2007.07.002

Messina A, Palmigiano A, Esposito F, Fiumara A, Bordugo A, Barone R, Sturiale L, Jaeken J, Garozzo D (2020) HILIC-UPLC-MS for high throughput and isomeric N-glycan separation and characterization in congenital disorders glycosylation and human diseases. Glycoconj J 38:201. https://doi.org/10.1007/s10719-020-09947-7

Mulkey SB, Ng BG, Vezina GL, Bulas DI, Wolfe LA, Freeze HH, Ferreira CR (2019) Arrest of fetal brain development in ALG11-congenital disorder of glycosylation. Pediatr Neurol 94:64–69. https://doi.org/10.1016/j.pediatrneurol.2018.12.009

Ng BG, Freeze HH (2018) Perspectives on glycosylation and its congenital disorders. Trends Genet. 34:466–476. https://doi.org/10.1016/j.tig.2018.03.002

Niehues R, Hasilik M, Alton G, Körner C, Schiebe-Sukumar M, Koch HG, Zimmer KP, Wu R, Harms E, Reiter K, von Figura K, Freeze HH, Harms HK, Marquardt T (1998) Carbohydrate-deficient glycoprotein syndrome type Ib. Phosphomannose isomerase deficiency and mannose therapy. J Clin Invest 101:1414–1420. https://doi.org/10.1172/JCI2350

Norata GD, Tsimikas S, Pirillo A, Catapano AL (2015) Apolipoprotein C-III: from pathophysiology to pharmacology. Trends Pharmacol Sci 36:675–687. https://doi.org/10.1016/j.tips.2015.07.001

Oda RP, Prasad R, Stout RL, Coffin D, Patton WP, Kraft DL, O'Brien JF, Landers JP (1997) Capillary electrophoresis-based separation of transferrin sialoforms in patients with carbohydrate-deficient glycoprotein syndrome. Electrophoresis 18:1819–1826. https://doi.org/10.1002/elps.1150181017

Ondruskova N, Cechova A, Hansikova H, Honzik T, Jaeken J (2021) Congenital disorders of glycosylation: still "hot" in 2020. Biochim Biophys Acta BBA – Gen Subj 1865:129751. https://doi.org/10.1016/j.bbagen.2020.129751

Parente F, Ah Mew N, Jaeken J, Gilfix BM (2010) A new capillary zone electrophoresis method for the screening of congenital disorders of glycosylation (CDG). Clin Chim Acta 411:64–66. https://doi.org/10.1016/j.cca.2009.10.004

Park JH, Mealer RG, Elias AF, Hoffmann S, Grüneberg M, Biskup S, Fobker M, Haven J, Mangels U, Reunert J, Rust S (2020) N-glycome analysis detects dysglycosylation missed by conventional methods in SLC39A8 deficiency. J Inherit Metab Dis 43:1370. https://doi.org/10.1002/jimd.12306

Parker WC, Bearn AG (1962) Studies on the transferrins of adult serum, cord serum, and cerebrospinal fluid. The effect of neuraminidase. J Exp Med 115:83–105. https://doi.org/10.1084/jem.115.1.83

Pedersen PS, Tygstrup I (1980) Congenital hepatic fibrosis combined with protein-losing enteropathy and recurrent thrombosis. Acta Paediatr Scand 69:571–574. https://doi.org/10.1111/j.1651-2227.1980.tb07136.x

Pérez B, Medrano C, Ecay MJ, Ruiz-Sala P, Martínez-Pardo M, Ugarte M, Pérez-Cerdá C (2013) A novel congenital disorder of glycosylation type without central nervous system involvement caused by mutations in the phosphoglucomutase 1 gene. J Inherit Metab Dis 36:535–542. https://doi.org/10.1007/s10545-012-9525-7

Poskanzer SA, Schultz MJ, Turgeon CT, Vidal-Folch N, Liedtke K, Oglesbee D, Gavrilov DK, Tortorelli S, Matern D, Rinaldo P, Bennett JT, Thies JM, Chang IJ, Beck AE, Raymond K,

Allenspach EJ, Lam C (2021) Immune dysfunction in MGAT2-CDG: a clinical report and review of the literature. Am J Med Genet A 185:213–218. https://doi.org/10.1002/ajmg.a.61914

Radenkovic S, Bird MJ, Emmerzaal TL, Wong SY, Felgueira C, Stiers KM, Sabbagh L, Himmelreich N, Poschet G, Windmolders P, Verheijen J, Witters P, Altassan R, Honzik T, Eminoglu TF, James PM, Edmondson AC, Hertecant J, Kozicz T, Thiel C, Vermeersch P, Cassiman D, Beamer L, Morava E, Ghesquière B (2019) The metabolic map into the pathomechanism and treatment of PGM1-CDG. Am J Hum Genet 104:835–846. https://doi.org/10.1016/j.ajhg.2019.03.003

Radenkovic S, Johnsen C, Schulze A, Lail G, Guilder L, Schwartz K, Schultz M, Mercimek-Andrews S, Boyer S, Morava E (2023) Novel insights into the phenotype and long-term D-gal treatment in PGM1-CDG: a case series. Ther Adv Rare Dis 4:26330040221150269. https://doi.org/10.1177/26330040221150269

Raynor A, Vincent-Delorme C, Alaix A-S, Cholet S, Dupré T, Vuillaumier-Barrot S, Fenaille F, Besmond C, Bruneel A (2021a) Normal transferrin patterns in congenital disorders of glycosylation with Golgi homeostasis disruption: apolipoprotein C-III at the rescue! Clin Chim Acta 519:285–290. https://doi.org/10.1016/j.cca.2021.05.016

Raynor A, Haouari W, Ng BG, Cholet S, Harroche A, Raulet-Bussian C, Lounis-Ouaras S, Vuillaumier-Barrot S, Pascreau T, Borgel D, Freeze HH, Fenaille F, Bruneel A (2021b) SLC37A4-CDG: new biochemical insights for an emerging congenital disorder of glycosylation with major coagulopathy. Clin Chim Acta 521:104–106. https://doi.org/10.1016/j.cca.2021.07.005

Raynor A, Haouari W, Lebredonchel E, Foulquier F, Fenaille F, Bruneel A (2024a) Biochemical diagnosis of congenital disorders of glycosylation. Adv Clin Chem 120:1–43. https://doi.org/10.1016/bs.acc.2024.03.001

Raynor A, Bruneel A, Vermeersch P, Cholet S, Friedrich S, Eckenweiler M, Schumann A, Hengst S, Tuncel AT, Fenaille F, Thiel C, Rymen D (2024b) "Hide and seek": misleading transferrin variants in PMM2-CDG complicate diagnostics. Proteomics Clin Appl 18:e2300040. https://doi.org/10.1002/prca.202300040

Rivinoja A, Hassinen A, Kokkonen N, Kauppila A, Kellokumpu S (2009) Elevated Golgi pH impairs terminal N-glycosylation by inducing mislocalization of Golgi glycosyltransferases. J Cell Physiol 220:144–154. https://doi.org/10.1002/jcp.21744

Rymen D, Peanne R, Millón MB, Race V, Sturiale L, Garozzo D, Mills P, Clayton P, Asteggiano CG, Quelhas D, Cansu A, Martins E, Nassogne M-C, Gonçalves-Rocha M, Topaloglu H, Jaeken J, Foulquier F, Matthijs G (2013) MAN1B1 deficiency: an unexpected CDG-II. PLoS Genet 9:e1003989. https://doi.org/10.1371/journal.pgen.1003989

Schachter H (2014) Complex N-glycans: the story of the "yellow brick road". Glycoconj J 31:1–5. https://doi.org/10.1007/s10719-013-9507-5

Schwartz NB, Domowicz MS (2018) Proteoglycans in brain development and pathogenesis. FEBS Lett 592:3791–3805. https://doi.org/10.1002/1873-3468.13026

Seta N, Barnier A, Hochedez F, Besnard MA, Durand G (1996) Diagnostic value of Western blotting in carbohydrate-deficient glycoprotein syndrome. Clin Chim Acta 254:131–140. https://doi.org/10.1016/0009-8981(96)06379-6

Shimada S, Ng BG, White AL, Nickander KK, Turgeon C, Liedtke KL, Lam CT, Font-Montgomery E, Lourenco CM, He M, Peck DS, Umana LA, Uhles CL, Haynes D, Wheeler PG, Bamshad MJ, Nickerson DA, Cushing T, Gates R, Gomez-Ospina N, Byers HM, UW Center for Mendelian Genomics, Scalco FB, Martinez NN, Sachdev R, Smith L, Poduri A, Malone S, Harris RV, Scheffer IE, Rosenzweig SD, Adams DR, Gahl WA, Malicdan MCV, Raymond KM, Freeze HH, Wolfe LA (2022) Clinical, biochemical and genetic characteristics of MOGS-CDG: a rare congenital disorder of glycosylation. J Med Genet:jmedgenet-2021-108177. https://doi.org/10.1136/jmedgenet-2021-108177

Stanley P (2011) Golgi glycosylation. Cold Spring Harb Perspect Biol 3:a005199. https://doi.org/10.1101/cshperspect.a005199

Stewart SH, Comte-Walters S, Bowen E, Anton RF (2010) Liver disease and HPLC quantification of disialotransferrin for heavy alcohol use: a case series. Alcohol Clin Exp Res 34:1956–1960. https://doi.org/10.1111/j.1530-0277.2010.01285.x

Sturiale L, Nassogne M-C, Palmigiano A, Messina A, Speciale I, Artuso R, Bertino G, Revencu N, Stephénne X, De Castro C, Matthijs G, Barone R, Jaeken J, Garozzo D (2021) Aberrant sialylation in a patient with a HNF1α variant and liver adenomatosis. iScience 24:102323. https://doi.org/10.1016/j.isci.2021.102323

Taday R, Park JH, Grüneberg M, DuChesne I, Reunert J, Marquardt T (2021) Mannose supplementation in PMM2-CDG. Orphanet J Rare Dis 16:359. https://doi.org/10.1186/s13023-021-01988-x

Tegtmeyer LC, Rust S, van Scherpenzeel M, Ng BG, Losfeld M-E, Timal S, Raymond K, He P, Ichikawa M, Veltman J, Huijben K, Shin YS, Sharma V, Adamowicz M, Lammens M, Reunert J, Witten A, Schrapers E, Matthijs G, Jaeken J, Rymen D, Stojkovic T, Laforêt P, Petit F, Aumaître O, Czarnowska E, Piraud M, Podskarbi T, Stanley CA, Matalon R, Burda P, Seyyedi S, Debus V, Socha P, Sykut-Cegielska J, van Spronsen F, de Meirleir L, Vajro P, DeClue T, Ficicioglu C, Wada Y, Wevers RA, Vanderschaeghe D, Callewaert N, Fingerhut R, van Schaftingen E, Freeze HH, Morava E, Lefeber DJ, Marquardt T (2014) Multiple phenotypes in phosphoglucomutase 1 deficiency. N Engl J Med 370:533–542. https://doi.org/10.1056/NEJMoa1206605

Ten Hagen KG, Fritz TA, Tabak LA (2003) All in the family: the UDP-GalNAc:polypeptide N-acetylgalactosaminyltransferases. Glycobiology 13:1R–16R. https://doi.org/10.1093/glycob/cwg007

Vaith P, Assmann G, Uhlenbruck G (1978) Characterization of the oligosaccharide side chain of apolipoprotein C-III from human plasma very low density lipoproteins. Biochim Biophys Acta 541:234–240. https://doi.org/10.1016/0304-4165(78)90396-3

van Eijk HG, van Noort WL, Kroos MJ, van der Heul C (1982) The heterogeneity of human serum transferrin and human transferrin preparations on isoelectric focusing gels; no functional difference of the fractions in vitro. Clin Chim Acta 121:209–216. https://doi.org/10.1016/0009-8981(82)90060-2

van Eijk HG, van Noort WL, Dubelaar ML, van der Heul C (1983) The microheterogeneity of human transferrins in biological fluids. Clin Chim Acta 132:167–171. https://doi.org/10.1016/0009-8981(83)90244-9

Van Schaftingen E, Jaeken J (1995) Phosphomannomutase deficiency is a cause of carbohydrate-deficient glycoprotein syndrome type I. FEBS Lett 377:318–320. https://doi.org/10.1016/0014-5793(95)01357-1

Van Scherpenzeel M, Timal S, Rymen D, Hoischen A, Wuhrer M, Hipgrave-Ederveen A, Grunewald S, Peanne R, Saada A, Edvardson S, Grønborg S, Ruijter G, Kattentidt-Mouravieva-A, Brum JM, Freckmann M-L, Tomkins S, Jalan A, Prochazkova D, Ondruskova N, Hansikova H, Willemsen MA, Hensbergen PJ, Matthijs G, Wevers RA, Veltman JA, Morava E, Lefeber DJ (2014) Diagnostic serum glycosylation profile in patients with intellectual disability as a result of MAN1B1 deficiency. Brain J Neurol 137:1030–1038. https://doi.org/10.1093/brain/awu019

Varki A (2017) Biological roles of glycans. Glycobiology 27:3–49. https://doi.org/10.1093/glycob/cww086

Verheijen J, Tahata S, Kozicz T, Witters P, Morava E (2020) Therapeutic approaches in congenital disorders of glycosylation (CDG) involving N-linked glycosylation: an update, genet. Med. Off. J. Am. Coll. Med Genet 22:268–279. https://doi.org/10.1038/s41436-019-0647-2

Vuillaumier-Barrot S, Le Bizec C, de Lonlay P, Barnier A, Mitchell G, Pelletier V, Prevost C, Saudubray JM, Durand G, Seta N (2002) Protein losing enteropathy-hepatic fibrosis syndrome in Saguenay-Lac St-Jean, Quebec is a congenital disorder of glycosylation type Ib. J Med Genet 39:849–851. https://doi.org/10.1136/jmg.39.11.849

Wada Y, Kadoya M, Okamoto N (2022) Mass spectrometry of transferrin and apolipoprotein CIII from dried blood spots for congenital disorders of glycosylation. Mass Spectrom (Tokyo Jpn) 11:A0113. https://doi.org/10.5702/massspectrometry.A0113

Willems AP, van Engelen BGM, Lefeber DJ (1860) Genetic defects in the hexosamine and sialic acid biosynthesis pathway. Biochim Biophys Acta 2016:1640–1654. https://doi.org/10.1016/j.bbagen.2015.12.017

Witters P, Edmondson AC, Lam C, Johnsen C, Patterson MC, Raymond KM, He M, Freeze HH, Morava E (2021) Spontaneous improvement of carbohydrate-deficient transferrin in PMM2-CDG without mannose observed in CDG natural history study. Orphanet J Rare Dis 16:102. https://doi.org/10.1186/s13023-021-01751-2

Wong SY-W, Gadomski T, van Scherpenzeel M, Honzik T, Hansikova H, Holmefjord KSB, Mork M, Bowling F, Sykut-Cegielska J, Koch D, Hertecant J, Preston G, Jaeken J, Peeters N, Perez S, Nguyen DD, Crivelly K, Emmerzaal T, Gibson KM, Raymond K, Abu Bakar N, Foulquier F, Poschet G, Ackermann AM, He M, Lefeber DJ, Thiel C, Kozicz T, Morava E (2017) Oral D-galactose supplementation in PGM1-CDG. Genet Med 19:1226–1235. https://doi.org/10.1038/gim.2017.41

Wopereis S, Grünewald S, Morava E, Penzien JM, Briones P, García-Silva MT, Demacker PNM, Huijben KMLC, Wevers RA (2003) Apolipoprotein C-III isofocusing in the diagnosis of genetic defects in O-glycan biosynthesis. Clin Chem 49:1839–1845. https://doi.org/10.1373/clinchem.2003.022541

Xia B, Zhang W, Li X, Jiang R, Harper T, Liu R, Cummings RD, He M (2013) Serum N-glycan and O-glycan analysis by mass spectrometry for diagnosis of congenital disorders of glycosylation. Anal Biochem 442:178–185. https://doi.org/10.1016/j.ab.2013.07.037

Yanagishita M (1993) Function of proteoglycans in the extracellular matrix. Acta Pathol Jpn 43: 283–293. https://doi.org/10.1111/j.1440-1827.1993.tb02569.x

Yen-Nicolaÿ S, Boursier C, Rio M, Lefeber DJ, Pilon A, Seta N, Bruneel A (2015) MALDI-TOF MS applied to apoC-III glycoforms of patients with congenital disorders affecting O-glycosylation. Comparison with two-dimensional electrophoresis. Proteomics Clin Appl 9: 787–793. https://doi.org/10.1002/prca.201400187

Zeevaert R, De Zegher F, Sturiale L, Garozzo D, Smet M, Moens M, Matthijs G, Jaeken J (2013) Bone dysplasia as a key feature in three patients with a novel congenital disorder of glycosylation (CDG) type II due to a deep intronic splice mutation in TMEM165. JIMD Rep 8:145. https://doi.org/10.1007/8904_2012_172

Zhang W, James PM, Ng BG, Li X, Xia B, Rong J, Asif G, Raymond K, Jones MA, Hegde M, Ju T, Cummings RD, Clarkson K, Wood T, Boerkoel CF, Freeze HH, He M (2016) A novel N-tetrasaccharide in patients with congenital disorders of glycosylation, including asparagine-linked glycosylation protein 1, phosphomannomutase 2, and mannose phosphate isomerase deficiencies. Clin Chem 62:208–217. https://doi.org/10.1373/clinchem.2015.243279

Zheng W, Kim H, Zhao Q (2000) Comparative toxicokinetics of manganese chloride and methylcyclopentadienyl manganese tricarbonyl (MMT) in Sprague-Dawley rats. Toxicol Sci 54:295–301. https://doi.org/10.1093/toxsci/54.2.295

Zheng W, Fu SX, Dydak U, Cowan DM (2011) Biomarkers of manganese intoxication. Neurotoxicology 32:1–8. https://doi.org/10.1016/j.neuro.2010.10.002

Glycosylation in Cancer

Henrique Oliveira Duarte, Celso Albuquerque Reis,
Veronique Blanchard, and Rudolf Tauber

Contents

Authors Celso Albuquerque Reis, Veronique Blanchard, and Rudolf Tauber shared senior authorship to this chapter.

H. O. Duarte
Instituto de Investigação e Inovação em Saúde, Universidade do Porto (i3S), Porto, Portugal

Institute of Molecular Pathology and Immunology of the University of Porto (IPATIMUP), Porto, Portugal

C. A. Reis
Instituto de Investigação e Inovação em Saúde, Universidade do Porto (i3S), Porto, Portugal

Institute of Molecular Pathology and Immunology of the University of Porto (IPATIMUP), Porto, Portugal

Instituto de Ciências Biomédicas Abel Salazar (ICBAS), Porto, Portugal

V. Blanchard
Medical School Berlin, Berlin, Germany

Institut für Laboratoriumsmedizin, Klinische Chemie und Pathobiochemie, Charité – Universitätsmedizin Berlin, Berlin, Germany

R. Tauber (✉)
Institut für Laboratoriumsmedizin, Klinische Chemie und Pathobiochemie, Charité – Universitätsmedizin Berlin, Berlin, Germany
e-mail: rudolf.tauber@charite.de

F. Pfrengle et al. (eds.), *Complex Carbohydrates in Health and Disease*, Handbook of Experimental Pharmacology 288, https://doi.org/10.1007/164_2025_751

Abstract

Structural alterations of the glycan chains attached to glycoproteins and glycolipids are present in all types of malignomas investigated to date, including adenocarcinomas, sarcomas and haematological malignancies. They occur in humans as well as in animals including experimental models of malignancy, regardless of the type, cause, or stage of the tumour. The biochemical and genomic characterization of the enzymatic machineries involved in glycan biosynthesis in cancer cells shows that tumour-associated glycosylation changes are a critical part of tumour initiation and progression. Experimental studies and epidemiological findings give clear evidence that tumour-associated glycans bear functional significance in the invasive and metastatic growth of malignancies, for immunological tumour defence and, hence, influence the clinical outcome and the prognosis of cancer patients. Tumour-associated glycan changes are, moreover, targets for new pharmacological and immunological therapy methods and serve as important clinical biomarkers for diagnosis, particularly for monitoring disease progression and therapeutic efficacy. This chapter provides an overview of the major types of changes of glycosylation, genetic and biochemical mechanisms contributing to cancer-associated glycosylation, functional consequences for tumour growth and the clinical significance in cancer diagnosis, monitoring and treatment.

Keywords

Anti-cancer vaccination · Cancer · Cell adhesion · Extracellular matrix · Glycan-specific receptors · Glycome · Glycosylation · Glycosyltransferase · Immune evasion · Metastasis · Tumour-associated glycan

1 Major Types of Changes of Glycosylation in Cancer

1.1 Brief Overview of Glycan Structures and Classes of Glycoconjugates

The mammalian cell glycome is characterized by astonishing structural complexity and functional diversity. By definition, the cellular glycome encompasses every complex carbohydrate or polysaccharide chain termed glycan existing either free or covalently attached, through a glycosidic bond, to other biological macromolecules, such as peptides, proteins and lipids, to form glycoconjugates (Lebrilla et al. 2022; Stowell et al. 2015). In the case of glycoproteins, for instance, the cellular process of co- and post-translational glycosylation consists in the stepwise or *en bloc* enzymatic addition of monosaccharide building blocks to specific amino acid residues of the peptidic backbone to form more elongated and branched glycan chains (Stowell et al. 2015). Complex glycan species can vary significantly in composition, length, branching degree and linkage specificity. Moreover, oligosaccharide chains can undergo further selective post-translational modifications (PTMs), including phosphorylation, sulfation and acetylation. The resulting glycan repertoire, or glycome, is thus inherently characterized by staggering structural complexity and diversity that actively ensure its fundamental role in a myriad of cellular functions and molecular pathways (Gagneux et al. 2022; Reily et al. 2019). Indeed, both in homeostatic and in pathological settings, through the fine tuning of the biochemical, structural and functional properties of their molecular carriers, glycans act as central regulators of virtually every cellular process, including signalling, extracellular communication, cell motility and differentiation. Glycan biosynthesis is accomplished by the highly regulated and well-coordinated action of endoplasmic reticulum (ER)- and Golgi-residing enzymes, i.e. glycosyltransferases and trimming glycosidases (Rini et al. 2022; Schjoldager et al. 2020). Each membrane-bound glycosyltransferase catalyses the selective, yet periodically redundant transfer of a monosaccharide from an activated nucleotide donor substrate to an acceptor molecule. Individual glycosyltransferases depict differential expression patterns that are often organ-, tissue- and cell-specific and are extremely sensitive to a cell's physiological state (Taniguchi et al. 2011).

There are three main classes of glycoconjugates, which differ in their non-glycan component: glycoproteins, glycolipids and proteoglycans. Glycoproteins consist of proteins to which one or more glycan chains have been co- or post-translationally attached at specific and sequence-determined sites (glycosites). Indeed, most secreted and membrane-anchored proteins undergo extensive co- and post-translational glycosylation that regulates, in turn, multiple aspects of a protein's molecular and functional features. There are several types of protein glycosylation, being *N*- and *O*-linked glycosylation, *O*-mannosylation and *O*-GlcNAcylation the most widely described (Lebrilla et al. 2022; Stowell et al. 2015). In the ER-initiated process of *N*-glycosylation, a 14-sugar lipid-like oligosaccharide precursor covalently bound to dolichol-phosphate (Dol-P) is transferred *en bloc* by the oligosaccharyltransferase (OST) multi-subunit protein complex to the nitrogen

(N) atom of asparagine (Asn) amino acids of nascent membrane or secreted proteins (Stanley et al. 2022). *N*-glycosylated Asn residues occur within the well-defined consensus sequence Asn-X-serine (Ser)/threonine (Thr), in which X can represent any amino acid with the exception of proline. Further *N*-glycan trimming (as part of quality control mechanisms of correct glycoprotein folding), elongation, branching and terminal capping occur progressively throughout the Golgi apparatus. All *N*-glycan structures share a common pentasaccharide core sequence composed of three mannose (Man) and two *N*-acetylglucosamine (GlcNAc) monosaccharides: Manα1,6(Manα1,3)Manβ1,4GlcNAcβ1,4GlcNAcβ1-Asn-X-Ser/Thr. There are three different classes of *N*-glycan chains, according to the monosaccharide composition attached to the pentasaccharide core: high-mannose or oligomannosidic *N*-glycans, composed solely of Man residues attached to the pentasaccharide core; complex-type *N*-glycans, in which the three mannose residues of the pentasaccharide core are free to support the formation of "antenna" arms by *N*-acetylglucosaminyltransferases (GnTs), which can then be further branched and terminally capped with galactose (Gal), fucose (Fuc) and *N*-acetylneuraminic acid or sialic acid (Neu5Ac) moieties; and hybrid *N*-glycans, in which mannose residues are linked to the Manα1–6 upper arm of the pentasaccharide core, and one or two antennae, which can undergo additional terminal modifications, can be attached to the Manα1–3 lower arm. The attachment of additional GlcNAc monosaccharides to the Man residues of the lower and upper arms of the *N*-glycan core by GnT4 and GnT5, respectively, can yield the formation of tri- and tetra-antennary *N*-glycan chains, a process known as *N*-glycan branching. Both complex and hybrid-type *N*-glycans can be modified with a bisecting GlcNAc residue attached to the central Man of the *N*-glycan core by GnT3, which, in turn, precludes these structures from being further modified with branching GlcNAc residues. Final maturation steps occurring in the trans-Golgi network comprise the transfer of a Fuc moiety to the asparagine-bound GlcNAc within the *N*-glycan core by the α1,6- fucosyltransferase 8 (FUT8) termed core fucosylation; and the outer arm addition of Fuc and Neu5Ac residues to cap *N*-glycan antennae, which impede further chain elongation and often determine a glycan species' biological functions including signalling, cell-cell and cell-matrix interactions (see below).

Mucin-type *O*-glycosylation takes place within the Golgi apparatus and is initiated with the covalent attachment of an *N*-acetylgalactosamine (GalNAc) monosaccharide to either a Ser/Thr residue of the peptidic backbone to form the Tn short *O*-glycan antigen (Brockhausen et al. 2022; Schjoldager et al. 2020). Contrarily to *N*-glycosylation, there is no known consensus sequence for this modification, which can be accomplished by one of twenty polypeptide GalNAc transferases (ppGalNAcTs) (Bennett et al. 2012; Clausen and Bennett 1996; Schjoldager et al. 2020). The cellular *O*-glycosylation landscape and glycosite occupancy are dictated, thus, by the highly regulated expression patterns of distinct ppGalNAcTs across different organs, tissues, and cell populations (Steentoft et al. 2013). Proteins harbouring domains rich in Ser/Thr repeats, such as mucins, are prone to be decorated by densely clustered *O*-glycan chains. As *O*-glycoproteins travel throughout the Golgi apparatus, the Tn antigen is further extended into one of four *O*-glycan core structures. The core 1 structure, or the Thomsen-Friedenreich (T) antigen, is formed following the modification of the Tn antigen with β1,3-Gal by the C1GalT1 glycosyltransferase. Core 1 structures can be further sialylated, giving rise to the sialyl-T or di-sialyl-T antigens, or have a β1,6-GlcNAc attached to their GalNAc moiety by C2GnT1, 2 or 3, which results in the formation of core 2 structures. The addition of one or two GlcNAc residues to the Tn antigen gives rise to core 3 and core 4 structures, respectively. The addition of GlcNAc-Gal (LacNAc) disaccharide repeating units to core structures allows the elongation, or extension, of *O*-glycan chains. These polysaccharide structures can be further decorated with sulfate groups or terminally capped with Fuc and Neu5Ac motifs, which prevents further *O*-glycan elongation. The linkage between Gal-GlcNAc in dimeric LacNAc structures can vary from β1,3 and β1,4, giving rise to type 1 and 2 precursor chains, respectively, which can present ABO blood group glycan antigens (Stanley et al. 2022; Yamamoto et al. 1990). The A, B and H blood group antigens can also exist within *O*-GalNAc glycan chains (type 3) and glycolipids (type 4). Their biosynthesis is initiated by the fucosyltransferase FUT1- or FUT2-catalyzed addition of α1,2Fuc to Gal to form the common blood group H glycan determinant within type 1 and 2 precursor chains, respectively. Complete A and B blood group glycan antigens are subsequently formed from H type-1:4 by additional glycosyltransferases.

Moreover, fucosylated tri- and tetrasaccharide structures, also known as Lewis antigens, can be found terminally decorating *N*- and *O*-glycan chains and are composed of a LacNAc disaccharide unit to which Fuc and Neu5Ac moieties can be attached (Stanley et al. 2022). In neutral Lewis antigens, which include Lewis a (Le^a) and Lewis x (Le^x), and the corresponding Lewis b (Le^b) and Lewis y (Le^y) isomers, one or two Fuc monosaccharides are attached to the GlcNAc moiety, or to both the Gal and GlcNAc residues, respectively (Oriol et al. 1986). Their sialylated counterparts, sialyl Lewis a (sialyl-Le^a) and sialyl Lewis x (sialyl-Le^x), harbour a Neu5Ac moiety attached to the Gal via α2,3-linkage.

Addition of oxygen-linked α-mannose (*O*-mannosylation) to Ser/Thr protein sites, a type of *O*-glycosylation highly conserved across evolution, is carried out by ER-residing *O*-mannosyltransferases POMPT1 and POMPT2 (Haltiwanger et al. 2022). *O*-mannosylation of proteins can partially compete with the mucin-type *O*-GalNAc modification. Interestingly, however, recent studies have demonstrated the existence of still unidentified players performing the *O*-mannosylation of specific families of protein substrates such as cadherins (Larsen et al. 2017a; Larsen et al. 2017b). Following the secretory pathway from the ER to the Golgi apparatus, protein-linked *O*-Man residues can be further elongated through the addition of other monosaccharides (GlcNAc, Gal, Fuc and Neu5Ac), giving rise to more complex and species-specific *O*-mannosylated glycans.

Nuclear, mitochondrial and cytoplasmic proteins are frequently modified with a single GlcNAc moiety at Ser/Thr residues by an *O*-GlcNAc transferase (OGT), a PTM known as *O*-GlcNAcylation (Zachara et al. 2022). The removal of *O*-GlcNAc motifs from proteins is accomplished by an *O*-GlcNAcase (OGA). This apparently simple single monosaccharide modification is a highly dynamic and reversible process, being strictly controlled by the concerted action of the OGT and OGA enzyme pair and depicts a similar subcellular distribution to that of protein phosphorylation (Hart et al. 2007; Hart et al. 2011). Importantly, protein-bound *O*-GlcNAc residues are not further elongated with additional monosaccharides and can be added and removed from the same glycosylation site multiple times throughout a protein's lifetime to actively regulate its substrate targeting, correct subcellular localization and activation state.

Membrane-bound glycosphingolipids (GSLs), ceramide lipid molecules containing a sphingoid base attached to one or more monosaccharide residues, represent the major class of mammalian glycolipids and depict a tissue-specific expression pattern (Schnaar et al. 2022). The often-saturated fatty acid component of ceramides depicts varying lengths, ranging from C14 to C30. In vertebrates, β-linked glucose (Glc) and Gal are usually the first monosaccharides linked to the ceramide molecule. GSL series classification is based on well-defined tetrasaccharide neutral sugar core sequences. For instance, Neu5Ac-containing GSLs, also known as gangliosides, occur predominantly in the brain. Unofficial names have been given to the most prominent GSL species. For instance, the ganglioside Galβ1,3GalNAcβ1,4(Neu5Acα2,3)Galβ1,4GlcβCer is designated GM1, in which G stands for the ganglioside series and M stands for mono-sialylated.

The carboxyl terminus (C-terminal) region of membrane-bound proteins can be covalently attached to the lipid bilayer via a glycosylphosphatidylinositol (GPI) anchor (Komath et al. 2022). GPI-anchored proteins are generally found in the outer leaflet of the lipid bilayer facing the extracellular space. In the ER, the GPI anchor is enzymatically assembled on a phosphatidylinositol lipid and covalently attached to a protein's C-terminal end. The GPI core structure consists of a PI moiety, one glucosamine and three Man residues, and a terminal phosphoethanolamine is attached through amide-bonding to the newly formed C-terminal of the protein during GPI attachment.

Finally, in proteoglycans, one or more glycosaminoglycan (GAG) chains are covalently attached to a core protein (Merry et al. 2022). GAGs consist, in turn, of linear polysaccharide chains of enzymatically breakable disaccharide units of amino sugars, GlcNAc or GalNAc, and uronic acids (glucuronic acid (GlcA) or iduronic acid (IdoA)) or Gal. Depending on the specific composition of their GAG chain, GAGs can be further grouped into chondroitin sulfate (CS), dermatan sulfate (DS), heparan sulfate (HS) and keratan sulfate (KS). Moreover, the high-molecular weight hyaluronan (HA) polysaccharide does not occur covalently attached to a protein core but is able to interact with proteoglycan receptors via HA-binding motifs. In addition to their GAG chain, proteoglycans are often decorated with *N*- and *O*-glycan chains. The precise composition of the GAG chains, as well as the conformational structure and distribution of the core protein, actively regulates the biological functions of proteoglycans (Couchman 2010).

1.2 Cancer-Associated Changes of Glycosylation

Structural alterations of the glycan moiety of glycoproteins and glycolipids are among the most frequently observed biochemical features of malignancy. They occur in both solid tumours and haematological malignancies. Alterations of glycosylation generated in malignant cells have to be distinguished from changes of glycosylation in secretory glycoproteins from non-malignant cells, namely immunoglobulins and acute-phase proteins reflecting systemic reactions of tumour-associated inflammation. Cancer-associated alterations may affect all types of glycosylation (*see* Subheading 1.1) and most notably include (i) highly branched *N*-glycans with increased sialylation, core and outer arm fucosylation and sialylated Lewis antigens, (ii) short-truncated *O*-glycan structures (e.g. Tn and STn), (iii) sialylated glycosphingolipids (gangliosides) and (iv) increased *O*-GlcNAcylation of intracellular proteins (Fig. 1).

N-Glycosylation

Structural changes in *N*-glycosylation in malignant cells and tissues include (1) increased branching of the *N*-glycans, (2) decreased bisecting GlcNAc residues in *N*-glycans, (3) increased fucosylation and sialylation and expression of sialyl-Le epitopes, and (4) either an elevated or decreased high-mannose-type *N*-glycans in distinct tumour entities.

(1) Increased branching of *N*-glycans leading to an elevated formation of tetra- and penta-antennary structures ranks among the most frequently observed types of alterations in cancer cells and tissues (Dennis et al. 1987; Pinho and Reis 2015). Increased *N*-glycan branching is caused by an elevated expression of GnT5 that transfers β1-6-linked GlcNAc from UDP-GlcNAc to the α1,6-linked Man of the pentasaccharide core (Kizuka and Taniguchi 2016). GnT5 has low affinity for its substrate and acceptor compared with other GlcNAc transferases such as GnT3, which underlies the enzyme's low activity in healthy tissues (Nishikawa et al. 1990; Osuka et al. 2024). On the opposite, GnT5 activity is higher in malignant tissues due to its elevated expression and to higher UDP-GlcNAc availability (Kizuka and Taniguchi 2016; Paneque et al. 2023). The GnT5 enzyme is encoded by the *MGAT5* gene, whose expression is regulated by diverse signalling cascades active during oncogenesis including RAS–RAF–MAPK (Galang et al. 1996). Elevated GnT5 expression acting on *N*-glycans located on cell surface receptors such as integrin $\alpha_1\beta_5$ triggers positive feedback loops, thereby amplifying tumourigenic intracellular signalling pathways (Dennis et al. 1999). Tumour growth and metastasis are suppressed in vitro and in vivo upon *MGAT5* knock-out (KO), which indicates the involvement of GnT5 in tumour progression (Granovsky et al. 2000; Guo et al. 2002, 2010) (see also Subheading 3). Increased *N*-glycan branching has also been observed in acute phase proteins and immunoglobulins in the plasma of patients with many forms of cancer including ovarian cancer (Biskup et al. 2013; Grzeski et al. 2022; Weiz et al. 2016), non–small cell lung cancer (Bartling et al. 2011), breast cancer (Benesova

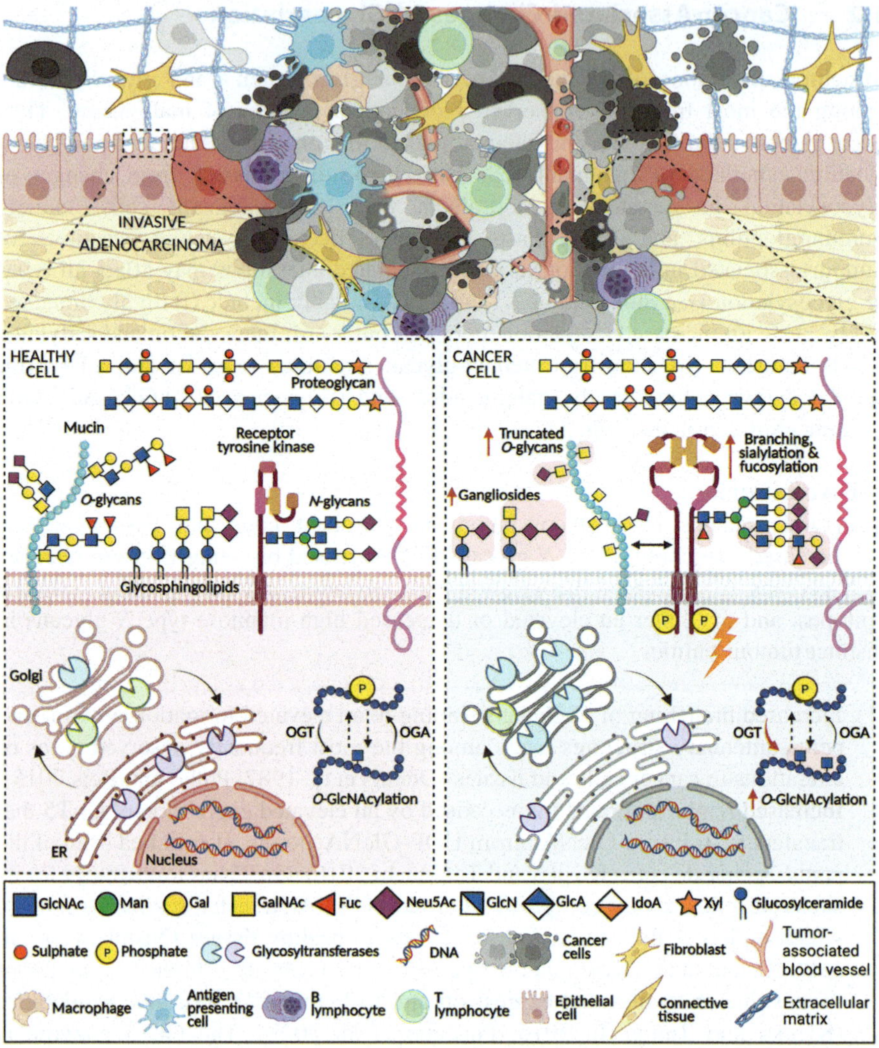

Fig. 1 Major glycosylation alterations occurring in cancer cells. The neoplastic transformation of epithelial tissues is accompanied by the emergence of a panoply of cell surface and intracellular cancer-associated glycoconjugates that regulate and actively support multiple malignant features of tumour cells. The disruption of the normal expression levels and subcellular localization patterns of specific glycosyltransferases leads to the upregulation of specific glycomic signatures and the de novo expression of glycan neoantigens. By impacting the tridimensional conformation, and the biochemical and interactomic properties of their molecular carriers, tumour-associated aberrant glycan structures impair the homeostatic functioning of molecules playing key roles in cancer cell malignancy. Moreover, cancer-associated glycans further protect malignant cells from anti-tumour immune responses by tightly controlling the activation threshold of immune cell effectors within the tumour milieu by engaging with glycan-binding protein (GBP) receptors. The most well-studied cancer-associated glycan signatures include short-truncated *O*-glycan structures (e.g. Tn and STn); highly branched *N*-glycans with increased sialylation, core and outer arm fucosylation and sialylated Lewis antigens; sialylated glycosphingolipids (gangliosides); increased *O*-GlcNAcylation of intracellular proteins

et al. 2024) and hepatocellular carcinoma (Goldman et al. 2009). As acute-phase proteins and immunoglobulins are secreted by hepatocytes and plasma cells, respectively, i.e. by non-malignant cells, these alterations most likely represent consequences of tumour-associated inflammation. β1,6-GlcNAc-branched *N*-glycans formed by GnT5 can be elongated by transfer of Gal residues β1,4 to GlcNAc by β1,4 Gal-T and by transfer of GlcNAc residues β1,3 to Gal by β1,3-GlcNAc-T to form poly-*N*-acetyllactosamine (Galβ1,4GlcNAcβ1,3 repeats) that serves as acceptor sites for additional Neu5Ac and Fuc residues and for the formation of sialyl-Le^x epitopes (see below) (Kizuka and Taniguchi 2016; Mitsui et al. 2012; Pinho and Reis 2015).

(2) *MGAT3*-encoded GnT3 catalyses the formation of so-called "bisecting" GlcNAc β1,4-linked to the innermost mannose residue (Kizuka and Taniguchi 2016). Addition of a bisecting GlcNAc inhibits the further action of GnT5, preventing *N*-glycan β1-6 branching and the formation of polylactosamine (LacNAc) repeats (Zhao et al. 2006). Depending on the cancer type, differences in the expression of GnT3 were observed and correlated with different signalling pathways, suggesting the possible role of *MGAT3* in modulating cancer-associated oncogenic signalling. In mouse models, loss of *MGAT3* increased mammary tumour progression, while high levels of GnT3 delayed their progression (Miwa et al. 2012, 2013; Potapenko et al. 2010; Song et al. 2010), suggesting a tumour-suppressive role of *MGAT3*. In addition, low levels of GnT3 were measured in colorectal cancer cells (Khare et al. 2014). Downregulation of *MGAT3* expression was associated with resistance to IFN-γ therapy in advanced colorectal carcinoma (Krug et al. 2023), correlating with low expression of the IFN-γ receptor. On the contrary, GnT3 activates Notch signalling and promotes tumour growth in ovarian cancer (Allam et al. 2017). Similarly, *MGAT3* expression was upregulated in ovarian cancer and intrahepatic cholangiocarcinoma in humans (Abbott et al. 2008; Dan et al. 2022). In mice lacking MGAT3 progression of chemically induced liver tumours was retarded (Yang et al. 2003). Increased *N*-glycan branching has been shown to be accompanied by a decrease of bisecting GlcNAc residues in blood sera of recurrent ovarian cancer patients (Alley et al. 2012). In addition, bisected biantennary *N*-glycans were found to be upregulated in ovarian cancer patients that were not responding to primary chemotherapy (Zahradnikova et al. 2021).

(3) Alterations in fucose metabolism rank among the earliest observed cancer-associated changes of glycosylation (Bauer et al. 1977). In *N*-glycans, L-Fuc can be attached α1,6 to the innermost GlcNAc of the pentasaccharide core (core fucosylation) by FUT8 (encoded by *FUT8*) or α1,3 to GlcNAc residues of the outer antennae by several other fucosyltransferases (Becker and Lowe 2003). FUT8 was found to be elevated in many forms of cancer including lung, liver, colorectal, ovarian, prostate, breast, thyroid, pancreatic and gastric cancer and melanoma (Bastian et al. 2021). Increased *FUT8* expression was associated with poor prognosis of non–small cell lung cancer patients (Park et al. 2020) and resistance to cisplatin treatment in ovarian cancer patients (Lv et al. 2019). In

hepatocellular carcinoma, *FUT8* was shown to be a p53 target gene (Okagawa et al. 2016). Moreover, *FUT8* upregulation promoted Wnt/β-catenin signalling (Yang et al. 2017) and the FAK/integrin pathway (Liu et al. 2019), thereby inducing epithelial-mesenchymal transition as well as the self-renewal of cancer stem cells.

(4) The relative ratio of high-mannose *N*-glycans as compared to complex and hybrid forms in human blood glycoproteins was found to be decreased in ovarian and gastric cancer (Biskup et al. 2013, 2014; Dědová et al. 2019; Hua et al. 2013; Kim et al. 2014; Ozcan et al. 2014). On the opposite, in tumour tissues, high-mannose *N*-glycans were proposed as malignant progression markers for colorectal cancer (Boyaval et al. 2022), lung adenocarcinoma (Ruhaak et al. 2015) and breast cancer (Benesova et al. 2024).

_O_-glycosylation

Immature *O*-glycosylation in the forms of the monosaccharide Tn antigen, the disaccharide T antigen and their sialylated counterparts Sialyl-Tn and sialyl-T is a hallmark of cancer. Tn and sialyl-Tn are highly expressed in pancreas, lung and breast and other types of cancer, whereas these glycans are unfrequently expressed in normal tissues (Kaur et al. 2013; Osako et al. 1993; Sanji et al. 2024). The predominance of Tn in the above-mentioned tumours was attributed to overexpression of *GALNT3* in lung and pancreatic cancer or of *GALNT6* in breast cancer (Bindeman and Fingleton 2022). Alternatively, truncated *O*-glycosylation can also result from the silencing of the *COSMC* gene, coding for a chaperone that is used by β1-3 galactosyltransferase (C1GalT1) to add Gal residues to nascent *O*-glycans (Radhakrishnan et al. 2014). Abundant expression of the sialyltransferase ST6GalNAc1 leads to the sialylation of Tn, which promotes tumour progression and poor prognosis (Munkley et al. 2015). Moreover, alterations in Golgi structure and mislocalization of glycosyltransferases due to changes in organelle pH may result in incomplete and disturbed *O*-glycosylation (Gill et al. 2010; Hassinen et al. 2011; Sanji et al. 2024). Sialyl-Tn antigen both in serum and in cancer tissues has been shown to be a prognostic marker in different types of cancer (Cotton et al. 2017; Hashiguchi et al. 2016; Pinho et al. 2007).

Sialylation and Lewis Epitopes

Sialylation of *N*- and *O*-glycans was found to be increased in various types of cancer, thereby promoting tumour progression and metastasis (Pinho and Reis 2015). Increased sialylation is due to the elevated activity of different sialyltransferases. The sialyltransferase ST6Gal1, catalysing the transfer of α1,6-linked Neu5Ac to Gal, can have different cancer-type-specific activities. On the one hand, *ST6GAL1* is overexpressed in many epithelial cancers including breast, cervical, gastric, liver, ovarian, pancreatic and prostate cancer (Bhalerao et al. 2023; Dall'Olio et al. 2004; Gretschel et al. 2003; Hsieh et al. 2017; Munkley et al. 2016; Recchi et al. 1998; Vazquez-Martin et al. 2005; Wang et al. 2003, 2005; Wei et al. 2016). Elevated activity of ST6Gal1 correlates with metastasis, resistance to treatment and poor prognosis in ovarian cancer (Wu et al. 2018). ST6Gal1 overexpression was shown to facilitate prostate cancer progression via the PI3K/Akt/GSK-3beta/beta-catenin signalling pathway (Wei et al. 2016). On the other hand, ST6Gal1 exerts tumour-suppressive functions for glioblastoma in cancer cell lines and in vitro experiments (Yamamoto et al. 1997, 2001). A similar inhibition was observed for colon cancer cell lines (Chiricolo et al. 2006; Park et al. 2012). In that case, loss of ST6Gal1 elevated EGF-induced EGFR phosphorylation, resulting in the activation of extracellular signal-regulated kinase (ERK) in colon cancer cells (Park et al. 2012). Moreover, ST6Gal1 site-specific modification of the Human Epidermal Growth Factor Receptor 2 (HER2/c-ErbB2) oncogenic receptor in gastric cancer protects HER2-addicted gastric carcinoma cells from the cytotoxicity mediated by anti-HER2 therapeutic monoclonal antibodies (mAbs) (Duarte et al. 2021). Similarly, ST3Gal1 has been shown to be an independent adverse prognostic factor for recurrence and survival of patients with clear cell renal cell carcinoma (Bai et al. 2015).

Increased levels of sialyl-Le^a and sialyl-Le^x epitopes were measured in patient blood and correlate with disease progression (Alley et al. 2012; Biskup et al. 2013; Biskup et al. 2014; Hua et al. 2013; Saldova et al. 2007). Their elevated expression is due to high levels of *ST3GAL4* and *ST3GAL6*, contributing to cell invasion and metastasis via binding to selectins. Selectins are vascular cell adhesion molecules of the C-lectin types that participate in the recruitment of myeloid-derived cells, promoting tumour cell extravasation in the metastatic cascade (Hauselmann et al. 2016; Trinchera et al. 2017).

_O_-GlcNAc

O-GlcNAc is a regulatory post-translational modification acting in essential cellular events such as signal transduction or carbohydrate metabolism, often by competing with phosphorylation. *O*-GlcNAc is transferred to proteins by the *O*-GlcNAc transferase OGT and cleaved by the hexosaminidase OGA. *O*-GlcNAc is able to inhibit phosphofructokinase 1, a central enzyme of the glycolysis, thereby favouring the pentose phosphate pathway that is profitable to cancer growth. *O*-GlcNAcylation regulates various signalling pathways such as NF-κB, NOTCH1, IL-6/STAT3, ERK1/2 contributing to cancer development as reviewed by He et al. (2023).

Glycosaminoglycans

Alterations in the composition of GAGs including the degree and the positioning of sulfate groups have been observed in different types of cancer, which result from changes in the complex enzymatic machinery of GAG biosynthesis and degradation (Marques et al. 2021; Soares da Costa et al. 2017). Due to their negative charge, GAGs interact with growth factors, growth factor receptors and cytokines participating in diverse cancer signalling pathways. As a result, GAGs contribute to cancer progression and metastasis (Afratis et al. 2012). For instance, the HS proteoglycan glypican-3 stimulates Wnt signalling in hepatocellular carcinoma and melanoma (Nakatsura et al. 2004; Nakatsura and Nishimura 2005). Moreover, HS chains on the syndecan-4 proteoglycan have been shown to contribute to the organotypic dissemination and internalization, by recipient cells, of extracellular vesicles secreted by gastric carcinoma cells (Pocas et al. 2023). Nevertheless, GAGs may also inhibit tumour growth as glypican-3 blocks metastatic spread in breast cancer (Guereno et al. 2020). Regarding CS- and DS-containing proteoglycans, sulfation patterns were associated with cancer progression, especially chondroitin-4,6-disulfate and dermatan-4,6-disulfate (Theocharis et al. 2006). In blood, changes in CS disaccharide sulfation were proposed as an ovarian cancer biomarker (Biskup et al. 2021). HA is a GAG abundantly found in the extracellular matrix (ECM) of healthy and malignant tissues. It has high affinity for CD44, which promotes tumour growth via the activation of the signalling pathways MAPK and PI3K/Akt (Misra et al. 2011). In addition, binding of HA accelerates epithelial-to-mesenchymal transition (EMT) by activating Akt signalling, which results in the formation of EMT-associated recurrent tumours (Misra et al. 2011). CD44 isoforms were shown to have an increased tumourigenic potential compared with CD44 (Eisterer et al. 2001). Moreover, CD44 isoforms have been used as prognostic markers as they correlate with the development of metastases (Tao et al. 2018; Wang et al. 2017).

Gangliosides

Located at the cell surface, gangliosides are highly expressed in many forms of tumours, whereby they enhance angiogenesis by increasing vascular endothelial growth factor-induced cell proliferation and migration (Furukawa et al. 2022; Groux-Degroote and Delannoy 2021; Schnaar 2023; Yoshida et al. 2020; Zeng et al. 2000). Their presence at the tumour surface can also lead to immune evasion due to binding with Siglec-7 on natural killer (NK) cells (Schnaar et al. 2022). Overexpression of gangliosides such as the bisialylated glycolipids GD2 and GD3 has been reported in some tumour tissues such as breast cancer or neuroblastoma as compared to control tissues (Lo Piccolo et al. 2001; Ruckhaberle et al. 2009). The transcription of the sialyltransferase ST8Sia1, which converts the glycolipid GM3 (Neu5Acα2,3Galβ1,4GlcβCer) to the disialylated GD3, is indeed increased in neuronal (Dae et al. 2009) and breast tumours (Bobowski et al. 2013). Moreover, epigenetic factors also play a role in ganglioside abundance. In fact, high expression of the GalNAc-transferase β4GalNAcT1, allowing the formation of GD2 from GD3, correlates with high histone acetylation levels in renal cell carcinoma (Banerjee et al. 2019). Neu5Ac *O*-acetylation constitutes another PTM of gangliosides that is increased in glioblastoma (Birks et al. 2011) or acute lymphoblastic leukaemia (Mukherjee et al. 2008).

Epigenetic Factors

As mentioned above, alterations of glycosylation in cancer result from multifactorial mechanisms that include genetic and epigenetic, transcriptional and posttranslational levels. Epigenetic factors, which were previously shown to regulate cancer glycosylation, may affect the expression level of glycosyltransferase genes as recently reviewed by Indellicato and Trinchera (2021). GalNAc-, Gal-, GlcNAc-, fucosyl-, and sialyltransferases can be modulated via promoter hypermethylation, post-transcriptional microRNA regulation and histone acetylation. For instance, the upregulation of *MGAT5* was associated with CpG island demethylation at the *nm23* promoter region (Chakraborty et al. 2006). Aberrant expression of microRNA affected intracellular signalling to mediate chemoresistance in ovarian cancer cells by driving cellular senescence, thereby reducing progression-free survival of patients. Nucleotide donor transporters were also reported to be downregulated by histone methylation (Huang et al. 2019). The mammalian sialidases Neu1-Neu4 cleaving glycolipids and glycoproteins also contribute to carcinogenesis. For instance, Neu1 is overexpressed in ovarian cancer tissues. Authors found that *NEU1* siRNA effectively inhibits the proliferation, apoptosis and invasion of human ovarian cancer cells by targeting lysosome and oxidative phosphorylation signalling (Ren et al. 2016).

2 Biological Consequences

Malignant growth is characterized by several central features: the ability of cancer cells (i) to evade cell division checkpoints and to escape apoptosis and immune surveillance, (ii) to grow invasively and to leave the cellular assembly of a tissue, (iii) to circulate with the lymphatics and the blood stream and (iv) to migrate to the sites of metastasis. Cancer-associated alterations of glycosylation (*see* Subheading 1) fundamentally contribute to these hallmarks of malignancy by several mechanisms at the molecular and cellular levels (Lin and Lubman 2024; Mereiter, Balmana et al. 2019a; Peixoto et al. 2019; Pinho and Reis 2015; Reily et al. 2019; Stowell et al. 2015; Thomas et al. 2021) (Fig. 1). These mechanisms are derived from the principal biological functions of glycosylation:

(a) The modulation of biophysical properties and biological activities of the proteins to which glycans are attached, e.g. receptors, ion channels, cell adhesion molecules, antibodies or matrix proteins. Modulation includes protein folding, conformation and bioactivity, solubility, stability and protection against proteolysis. The modulatory functions depend on the interactions of the glycan chains with the amino acids of the polypeptide chain mediated by hydrogen bonds, electrostatic bonds or dipole attraction.
(b) The specific interaction of distinct glycan epitopes with glycan-binding proteins (GBPs) and receptors involved in inter- and intracellular signalling, cell-cell and cell-matrix adhesion, immune function, endocytosis and intracellular trafficking and other functions.
(c) The impact of glycosphingolipids (GSL) on the structure of glycolipid-enriched membrane microdomains influencing localization and organization of cell surface signalling receptors (Hanafusa et al. 2020).
(d) The binding of growth factors, chemokines and cytokines by GAGs/proteoglycans of the ECM and the cell surface (Theocharis and Karamanos 2019).

In the following sections, the roles of glycosylation in cancer cell biology, cancer systems biology and clinical consequences for cancer patients will be discussed, although only a few examples are cited from the extensive literature on these topics.

2.1 Invasive Growth and Metastasis

Invasive growth into neighbouring tissues and organs, invading lymphatic and blood vessels and forming metastases in lymph nodes and distant organs is a key feature of cancer. Cancer-associated alterations of glycosylation contribute to the aggressive capability of cancer cells by various mechanisms on different levels (Bindeman and Fingleton 2022; Josic et al. 2019; Peng et al. 2024):

- Interaction and crosstalk with the surrounding ECM
- Modification of cell-cell adhesion
- Impact on inter- and intracellular signalling
- Facilitated circulation of cancer cells in the lymph and blood stream
- Mediating extravasation (diapedesis) and migration of cancer cells to sites of metastasis

Interaction and Crosstalk with the Surrounding Extracellular Matrix (ECM)
Binding of glycoproteins and proteoglycans of the ECM to receptors and adhesion molecules of cancer cells particularly to integrins plays a pivotal role in shaping invasiveness, mobility and migration. This crosstalk is modified by glycosylation of both cell surface and ECM glycoproteins. Integrins, heterodimeric transmembrane receptors coupling cells to the ECM by formation of cell adhesion complexes and modulation of signalling pathways, play a key role in this interaction. Glycosylation may modify integrin binding to its ligands, the assembly of integrins in the plasma membrane and the interaction with signalling molecules, among others (Cai et al. 2017; Marsico et al. 2018; Singh et al. 2018; Zhao et al. 2008), most likely by affecting integrin protein conformation (Luo et al. 2003). One of the most frequently observed cancer-associated alterations of glycan structures is the increase in β1-6 GlcNAc branching of *N*-glycans catalysed by upregulated glycosyltransferase GnT-5 (de-Souza-Ferreira et al. 2023; Nagae et al. 2018). While the GnT3, catalysing the formation of bisecting *N*-glycan structures, was shown to inhibit cell migration, GnT5 catalysing β-1,6-GlcNAc residues promotes migration (Zhao et al. 2008). This effect was, for example, observed in human fibrosarcoma cells with induced expression of GnT5 that stimulated cell migration toward fibronectin and invasion through Matrigel probably by inhibition of α5β1 integrin clustering on the cell surface (Guo et al. 2002). Introduction of bisecting GlcNAc residues into integrin α5β1, on the other hand, was reported to reduce ligand binding to fibronectin and to downregulate cell adhesion and cell migration of melanoma B16 cells (Isaji et al. 2004). Likewise, GnT3 was shown to antagonize the effect of GnT5 on α3β1 integrin-mediated cell migration on the ECM glycoprotein laminin 5 (Zhao et al. 2006).

Apart from GnT5, GnT4a, catalysing the formation of (β1,4 GlcNAc) branches on the mannose core of *N*-glycans, was reported to be upregulated in hepatocellular carcinoma and to enhance cell migration, invasion and adhesion (Yang et al. 2023). The effect was most likely caused by the increase of β1,4 GlcNAc branching of *N*-glycans on integrin β1, thereby enhancing the interaction with the intermediate filament protein vimentin. Interruption of the formation of β1,4 GlcNAc branches on integrin β1 suppressed the interaction with vimentin and inhibited cell motility (Yang et al. 2023). As was shown for ovarian cancer cells derived from tissue and ascites, *N*- and *O*-glycosylation, and particularly α2-6 sialylation, of the integrin α2 is essential for binding to collagen and laminin and to protect the integrin from proteasomal degradation (Huang et al. 2021).

Apart from changes of *N*-glycosylation, integrins are also prone to alterations of *O*-glycosylation. For instance, pancreatic adenocarcinomas were shown to overexpress core 1 β3-galactosyltransferase (C1GalT1) controlling the elongation of GalNAc-type *O*-glycosylation. Overexpression of the enzyme was associated with poor disease-free and overall survival, and with enhanced cell migration and invasion involving integrin α_v (Kuo et al. 2021). Knockdown of the enzyme decreased tumour growth and metastasis in subcutaneous and pancreatic orthotopic injection models, and suppressed cell adhesion to the ECM (Kuo et al. 2021).

Not only is the function of integrins altered by glycosylation, but certain integrins also appear to influence the glycosylation of cancer cells. For example, overexpression of both integrin α3β1 and phosphatidylinositol 4-kinase IIα (PI4KIIα) induced hypersialylation in several cell lines (Isaji et al. 2019). This appears to be important since hypersialylation is another hallmark of cancer-associated glycosylation (*see* Subheading 1) and was shown to affect the structure and function of integrins. For example, in vitro studies demonstrated that, unlike the α3β1 integrin from normal epithelium, this integrin from a human bladder carcinoma cell line carried sialylated tetra-antennary *N*-glycans (Pochec et al. 2006). Desialylation diminished binding of the integrin to the ECM glycoprotein fibronectin indicating that upregulated sialylation of integrins may modulate anchoring of cancer cells to the ECM (Pochec et al. 2006).

Another mechanism that influences the function of integrins in cancer cells is the expression of bulky glycoproteins of the cancer cell glycocalyx (Paszek et al. 2014). Expression of the bulky cancer-associated glycoprotein MUC1 fostered the assembly of focal adhesions and supported integrin-dependent growth factor signalling leading to enhanced cell growth and survival. Moreover, it was shown that metastatic tumours upregulate expression of bulky glycoproteins. These findings indicate that upregulated expression of bulky glycoproteins in cancer cells generates a bulky glycocalyx that facilitates integrin clustering by funnelling active integrins into focal adhesions and enhance growth factor signalling (Paszek et al. 2014).

Apart from integrins, the HA receptor CD44 is intimately involved in the crosstalk of cancer cells with the ECM influencing cell migration, invasiveness and metastasis, among others (Senbanjo and Chellaiah 2017). In vitro experiments demonstrated that alterations in the glycosylation of CD44, particularly desialylation, can affect HA-binding ability (English et al. 1998; Mereiter et al.

2019b). Studies on the effects of individual glycosylation sites of CD44 in hepatocellular carcinoma cells revealed that *N*-glycans at three distinct glycosylation sites stabilized CD44 against degradation and as a consequence promoted metastasis and invasion (Cheng et al. 2023). Moreover, another frequently observed type of cancer-associated glycosylation, the truncation of *O*-glycosylation with an enrichment of Tn carbohydrate antigen, was shown to affect CD44 in pancreatic cancer and to increase activation of the homeobox domain transcription factor NANOG that mediates increased cancer stem cell activation (Leon et al. 2022).

Apart from glycosylation of receptors that bind to ECM components, glycosylation of ECM proteins such as collagen, laminin and fibronectin may contribute to malignant growth. For instance, culturing of lung cancer cells on neoglycosylated collagen was demonstrated to determine a glycosylation-dependent positive selection of cancer stem cells and to trigger their expansion (Gardelli et al. 2021). Oncofetal fibronectin defined by addition of an *O*-glycan to a specific threonine residue has been shown to be involved in the EMT process (Ding et al. 2012). Moreover, glycosylation of laminin, a glycoprotein of the basement membrane promoting tumour invasion and cell survival (Marinkovich 2007), appears to influence cell migration (Kariya et al. 2010). The role of glycosylation of ECM proteins in malignant growth, however, remains to be examined in more detail. Moreover, alterations of the glycan moiety of proteoglycans both of the ECM and the cell surface appear to play a crucial role in invasion and metastasis, but are not covered in this review.

Cell-Cell Adhesion Cell-cell adhesion is mediated by the members of different families and superfamilies of cell adhesion molecules (CAMs) including cadherins, integrins, selectins and immunoglobulin-like cell adhesion molecules (IgSF-CAMs). Apart from tying cells to each other by the formation of hetero- or homotypic bonds on the cell surface, CAMs, on the cytoplasmic side of the plasma membrane, are connected to the cellular cytoskeleton and to signalling pathways intimately involved in the control of differentiation and proliferation. There is clear evidence that cancer-associated alterations of glycosylation affect structure and function of CAMs, which in turn leads to disturbance and attenuation of cell-cell bonds favouring invasiveness and metastasis. This has exemplarily been demonstrated for cadherins (Carvalho et al. 2016b). For instance, N-cadherin expressed by neural cells and multiple other tissues including cardiac muscle not only mediates cell-cell adhesion but also activates outside-in signal transduction pathways and is involved in the regulation of differentiation and development (Garcia-Castro et al. 2000). Enhancement of both poly-*N*-acetyllactosamine expression and β1,6-branching on *N*-glycans catalysed by GnT5 on N-cadherin affected the outside-in signal transduction pathway of ERK mediated by N-cadherin, sensitized stimulation of tyrosine phosphorylation of catenins by growth factors and expression of v-src consistent with reduction of cell-cell adhesion (Guo et al. 2003). Moreover, overexpression of GnT5 caused increased cell motility supporting the assumption that aberrant *N*-linked β1,6 branching can lessen cell-cell adhesion and lead to increased cellular motility and invasiveness during oncogenesis (Guo et al. 2003). This was further supported by the finding that decreased expression of *N*-linked β1,6-branched glycans on N-cadherin achieved by knockdown of GnT5 caused enhanced N-cadherin-mediated cell-cell adhesion (Guo et al. 2009). Studies in a gastric cancer model revealed that, similar to N-cadherin, *N*-linked β1,6 branching catalysed by GnT5 also destabilized epithelial E-cadherin and impaired cell-cell adhesion (Carvalho et al. 2016a, b; Pinho et al. 2013). By contrast, GnT3 catalysing the formation of bisecting GlcNAc residues stabilized E-cadherin on the cell surface and contributed to the maintenance of stable and functional adherens-junctions (Pinho et al. 2013). These findings are consistent with the general observation that loss of cadherin function in malignant cells leads to an increase in cellular motility and invasive growth (Mendonsa et al. 2018). Since similar effects were observed for integrins (Zhao et al. 2006, 2008), GnT3 and GnT5 appear to either impair or induce invasive and metastatic growth by different mechanisms of action.

Aberrant cancer-associated glycosylation may also affect members of the Ig-CAM superfamily as was shown for the neural cell adhesion molecule (NCAM, CD56) that is expressed on the cell surface of neurons, astrocytes, oligodendrocytes and myotubes and of other tissues and mediates homo- and heterotypic adhesion (Walmod et al. 2004). Glycosylation of NCAM is characterized by the presence of different levels of polysialic acid (PSA) residues that are synthesized by two α2,8-polysialyltransferases, ST8Sia2 and ST8Sia4 (Angata and Fukuda 2003; Mühlenhoff et al. 2013). Polysialylated NCAM is present in several metastatic types of human cancer such as Wilm's tumour (Roth et al. 1988) and neuroblastoma (Livingston et al. 1988). PSA controlled tumour cell growth and differentiation by interfering with NCAM signalling at cell-cell contacts of human neuroblastoma cell lines (Seidenfaden et al. 2003). In clear cell renal cell carcinoma, NCAM polysialylation promoted tumourigenesis and progression via the Hedgehog and Wnt/β-catenin signalling pathways (Jian et al. 2023). In human breast cancer cells, polysialylation enhanced NCAM-mediated cell migration, but not cell proliferation or motility (Guan et al. 2015). In addition, polysialylation catalysed by different polysialyltransferases affected the adhesion of NCAM to different components of the ECM (Guan et al. 2015).

Epithelial-Mesenchymal Transition (EMT) EMT is a developmental regulatory program by which transformed epithelial cell are enabled to grow invasively, to disseminate and to metastasize by the loss of cell polarity and cell-cell adhesion and by other capacities (Hanahan and Weinberg 2011). The regulation of *N*-glycan branching appears to play a major role in EMT (Taniguchi et al. 2021). Apart from *N*-acetylglucosaminyltransferases GnT3 and GnT5 and α1,6 fucosyltransferase (FUT8), *N*-acetylglucosaminyltransferase-1 (GnT1) has been identified as a regulator of EMT (Zhang, Isaji et al. 2019).

Circulation of Cancer Cells in the Blood Stream and Extravasation While circulating with the blood stream, cancer cells may adhere to platelets by means of the specific binding of cell surface glycans to selectins expressed on the platelet surface (Shu et al. 2024). As a consequence, platelets form a shield both protecting circulating cancer cells from encountering immune cells and mediating adhesion to the vascular endothelium at sites of extravasation. For example, it has been shown that P-selectin strongly expressed on platelets mediates adhesion of platelets to sialyl-Le^a and sialyl-Le^x on malignant myeloma cells protecting them from NK cell-mediated cytotoxicity (Natoni et al. 2023). Moreover, human HT-29 colon cancer cells were found to roll and tether on an E- and P-selectin matrix in a fashion similar to leukocytes. This seems to suggest that platelet cloak-mediated extravasation of circulating cancer cells to metastatic sites proceeds in a way similar to extravasation of leukocytes in contexts of local or systemic inflammation (Kohler et al. 2010). This assumption is supported by the identification of several glycoprotein ligands of endothelial selectins, i.e. leukocyte E-selectin ligands, P-selectin glycoprotein ligand-1 and E-selectin ligand-1, on human metastatic prostate tumour cells (Dimitroff et al. 2005).

Beyond extravasation, interaction of selectins expressed by endothelial cells with cancer cell surface glycans can reprogram circulating cancer cells as was demonstrated by the finding that bone vascular niche E-selectin by direct binding to cancer cells may promote bone metastasis by inducing mesenchymal-epithelial transition and activating Wnt signalling (Esposito et al. 2019).

2.2 Impact on Signalling Pathways

Fundamental characteristics of malignancy such as loss of growth control, impaired cell-cell adhesion, or EMT are causally linked to aberrations of signalling receptors and pathways. Cancer-associated glycosylation contributes to these aberrations as was shown for numerous signalling pathways such as Notch receptors and the Notch signalling pathway, receptor tyrosine kinases (RTKs) and their downstream signalling pathways PI3K/Akt, JAK/STATand MAPK, TGF beta signalling pathway, death receptors, receptors for neurotropic factors or the Wnt-signalling pathway (Gao et al. 2021). Only a few examples from the large number of signalling pathways can be discussed here.

The Notch signalling pathway plays a fundamental role in the regulation of differentiation, proliferation and apoptotic programs (Artavanis-Tsakonas et al. 1999) and depends on the glycosylation of Notch receptors with *O*-glucose and *O*-fucose (Luca et al. 2015; Pandey et al. 2021). Up- or downregulation of the *O*-glucosyltransferase may result in enhancement or impairment of Notch signalling in human myeloid leukemia U937 cells and may be involved in the pathogenesis of leukemia (Chu et al. 2013; Ma et al. 2011). Overexpression of protein *O*-fucosylation was shown to accelerate hepatocellular carcinoma progression via the Notch signalling pathway (Ma et al. 2016).

Dysregulation of oncogenic RTKs and their downstream signalling pathways is a detrimental factor in malignant growth. *N*- and *O*-glycosylation of RTKs may affect receptor conformation, ligand binding, oligomerization and activation of signalling cascades (Gao et al. 2021). As was shown for members of the ErbB family of receptors, aberration of both *N*- and *O*-glycosylation may modify binding of epidermal growth factors and may lead to hyperactivation of downstream signalling pathways (reviewed in Duarte et al. (2022)). For example, in oral squamous cell carcinoma cells, expression of Le^y on the epidermal growth factor (EGF) receptor was increased. Inhibition of Le^y biosynthesis weakened phosphorylation of Akt and ERK in response to EGF and reduced EGF-triggered cell migration (Lin et al. 2015). Core 1 β1,3-galactosyltransferase catalyzing the transfer of D-galactose from UDP-galactose to Tn antigen to form Thomsen-Friedenreich (T) antigen was shown to modify *O*-glycosylation of the EGF receptor in head and neck squamous cell carcinoma. Knockdown of core 1 β1,3-galactosyltransferase decreased EGF-EGFR binding affinity and inhibited EGFR signalling (Lin et al. 2018). This is in line with the finding that knockdown of *N*-acetylgalactosaminyltransferase III (β4GalNAcT3) catalysing the transfer of *N*-acetylgalactosamine to Ser/Thr residues caused alterations of *O*-glycosylation of HER2 with enhanced phosphorylation and augmented growth, motility and tumour–endothelium adhesion in pancreatic ductal adenocarcinoma (Chugh et al. 2016). Another example of the ample evidence that glycosylation modifies EGF receptor activity is the finding that sialylation of EGFR by the ST6Gal1 sialyltransferase promotes EGFR activation and resistance to gefitinib- and cetuximab-mediated cell death (Britain et al. 2018; Rodrigues et al. 2021).

Transforming growth factor-β receptors (TGFR) are Ser/Thr kinase receptors that activate SMAD pathways upon ligand binding. TGFR are involved in the regulation of proliferation and differentiation, tissue repair and immune responses (Gao et al. 2021). *N*-Glycosylation is necessary for the function of the three known TGFR and is subject to changes during carcinogenesis. For instance, FUT8 catalysing the addition of L-fucose from GDP-fucose α-1,6 to the core *N*-acetylglucosamine residue of *N*-glycans was found overexpressed in breast carcinoma cell lines upon induction of EMT with TGF-β and was associated with enhanced migration, invasiveness and metastasis, while pharmacological inhibition of FUT8 impaired these processes (Tu et al. 2017). Secondly, the dualism of GnT3 and GnT5 appears to play another fundamental role in carcinogenesis suppressing or enhancing TGF-β-induced EMT in cancer cells (Taniguchi et al. 2021). Moreover, I-branching β-1, 6-*N*-acetylglucosaminyl transferase 2 (GCNT2) was found to be required for TGF-β1–induced EMT enhancing cell detachment, adhesion to endothelial cells, cell migration and invasion in vitro, as well as lung metastasis of breast cancer cells in vivo (Zhang et al. 2011). Vice versa, TGF-β upregulated expression of the genes of GnT5 and GCNT2 in mouse melanoma and breast cancer cells, respectively, evidencing the close interaction of *N*-glycan branching and the TGF-β signalling pathways (Miyoshi et al. 1995).

2.3 Immune Evasion

Evasion of anti-cancer immunity remains one of the key hallmarks of malignancy (Gupta et al. 2023; Park, Colville et al. 2024; Rodriguez et al. 2018). Under normal conditions, the immune system is able to recognize and eliminate newly transformed cancer cells (Ribatti 2017). Glycan structures expressed on glycoproteins, proteoglycans and glycolipids of cancer cell surfaces associated with malignancy and also on glycoproteins secreted or shed by cancer cells may suppress immune tumour surveillance by at least four different ways (Ghasempour and Freeman 2023; Gupta et al. 2023; Kim and Cho 2022; Mereiter et al. 2019a; Rodriguez et al. 2018): (1) the interaction with glycan-binding receptors also referred to as mammalian lectins with affinity for distinct structural glycan epitopes; (2) the modulation of immune checkpoint molecules, e.g. programmed cell death ligand-1 (PD-L1); (3) masking of cancer cell surface antigens by the glycocalyx; (4) masking circulating cancer cells by adhesion of platelets forming a protective shield.

Interaction with Glycan-Specific Receptors Interaction with glycan-specific receptors contributing to immune evasion involves four or even more different families of receptors, i.e. sialic acid-binding immunoglobulin-like lectins (Siglecs) (Coccimiglio et al. 2024; Duan and Paulson 2020; Macauley et al. 2014; Stanczak and Laubli 2023), dendritic cell-specific intercellular adhesion molecule-3 grabbing non-integrin-1 (DC-SIGN) (Garcia-Vallejo and van Kooyk 2013), macrophage galactose-specific lectin (MGL) (Denda-Nagai et al. 2010; van Kooyk et al. 2015; van Vliet et al. 2008) and galectins (Cummings et al. 2022b; Elola et al. 2018; Verkerke et al. 2022). Each of these receptor families comprises several members that are distinguished by structural and functional characteristics. The immunosuppressive role of these receptors in malignancy has been widely observed in human as well as in animal cancer models. Members of these receptor families may exhibit species-specific differences that have to be taken into account when comparing results obtained in animal models and in humans (Garcia-Vallejo and van Kooyk 2013; RodrIguez et al. 2018). Interactions of cancer-associated glycans with these lectin receptors may affect T-cell differentiation, the activity of NK cells, the differentiation of tumour-associated or anti-inflammatory M2 macrophages and antigen-presenting cells resulting in immune evasion, among others (Rodriguez et al. 2018).

Siglecs are expressed on the surface of various immune cells, e.g. macrophages, monocytes, dendritic cells, tumour-infiltrating T cells and NK cells, may bind to sialylated glycan epitopes on cancer cells such as sialyl-Tn, sialyl-T, sialyl-Le^a, sialyl-Le^x, or the gangliosides GD3 and GM3, and recognize cancer cells by their elevated sialylation of glycoproteins and glycolipids, a feature that ranges among the most widely observed types of cancer-associated glycosylation (Bull et al. 2014; Duan and Paulson 2020; Pearce and Laubli 2016; Stanczak and Laubli 2023; Zhou et al. 2023). Siglecs were identified as critical immune suppressors that among others can induce the differentiation of monocytes into tumour-associated macrophages inhibiting the function of T cells (Beatson et al. 2020), suppress antigen-specific T-cell responses (Wang, Sun et al. 2019), or inhibit T-cell proliferation and induce regulatory T cells (Perdicchio et al. 2016). Recently, Siglec-9 was shown to act as an immune-checkpoint molecule on macrophages in glioblastoma restricting T-cell priming (Mei et al. 2023).

Members of the DC-SIGN family are expressed on dendritic cells and macrophages (Geijtenbeek et al. 2000; RodrIguez et al. 2018) and bind to distinct glycan epitopes including Le^x and Le^y structures (van Gisbergen et al. 2005) and high mannose *N*-glycans (Valle-Argos et al. 2021) on malignant cells. As was observed for colorectal cancer, dendritic cells recognize CEA on cancer cells through DC-SIGN binding (van Gisbergen et al. 2005). Moreover, binding of DC-SIGN was shown to drive follicular T-helper cell (TFH) differentiation via IL-27 expression (Gringhuis et al. 2014). Interaction of fucosylated glycans with DC-SIGN was reported to enhance IL-10 secretion and decrease T-cell proliferation in a DC-SIGN-, glycosylation-, and Raf1-dependent manner (Garcia-Vallejo et al. 2014). Moreover, as was shown for pancreatic ductal adenocarcinoma, fucosylated *O*-glycans can serve as ligands for DC-SIGN positive tumour-associated macrophages, modulating their activation and inducing the production of anti-inflammatory IL-10 (Rodriguez et al. 2022).

Human MGL is expressed on antigen-presenting cells, has binding specificity for glycans containing terminal GalNAc residues, such as Tn antigen or GlcNAc-GalNAc disaccharide (LacdiNAc) repeats, and also sialylated Tn antigens and favours immunosuppressive effects (Szczykutowicz 2023). For instance, interaction of MGL with its cognate glycan ligands has been observed to promote an immunosuppressive microenvironment in a murine lung cancer model (da Costa et al. 2021) to enhance IL-10 secretion (van Vliet et al. 2013) and to both decrease T-cell proliferation and increase T-cell death (van Vliet et al. 2006). Moreover, expression of MGL ligands has been shown to be associated with poor survival of stage III colon cancer patients (Lenos et al. 2015).

Galectins constitute a large family of soluble GBPs expressed by various cell types including cancer cells that may be present in cytosol, nucleus, outer plasma membrane and the extracellular matrix and have binding affinity for β-Gal-containing glycans (Cummings et al. 2022b; Elola et al. 2018; Verkerke et al. 2022). Among the numerous functions galectins have, members of the galectin family secreted by cancer cells may bind to glycan epitopes on immune cells triggering immunosuppressive effects. Galectin-1 was shown to induce dendritic cells to acquire an interleukin 27-dependent regulatory function and to promote IL-10-mediated T-cell tolerance (Ilarregui et al. 2009). With respect to immune evasion, for example, galectin-3 ligand complexes of extracellular vesicles in the plasma from patients with triple negative breast cancer were shown to induce a significant increase in T-regulatory cells in peripheral blood mononuclear cells that correlated with a significant increase in immunosuppressive interleukins 10 and 35 (Raiter et al. 2023). Moreover, extracellular galectin-4 was reported to drive immune evasion and to promote T-cell apoptosis in human pancreatic cancer (Lidstrom et al. 2023). The recent finding that the C-type lectins CLEC10A (CD301) and CLEC4G (LSECtin) are exclusively expressed in distinct subpopulations, particularly in dendritic cells and macrophages of human breast cancer infiltrating immune cells, indicates that additional lectins are involved in cancer immune response (Wegscheider et al. 2024).

Apart from the activation of these GBPs, cancer-associated glycosylation affects the functional efficiency of immune checkpoint molecules. PD-1 was shown to be extensively *N*-glycosylated in T cells with intensities of its specific glycoforms being altered upon T-cell receptor (TCR) activation (Sun et al. 2020). Glycosylation was critical for maintaining PD-1 protein stability and cell surface localization (Sun et al. 2020). Moreover, glycosylation of PD-1 was essential for mediating its interaction with PD-L1 (Sun et al. 2020). In addition, glycosylation of PD-L1 was demonstrated to be essential for PD-L1/PD-1 interaction and for immunosuppression in triple-negative breast cancer (Li et al. 2018) offering potential targets for immunotherapy (Lee et al. 2024).

Very recently it has been shown that the glycosylation of CD155, an adhesion molecule belonging to the Nectin/Nectin-like family that may be overexpressed on tumour cells is involved in binding to both activating and inhibitory receptors either up- or down-regulating anti-cancer immunity (Molfetta et al. 2020; Tahara et al. 2024).

In addition to the specific interactions of cancer-associated glycosylation with immunosuppressive receptors, the glycocalyx of cancer cell has been proposed to build a repulsive barrier that masks tumour neoantigens and inhibits phagocytosis by myeloid cells (Ghasempour and Freeman 2023).

3 Clinical Significance

3.1 Cancer Diagnosis and Monitoring

Altered glycosylation remains a fundamental and universal molecular feature of cancer cells (Bellis et al. 2022; Mereiter et al. 2019a; Pinho and Reis 2015). Indeed, the neoplastic transformation of healthy cells into their malignant counterparts is characterized by the dysregulation or disruption of specific glycosylation pathways, namely via the altered expression levels, activity and distribution patterns of specific glycosyltransferases, or the unbalanced levels or compromised availability of sugar acceptor/donor substrates. The aberrant functioning of the glycosylation machinery leads to the cell surface presentation of an array of cancer-associated glycomic profiles, or tumour-associated carbohydrate antigens (TACAs), that include the excessive or insufficient levels of specific glycan structures, the emergence of incomplete/truncated *O*-glycan chains and, although less frequent, the de novo expression of glycan epitopes, as described in the previous sections. Several classical serological assays have been established for the non-invasive detection of certain cancer types, as well as the clinical monitoring of tumour progression, therapeutic response and disease relapse (Adamczyk et al. 2012; Reis et al. 2010). The CA 19-9 serological assay, which detects circulating sialyl-Le^{a}, has been widely applied in the clinical management of gastrointestinal malignancy, including gastric, colorectal and pancreatic cancers. Two other well-established glycan-based serological biomarkers are the carcinoembryonic antigen (CEA) glycoprotein, used for the early detection and monitoring of recurrent or metastatic colorectal cancer, and the

CA 125, which detects cleaved/shed MUC16 glycoprotein for the clinical monitoring of ovarian cancer. Moreover, core fucosylated alpha fetoprotein (AFP-L3) glycoforms have been used as clinical biomarkers in the monitoring of hepatocellular carcinoma progression and recurrence (Park et al. 2017). However, since most of these glycan-based serological markers mostly rely on the overall quantification of a single glycoprotein, while ignoring target-specific glycosylation features, they often lack, therefore, clinical specificity. In particular, the targeted detection and quantification of specific cancer-associated protein glycoforms, which may not be reflected in the overall quantification of the glycoprotein alone, may provide invaluable insights on the glycomic alterations accompanying disease progression, improve diagnostic specificity, and prognostic and therapeutic stratification (de Haan et al. 2020; Mereiter et al. 2019a). In that sense, efforts have been made towards the clinical implementation of the glycomic profiling of glycoproteins present in the serum or other body fluids, either free as part of the cargo of extracellular vesicles (EVs). Indeed, the non-invasive longitudinal detection of circulating glycoconjugates at key moments of a patient's clinical history, such as diagnosis, completion of neoadjuvant chemotherapy and curative surgical resection and post-treatment follow-up, may provide invaluable insights on disease progression and the host therapeutic response. Moreover, glycans whose altered expression is a recognized feature of cancer are currently being explored as novel agents for real-time intraoperative tumour imaging (Cheng et al. 2021).

The tissue-specific glycosylation patterns of normal epithelia have been extensively characterized with the aid of hybridoma-produced glycan-specific mAbs and plant-derived lectins (Croce et al. 2007; Gagneux et al. 2022). In the non-cancer setting, glycan antigens frequently depict a highly regulated spatially resolved distribution that often accompanies that one of well-defined sets of molecular carriers. This is well illustrated by the organ-specific and gradient-like glycosylation profile of mucins carrying specific Lewis structures, which is accompanied by the highly restricted expression of genes coding for glycosyltransferases catalysing key steps in their biosynthesis (Reis et al. 1999). The neoplastic transformation of cells and tissues disrupts this highly regulated glycomic landscape, and leads to the emergence, at the cell surface, of an array of abnormal cancer-associated glycan antigens that represent an appealing source of tissue-based biomarkers and promising molecular targets for antibody- and cell-based therapeutic approaches. Indeed, while the expression of prematurely terminated *O*-glycan Tn and the T antigens is absent in healthy epithelial mucosae, their de novo cell surface expression constitutes a well-established glycosylation feature of several types of epithelial-derived carcinomas, mainly due to the overexpression of ST6GalNAc1 or inactivating mutations in the *COSMC* gene (Ju et al. 2008; Marcos et al. 2011). Expression of short-truncated *O*-glycan determinants further associates with tumour clinical aggressiveness and poor patient prognosis (Bellis et al. 2022; Brockhausen et al. 2022; Mereiter et al. 2019a; Romer et al. 2021). Moreover, extensive glycosylation of proteins may hinder their tissue-based diagnostic detection through epitope masking, as demonstrated for PD-L1 (Lee et al. 2019), and compromise their use as reliable predictive biomarkers to determine patient prognosis and therapeutic eligibility.

More recently, tissue-based proximity ligation assay (PLA) has been employed towards the in situ validation of specific protein glycoforms, which becomes particularly relevant in the absence of glycoform-sensitive diagnostic mAbs (reviewed in (Duarte et al. 2023)). This approach has generated invaluable mechanistic data on the impact that specific glycosylation signatures may exert on the oncogenic or tumour suppressive functions of their molecular carriers. In situ PLA has been successfully applied in the detection of cancer-associated proteins modified with tumourigenic glycan signatures, including MUC2-sialyl-Tn, CD44v6-sialyl-Tn, E-cadherin-branched *N*-glycans, HER2-α2,6Neu5Ac (Campos et al. 2015; Carvalho et al. 2016b; Conze et al. 2010; Duarte et al. 2021).

Additionally, matrix-assisted laser desorption/ionization mass spectrometry imaging (MALDI-MSI) has been extensively explored for the spatially resolved semi-quantitative *N*-glycomic profiling of formalin-fixed paraffin-embedded (FFPE) normal and cancer tissues, following the on-tissue enzymatic release of *N*-glycans from their glycoprotein carriers with the PNGase F endoglycosidase (Drake et al. 2017; Grzeski et al. 2022). MALDI-MSI has become an increasingly appealing tool in both pharmaceutical and clinical settings, by providing enhanced spatial resolution, as well as improved reproducibility, throughput and depth. Within the complex tumour microenvironment, MALDI-MSI-based in situ analysis allows the disclosure of the region-specific profiles of cancer-associated glycans in pathology-guided well-defined histological regions, namely normal adjacent mucosa, tumour invasive front, immune infiltrate and stromal compartment. Moreover, the miniaturization of large cohorts of cancer clinical specimens within tissue microarrays (TMAs) allows for the high throughput analysis of a large number of samples representative of the histological and molecular diversity of tumour subtypes and the association of region-specific glycomic profiles to relevant clinicopathological features (Powers et al. 2014). The compatibility of MALDI-MSI analytical pipelines with FFPE tissues is of significant relevance, since this represents the most widely available clinical material from cancer patients archived in repositories worldwide.

3.2 Glycan-Based Immunogens for Effective Anti-cancer Vaccination

There is extensively reported evidence that the immune system is able to mount immunological responses against glycoconjugates (Anderluh et al. 2022). In fact, since several cancer-associated glycan antigens elicit robust immune responses in their cancer hosts, they have been extensively studied as promising immunogens for the generation of effective anti-cancer vaccine formulations. Moreover, cancer-associated glycans, absent from non-transformed healthy cells, are often presented at the tumour cell surface and can, thus, be readily engaged by antigen presenting immune cell effectors (Angata et al. 2022). Indeed, extensive glycomic and glycoproteomic profiling of cancer cells and tissues is invaluable for the identification of unique aberrant glycosylation signatures restricted to the cancer cell population as promising targets for the development of effective and non-toxic glycovaccine formulations. In particular, ganglioside-based glycan antigens and glycopeptides carrying the short-truncated *O*-glycan antigens Tn and sialyl-Tn, often conjugated with immunogenic carriers, such as dendrimers, viral-like particles and nanoparticles, as well as chemically modified carbohydrate epitopes have been widely explored to bypass immunological self-tolerance and the low immunogenicity inherent to glycan-based antigens and to avoid in vivo enzymatic degradation of administered immunogens. Besides mucin glycopeptides enriched in short-truncated *O*-glycans (e.g. MUC1-Tn/sialyl-Tn), anti-cancer vaccine formulations based on gangliosides and Lewis antigens are currently under pre-clinical and clinical development (Anderluh et al. 2022; Duinkerken et al. 2019).

3.3 Therapeutic Targeting of Cancer-Associated Glycans

Over the last decades, rapid developments in the fields of antibody engineering and adoptive cell-based therapies have consolidated immunotherapy as one of the main therapeutic avenues in the clinical oncology setting (Riley et al. 2019). In particular, three main strategies have achieved significant clinical success and mostly rely on the development of i) mAbs targeting cancer-specific molecular targets, ii) immune checkpoint inhibitors (ICIs), iii) chimeric antigen receptor (CAR) T lymphocytes (Smith and Bertozzi 2021) (Fig. 2). The tumour-confined expression and cell surface presentation of cancer-associated glycan antigens, embedded within the dense glycocalyx layer, make them ideal molecular candidates for the development of targeted therapeutic strategies. However, despite having their expression highly restricted to cancer cell populations and neoplastic tissues, most TACAs are poorly immunogenic and require further chemical modifications or conjugation to protein carriers for the triggering of robust anti-cancer immune responses. Passive immunotherapeutic strategies based on unconjugated glycan-specific mAbs rely on their intrinsic cytolytic activity, which includes antibody-dependent cellular cytotoxicity (ADCC), antibody-dependent cellular phagocytosis (ADCP), and complement-dependent cytotoxicity (CDC). The GSL antigens GD2 and Globo H, the short-truncated *O*-glycans Tn and sialyl-Tn, and the tetrasaccharide Lewis antigens Le^y and sialyl-Le^a are amongst the most promising TACAs against which therapeutic mAbs have been generated and are under clinical evaluation (Berois et al. 2022). Another strategy is to stimulate the immune-mediated destruction of cancer cells by increasing their proximity to effector immune cells. To achieve this goal, bispecific mAbs capable of the simultaneous binding of TACAs, expressed by target cancer cells, and immune receptors, such as CD3 and $F_c\gamma RI$, expressed by immune cells, are under clinical investigation.

The inhibition or blockade of glycan-mediated tumour-supportive molecular interactions constitutes another promising therapeutic strategy under clinical development. For instance, sialyl-Lex and sialyl-Lea are glycan isomeric species are the cognate ligands of C-calcium-dependent (C-type) selectin receptors expressed by

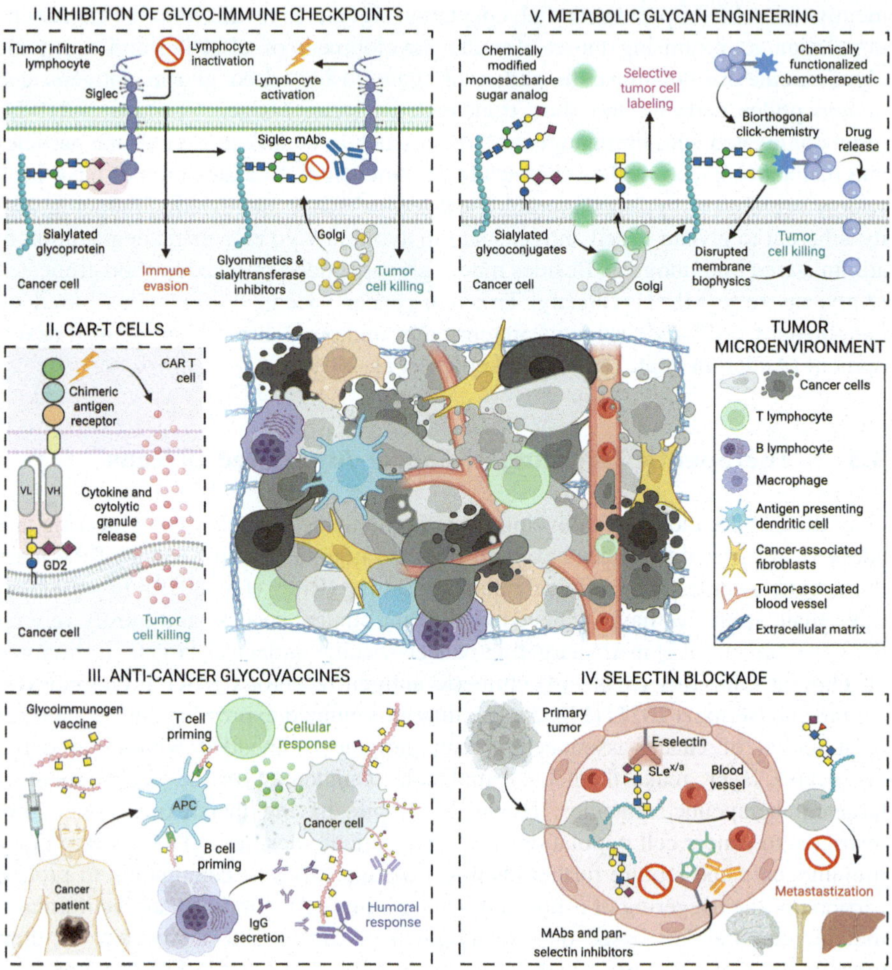

Fig. 2 Strategies for the therapeutic targeting of tumour-associated glycans in cancer. **I.** The targeted inhibition of glyco-immune checkpoints boosts anti-cancer immune responses by limiting the cell surface exposure of sialylated glycans on tumour cells or directly blocking their interaction with cognate immune receptors (e.g. Siglecs) with specific antibodies. **II.** Chimeric antigen receptor (CAR) T cells directed at cancer-specific glycoconjugate-based epitopes efficiently and selectively eliminate aberrantly glycosylated tumour cells, while sparing healthy cells. **III.** Vaccines based on chemically modified tumour-associated glycan antigens, whose aberrant expression is specifically confined to malignant cells, elicit robust humoral and cellular anti-cancer immune responses. **IV.** The efficient blockade of glycan-dependent and selectin-mediated extravasation of circulating tumour cells across the endothelium represents a promising therapeutic approach to impede cancer cell metastasis. **V.** The metabolic labelling of glycans harbours a dual anti-cancer effect: the specific delivery of chemotherapeutic agents through highly efficient biorthogonal click chemistry reactions, and the compromising of tumour cell integrity through disrupted cell membrane biophysics

endothelial cells (E-selectins), and mediate, therefore, the adhesion and endothelial extravasation of immune cells to secondary lymphoid organs and inflammation sites (Cummings et al. 2022a). In the cancer setting, sialyl-Le$^{x/a}$-expressing tumour cells rely on a similar selectin-mediated adhesion mechanism to facilitate their removal from circulation to secondary metastatic sites. Indeed, E-selectin has been established as a crucial bone marrow homing receptor within the complex tumour microenvironment, and a molecular driver of bone metastasis in sialyl-Le$^{x/a}$-expressing both solid and liquid tumours (Dimitroff et al. 2004; Esposito et al. 2019; Katayama et al. 2003). These observations have consolidated the rationale for the therapeutic targeting of glycan-selectin interactions to undermine tumour cell systemic and organotypic dissemination. Crystallographic and nuclear magnetic resonance (NMR)-based elucidation of selectin-glycan molecular interactions has provided a structural blueprint for the design of high-affinity glycomimetic pan-selectin inhibitors and recombinant ligands (Smith and Bertozzi 2021). Finally, selectin-targeting antagonist mAbs have also garnered interest in the oncological setting, by efficiently blocking selectin-glycan molecular interactions and ultimately disrupting a mostly immunosuppressive tumour microenvironment. Selectin-specific mAbs may be used for the therapeutic blockade of selectin-dependent immune checkpoints such as the one mediated by P-selectin glycoprotein ligand-1 (PSGL-1) to dampen the activity of antigen-presenting cells (APCs) and effector T cells.

The threshold of immune cell activation is finely regulated by the overall balance between stimulatory and inhibitory molecular signals, which are, in turn, governed by a set of cell surface immune checkpoint receptors (Manni and Laubli 2021). Illustratively, the activation of molecular immune checkpoints in the homeostatic setting prevents autoimmune responses. Within the complex tumour milieu, cancer cells harness the immunosuppressive properties of selected glycan epitopes presented at the cell surface to evade immunosurveillance and dampen anti-tumour immune responses. More specifically, aberrant glycan structures foster an immunologically permissive tumour microenvironment through the engagement of carbohydrate-specific lectin-based receptors, or glyco-immune checkpoints, such as the Siglec I-type lectin receptors, and galectins expressed by most immune cell types (Angata et al. 2022; Boelaars and van Kooyk 2024; Cummings et al. 2022a). Membrane-anchored Siglec receptors engage with cell surface proteins and lipids decorated with Neu5Ac-containing glycans and maintain tolerance to self by activating inhibitory pathways through cytoplasmic signalling motifs. Cell surface hypersialylation, a well-established glycophenotype of cancer cells, promotes immune evasion through Siglec-receptor mediated inhibition of anti-tumour immunity (Ibarlucea-Benitez et al. 2021). Therapeutic mAbs designed to block the engagement of CD33-related Siglecs receptors are under clinical development. They are being tested either alone or combined with potent cytotoxic agents in the format of antibody-drug conjugates (ADCs) with cytotoxic agents. Moreover, the highly regulated organ- and tissue-specific expression pattern of different Siglec receptors makes them appealing molecular candidates for the targeted delivery of chemotherapeutic agents (Bull et al. 2021). Siglec-targeting mAbs can be directed at either the Siglec receptor or its cognate ligand. Unfortunately, since few Siglec ligands have been unequivocally identified, the former strategy has been more widely explored in the clinical setting. Importantly, complementary immunoglobulin design and engineering ensure that the developed mAb does not act as an agonist upon Siglec binding. Regarding the therapeutic targeting of Siglecs with unknown physiological ligands, alternative approaches to achieve inhibition of sialoglycan-mediated immunosuppression have been developed and include Neu5Ac glycomimetics and sialyltransferase small molecule inhibitors (SMIs). Alternatively, the conjugation of a sialidase with mAbs directed at HER2 oncogenic RTK, overexpressed in breast and gastric adenocarcinomas, has been shown to promote the targeted cell surface desialylation of HER2-positive and neighbouring tumour cells, to dampen the activation of inhibitory Siglecs and, consequently, to enhance local anti-tumour immunity (Xiao et al. 2016). Galectins constitute another family of endogenous lectins that bind preferentially to glycoproteins decorated with glycan chains terminated with uncapped Gal moieties. Galectins can facilitate drive cancer cell evasion from immune surveillance by actively regulating T-cell activation and exhaustion. In this sense, the therapeutic efficacy of mAbs targeting galectins and their ligands is being assessed in clinical trials (Marino et al. 2023).

Given that numerous aberrantly glycosylated antigens are almost exclusively limited to the tumour cell surface, while depicting absent expression in non-transformed cells, mAbs targeting cancer-associated glycans have been additionally engineered into CAR-bearing T cells (Abrantes et al. 2022; Raglow et al. 2022). In particular, advances in the structural and glycoproteomic elucidation of cancer-specific carbohydrate chains have expanded the available repertoire of putative glycosylated targets against which effective CAR T formulations can be developed. Importantly, glycan-based target antigens may contribute to overcome current limitations hindering the translation of CAR Ts to the solid tumour setting, characterized by the lack of tumour-specific antigens with neglectable expression in normal tissues. Moreover, CAR target engagement is independent from major histocompatibility complex (MHC)-dependent presentation. Several CAR T formulations targeting cancer-specific aberrant glycosylation signatures are under preclinical and clinical development in both liquid and solid tumour settings. Target glycoconjugates include MUC1-Tn epitopes (Posey et al. 2016), whose expression has been widely reported across several types of both liquid and solid malignancies; the disialoganglioside GD2 (Majzner et al. 2022; Prapa et al. 2021), whose expression is residual in non-transformed tissues and confined to the peripheral nerves, brain and melanocytes; the Le^y tetrasaccharide (Meyran et al. 2020; Ritchie et al. 2013), whose membrane expression can be detected in the majority of carcinomas of epithelial origin. On the other hand, densely packed *N*-glycan chains decorating membrane-spanning glycoproteins can shield their molecular carriers from engaging with CAR T cells and, thus, reduce CAR T overall therapeutic efficacy (Greco et al. 2022). With the aim of improving CAR engagement with cancer-associated glycan antigens on the surface of target cells, novel CAR formulations have been designed to integrate highly specific glycan-binding motifs of mammalian selectins (e.g. E-selectin (Esposito et al. 2019)), and bacteria- and plant-derived lectins.

Finally, the tumour-confined expression patterns of aberrant glycan antigens make them ideal molecular candidates for the targeted delivery of anti-cancer therapeutics (Berois et al. 2022). In glycan-directed ADCs, a glycan-specific mAb, acting as the targeting component, is conjugated to a cytotoxic molecule, such as cytotoxins or chemotherapeutic drugs (e.g. anti-mitotic agents, inhibitors of microtubule polymerization). Moreover, glycan-specific mAbs have been successfully used for surface functionalization and targeted delivery of nanoparticles encapsulating anti-cancer therapeutics to the tumour site, while reducing off-target toxicity (Berois et al. 2022).

In metabolic oligosaccharide engineering (MOE) or metabolic glycan labelling approaches, unnatural, chemically modified monosaccharide sugar analogues are selectively fed to target cancer cells, subsequently integrated within specific metabolic pathways and, finally, biosynthetically incorporated into the cell surface glycocalyx for the functional tuning of cell behaviour (Dube and Bertozzi 2003; Liu et al. 2021). These cell surface chemical tags can then be leveraged for the targeted delivery of chemotherapeutics and anti-cancer drugs via highly selective biorthogonal click-chemistry reactions with predetermined functional groups. To avoid off-target cytotoxicity and improve selective tumour cell labelling, several strategies based on rational chemical design and the use of delivery vehicles are currently under development to restrict the in vivo delivery and incorporation of the above-referred analogues to cancer cells. Following the successful labelling of cancer cells with chemical tags, therapeutic agents modified with the chemical tag's complementary functional group can be efficiently delivered to target cancer cells. Beyond acting as molecular targets, metabolically incorporated cell surface chemical tags profoundly reshape the composition and biophysical properties of cancer cell membranes, which can both dampen the activation of oncogenic signalling pathways by membrane-anchored glycosylated receptors and sensitize labelled cells to chemoradiation-and immune-based therapeutic modalities (Han et al. 2023). Indeed, the chemical labelling of glycoproteins essential in cancer cell recognition and intercellular interactions undermines several glycan-mediated malignant features of tumour cells, such as E-selectin binding during metastatic dissemination. By actively reshaping the cell surface glycome and glycoproteome, MOE-based strategies have the potential to improve the therapeutic efficacy of immunotherapeutic modalities. For instance, the incorporation of unnatural monosaccharide analogues by cancer cells increases their immunogenicity and boosts anti-tumour immune responses, namely ADCC and CDC.

3.4 Clinical Studies: Consequences for the Outcome of Patients

Cancer-associated alterations of glycosylation are associated with clinical characteristics reflecting poor patient prognosis such as growth rate, rate of metastasis, invasiveness or non-response to therapy. In clinical studies, it could be shown that distinct types of cancer-associated glycosylation are concomitant with poor overall survival or progression-free survival. Representative examples from the extensive literature are listed below.

For instance, expression of the *GCNT2* gene encoding the I-branching enzyme, a β1,6-*N*-acetylglucosaminyltransferase responsible for formation of the blood group I antigen with branched poly-LacNAcylated glycans is downregulated during clinical progression of melanoma (Sweeney et al. 2018). Moreover, expression of PSA is correlated with tumour progression in non–small cell lung cancer (Tanaka et al. 2000). For human hepatocellular carcinoma, a signature of five glycosylation-related genes identified with bioinformatics analysis allowed to identify patients with a worse prognosis associated with clinical features, such as the tumour grade and stage (Shi et al. 2022). Likewise, a glycosylation signature was introduced as a promising tool for predicting the prognosis and immunotherapeutic response of prostate cancer (Sun et al. 2023). Kaplan–Meier survival analysis of 99 clinical cases showed that high expression of C1GalT1 controlling the elongation of GalNAc-type *O*-glycosylation in pancreatic adenocarcinoma correlated with poor disease-free and overall survival of (Kuo et al. 2021). High expression levels of core 1 β1,3-galactosyltransferase catalysing the transfer of D-galactose from UDP-galactose to Tn antigen to form the T antigen predicted poor disease-free and overall survivals of patients with head and neck squamous cell carcinoma (Lin et al. 2015). Moreover, expression of glycans on cancer cells containing terminal GalNAc residues such as Tn antigen or LacdiNAc but also sialylated Tn antigens that bind to human macrophage galactose-type lectin (MGL) has been shown to be associated with poor survival of stage III colon cancer patients (Lenos et al. 2015).

3.5 Animal Models

The mechanistic basis underlying the biological role of aberrant glycosylation signatures in the activation of specific cellular pathways and the sustenance of cancer cell malignant phenotypes can be systematically probed with precise genome editing pipelines in tumour cell in vitro models with selected glycosylation capacities. Indeed, the genomic silencing of glycosyltransferases, glycosidases, transporters of nucleotide sugar donors and molecular chaperones allow for the unequivocal functional dissection of the biological role played by individual glycogenes in relatively simplistic glycoengineered cellular systems free from the influence of systemic confounding factors. However, the elucidation of the structure/function relationship for individual glycogenes, glycan determinants or glycosylation sites requires biochemical and genetic combinatorial strategies within complete model organisms of disease (e.g. mice) (Stanley 2016). Moreover, in the clinical oncology setting, in vivo studies in animal models become of particularly importance when assessing the clinical utility of a given glycan determinant as a potential predictive/stratifying biomarker or a novel therapeutic target. However, such in vivo studies are technically challenging, since the KO of selected glycogenes may be deleterious to embryonic and postnatal development. In addition, the biological interpretation of generated phenotypes may be hindered by the partial functional redundancy of related glycosyltransferases, the wiring of "rescue" or compensatory glycosylation pathways, and the vast repertoire of protein and lipid substrates modified by a single enzyme. Over the last two decades, the Consortium for Functional Glycomics (CFG) has allocated significant resources toward the comprehensive phenotypical characterization of over 30 mutant mouse strains deficient in genes codifying for individual glycosyltransferases or GBPs (Orr et al. 2013).

When studying the tumourigenic role of glycan determinants in cancer onset, progression, phenotypical aggressiveness and therapeutic resistance, organ-specific conditional KO represents an indispensable tool. The organ- or tissue-confined conditional deletion or downregulation of a target glycogene produces invaluable functional insights regarding its role in the process of neoplastic transformation of that specific organ. Using conditional KO mice models, the oncogenic and tumour suppressive functions of several glycogenes have been elucidated in the context of organ-specific carcinogenesis. The abrogation of bisecting GlcNAc through the conditional, breast gland-confined deletion of *Mgat3* promotes the spontaneous formation of mammary tumours and lung metastasis in mice (Song et al. 2010). On the other hand, and in support of the opposing roles played by *N*-glycan branching and bisection, *Mgat5*-deficient mice are less susceptible to spontaneous mammary tumour formation and metastatic dissemination (Granovsky et al. 2000). The conditional KO of the *O*-glycan core 1 synthetase *C1galt1* in gastric epithelial cells leads to spontaneous gastritis rapidly evolving to gastric adenocarcinoma characterized by the de novo expression of short-truncated *O*-glycan determinants and gastric mucin displacement in the gastric epithelium (Liu et al. 2019). Moreover, *Fut2*-null mice have been used to characterize the binding dependency of BabA-expressing *Helicobacter pylori* strains, which constitute main drivers of the gastric multi-step carcinogenesis cascade, on the presence of α1,2-fucosylated glycan structures in the gastric mucosa (Magalhaes et al. 2009). The inducible, pancreas-restricted expression of *Fut3* and *β3galt5* was sufficient to reconstitute sialyl-Le^a expression and hyperactivation EGFR-mediated signalling, which consequently leads to acute and severe pancreatitis and the development of aggressive *KRAS*-driven pancreatic adenocarcinoma (Engle et al. 2019). Furthermore, the pancreas-confined deletion of *ST6Gal1* delayed the spontaneous development of malignant pancreatic lesions (Chakraborty et al. 2018).

Acknowledgements RT was partly supported by the Deutsche Forschungsgemeinschaft (DFG, German Research Foundation) SFB 1340/1 2018 "Matrix in Vision", No. 372486779, Project A06. CAR and HOD were partly supported by Portuguese funds through Fundação para a Ciência e a Tecnologia (FCT)/Ministério da Ciência, Tecnologia e Inovação through the research projects PTDC/MEC-ONC/0491/2021 and 2022.04138.PTDC. This work was in part funded by Programa Operacional Regional do Norte and co-funded by European Regional Development Fund under the project "The Porto Comprehensive Cancer Center" with the reference NORTE-01-0145-FEDER-072678—Consórcio PORTO.CCC.

Conflict of Interests There are no conflicts of interests to declare.

References

Abbott KL, Nairn AV, Hall EM, Horton MB, McDonald JF, Moremen KW, Dinulescu DM, Pierce M (2008) Focused glycomic analysis of the N-linked glycan biosynthetic pathway in ovarian cancer. Proteomics 8:3210–3220. https://doi.org/10.1002/pmic.200800157

Abrantes R, Duarte HO, Gomes C, Walchli S, Reis CA (2022) CAR-Ts: new perspectives in cancer therapy. FEBS Lett 596:403–416. https://doi.org/10.1002/1873-3468.14270

Adamczyk B, Tharmalingam T, Rudd PM (2012) Glycans as cancer biomarkers. Biochim Biophys Acta 1820:1347–1353. https://doi.org/10.1016/j.bbagen.2011.12.001

Afratis N, Gialeli C, Nikitovic D, Tsegenidis T, Karousou E, Theocharis AD, Pavao MS, Tzanakakis GN, Karamanos NK (2012) Glycosaminoglycans: key players in cancer cell biology and treatment. FEBS J 279:1177–1197. https://doi.org/10.1111/j.1742-4658.2012.08529.x

Allam H, Johnson BP, Zhang M, Lu Z, Cannon MJ, Abbott KL (2017) The glycosyltransferase GnT-III activates Notch signaling and drives stem cell expansion to promote the growth and invasion of ovarian cancer. J Biol Chem 292:16351–16359. https://doi.org/10.1074/jbc.M117.783936

Alley WR Jr, Vasseur JA, Goetz JA, Svoboda M, Mann BF, Matei DE, Menning N, Hussein A, Mechref Y, Novotny MV (2012) N-linked glycan structures and their expressions change in the blood sera of ovarian cancer patients. J Proteome Res 11:2282–2300. https://doi.org/10.1021/pr201070k

Anderluh M, Berti F, Bzducha-Wrobel A, Chiodo F, Colombo C, Compostella F, Durlik K, Ferhati X, Holmdahl R, Jovanovic D, Kaca W, Lay L, Marinovic-Cincovic M, Marradi M, Ozil M, Polito L, Reina JJ, Reis CA, Sackstein R, Silipo A, Svajger U, Vanek O, Yamamoto F, Richichi B, van Vliet SJ (2022) Recent advances on smart glycoconjugate vaccines in infections and cancer. FEBS J 289:4251–4303. https://doi.org/10.1111/febs.15909

Angata K, Fukuda M (2003) Polysialyltransferases: major players in polysialic acid synthesis on the neural cell adhesion molecule. Biochimie 85:195–206. https://doi.org/10.1016/s0300-9084(03)00051-8

Angata T, von Gunten S, Schnaar RL, Varki A (2022) I-Type Lectins. In: Varki A, Cummings RD, Esko JD, Stanley P, Hart GW, Aebi M, Mohnen D, Kinoshita T, Packer NH, Prestegard JH, Schnaar RL, Seeberger PH (eds) Essentials of glycobiology, 4th edn. Cold Spring Harbor Laboratory Press, Cold Spring Harbor, pp 475–490

Artavanis-Tsakonas S, Rand MD, Lake RJ (1999) Notch signaling: cell fate control and signal integration in development. Science 284:770–776. https://doi.org/10.1126/science.284.5415.770

Bai Q, Liu L, Xia Y, Long Q, Wang J, Xu J, Guo J (2015) Prognostic significance of ST3GAL-1 expression in patients with clear cell renal cell carcinoma. BMC Cancer 15:880. https://doi.org/10.1186/s12885-015-1906-5

Banerjee A, Mahata B, Dhir A, Mandal TK, Biswas K (2019) Elevated histone H3 acetylation and loss of the Sp1-HDAC1 complex de-repress the GM2-synthase gene in renal cell carcinoma. J Biol Chem 294:1005–1018. https://doi.org/10.1074/jbc.RA118.004485

Bartling B, Vanhooren V, Dewaele S, Libert C, Hofmann HS, Haerting J, Nuding S, Silber RE, Simm A, Chen CC (2011) Altered desialylated plasma N-glycan profile in patients with non-small cell lung carcinoma. Cancer Biomark 10:145–154. https://doi.org/10.3233/CBM-2012-0239

Bastian K, Scott E, Elliott DJ, Munkley J (2021) FUT8 Alpha-(1,6)-Fucosyltransferase in cancer. Int J Mol Sci 22. https://doi.org/10.3390/ijms22010455

Bauer CH, Vischer P, Grunholz H, Reutter W (1977) Glycosyltransferases and glycosidases in Morris hepatomas. Cancer Res 37:1513–1518

Beatson R, Graham R, Grundland Freile F, Cozzetto D, Kannambath S, Pfeifer E, Woodman N, Owen J, Nuamah R, Mandel U, Pinder S, Gillett C, Noll T, Bouybayoune I, Taylor-Papadimitriou J, Burchell JM (2020) Cancer-associated hypersialylated MUC1 drives the differentiation of human monocytes into macrophages with a pathogenic phenotype. Commun Biol 3:644. https://doi.org/10.1038/s42003-020-01359-5

Becker DJ, Lowe JB (2003) Fucose: biosynthesis and biological function in mammals. Glycobiology 13:41R–53R. https://doi.org/10.1093/glycob/cwg054

Bellis SL, Reis CA, Varki A, Kannagi R, Stanley P (2022) Glycosylation changes in cancer. In: Varki A, Cummings RD, Esko JD, Stanley P, Hart GW, Aebi M, Mohnen D, Kinoshita T, Packer NH, Prestegard JH, Schnaar RL, Seeberger PH (eds) Essentials of glycobiology, 4th edn. Cold Spring Harbor Laboratory Press, Cold Spring Harbor, pp 631–644

Benesova I, Nenutil R, Urminsky A, Lattova E, Uhrik L, Grell P, Kokas FZ, Halamkova J, Zdrahal Z, Vojtesek B, Novotny MV, Hernychova L (2024) N-glycan profiling of tissue samples to aid breast cancer subtyping. Sci Rep 14:320. https://doi.org/10.1038/s41598-023-51021-3

Bennett EP, Mandel U, Clausen H, Gerken TA, Fritz TA, Tabak LA (2012) Control of mucin-type O-glycosylation: a classification of the polypeptide GalNAc-transferase gene family. Glycobiology 22:736–756. https://doi.org/10.1093/glycob/cwr182

Berois N, Pittini A, Osinaga E (2022) Targeting tumor Glycans for cancer therapy: successes, limitations, and perspectives. Cancers (Basel) 14. https://doi.org/10.3390/cancers14030645

Bhalerao N, Chakraborty A, Marciel MP, Hwang J, Britain CM, Silva AD, Eltoum IE, Jones RB, Alexander KL, Smythies LE, Smith PD, Crossman DK, Crowley MR, Shin B, Harrington LE, Yan Z, Bethea MM, Hunter CS, Klug CA, Buchsbaum DJ, Bellis SL (2023) ST6GAL1 sialyltransferase promotes acinar to ductal metaplasia and pancreatic cancer progression. JCI Insight 8. https://doi.org/10.1172/jci.insight.161563

Bindeman WE, Fingleton B (2022) Glycosylation as a regulator of site-specific metastasis. Cancer Metastasis Rev 41:107–129. https://doi.org/10.1007/s10555-021-10015-1

Birks SM, Danquah JO, King L, Vlasak R, Gorecki DC, Pilkington GJ (2011) Targeting the GD3 acetylation pathway selectively induces apoptosis in glioblastoma. Neuro-Oncology 13:950–960. https://doi.org/10.1093/neuonc/nor108

Biskup K, Braicu EI, Sehouli J, Fotopoulou C, Tauber R, Berger M, Blanchard V (2013) Serum glycome profiling: a biomarker for diagnosis of ovarian cancer. J Proteome Res 12:4056–4063. https://doi.org/10.1021/pr400405x

Biskup K, Braicu EI, Sehouli J, Tauber R, Blanchard V (2014) The serum glycome to discriminate between early-stage epithelial ovarian cancer and benign ovarian diseases. Dis Markers 2014: 238197. https://doi.org/10.1155/2014/238197

Biskup K, Stellmach C, Braicu EI, Sehouli J, Blanchard V (2021) Chondroitin sulfate disaccharides, a serum marker for primary serous epithelial ovarian cancer. Diagnostics (Basel) 11. https://doi.org/10.3390/diagnostics11071143

Bobowski M, Vincent A, Steenackers A, Colomb F, Van Seuningen I, Julien S, Delannoy P (2013) Estradiol represses the G(D3) synthase gene ST8SIA1 expression in human breast cancer cells by preventing NFkappaB binding to ST8SIA1 promoter. PLoS One 8:e62559. https://doi.org/10.1371/journal.pone.0062559

Boelaars K, van Kooyk Y (2024) Targeting myeloid cells for cancer immunotherapy: Siglec-7/9/10/15 and their ligands. Trends Cancer 10:230–241. https://doi.org/10.1016/j.trecan.2023.11.009

Boyaval F, Dalebout H, Van Zeijl R, Wang W, Farina-Sarasqueta A, Lageveen-Kammeijer GSM, Boonstra JJ, McDonnell LA, Wuhrer M, Morreau H, Heijs B (2022) High-mannose N-Glycans as malignant progression markers in early-stage colorectal cancer. Cancers (Basel) 14. https://doi.org/10.3390/cancers14061552

Britain CM, Holdbrooks AT, Anderson JC, Willey CD, Bellis SL (2018) Sialylation of EGFR by the ST6Gal-I sialyltransferase promotes EGFR activation and resistance to gefitinib-mediated cell death. J Ovarian Res 11:12. https://doi.org/10.1186/s13048-018-0385-0

Brockhausen I, Wandall HH, Hagen KGT, Stanley P (2022) O-GalNAc glycans. In: Varki A, Cummings RD, Esko JD, Stanley P, Hart GW, Aebi M, Mohnen D, Kinoshita T, Packer NH, Prestegard JH, Schnaar RL, Seeberger PH (eds) Essentials of glycobiology, 4th edn. Cold Spring Harbor Laboratory Press, Cold Spring Harbor, pp 117–128

Bull C, den Brok MH, Adema GJ (2014) Sweet escape: sialic acids in tumor immune evasion. Biochim Biophys Acta 1846:238–246. https://doi.org/10.1016/j.bbcan.2014.07.005

Bull C, Nason R, Sun L, Van Coillie J, Madriz Sorensen D, Moons SJ, Yang Z, Arbitman S, Fernandes SM, Furukawa S, McBride R, Nycholat CM, Adema GJ, Paulson JC, Schnaar RL, Boltje TJ, Clausen H, Narimatsu Y (2021) Probing the binding specificities of human Siglecs by cell-based glycan arrays. Proc Natl Acad Sci USA 118. https://doi.org/10.1073/pnas.2026102118

Cai X, Thinn AMM, Wang Z, Shan H, Zhu J (2017) The importance of N-glycosylation on beta (3) integrin ligand binding and conformational regulation. Sci Rep 7:4656. https://doi.org/10.1038/s41598-017-04844-w

Campos D, Freitas D, Gomes J, Magalhaes A, Steentoft C, Gomes C, Vester-Christensen MB, Ferreira JA, Afonso LP, Santos LL, Pinto de Sousa J, Mandel U, Clausen H, Vakhrushev SY, Reis CA (2015) Probing the O-glycoproteome of gastric cancer cell lines for biomarker discovery. Mol Cell Proteomics 14:1616–1629. https://doi.org/10.1074/mcp.M114.046862

Carvalho S, Catarino TA, Dias AM, Kato M, Almeida A, Hessling B, Figueiredo J, Gartner F, Sanches JM, Ruppert T, Miyoshi E, Pierce M, Carneiro F, Kolarich D, Seruca R, Yamaguchi Y, Taniguchi N, Reis CA, Pinho SS (2016a) Preventing E-cadherin aberrant N-glycosylation at Asn-554 improves its critical function in gastric cancer. Oncogene 35:1619–1631. https://doi.org/10.1038/onc.2015.225

Carvalho S, Reis CA, Pinho SS (2016b) Cadherins glycans in cancer: sweet players in a bitter process. Trends Cancer 2:519–531. https://doi.org/10.1016/j.trecan.2016.08.003

Chakraborty AK, Sousa Jde F, Chakraborty D, Funasaka Y, Bhattacharya M, Chatterjee A, Pawelek J (2006) GnT-V expression and metastatic phenotypes in macrophage-melanoma fusion hybrids is down-regulated by 5-Aza-dC: evidence for methylation sensitive, extragenic regulation of GnT-V transcription. Gene 374:166–173. https://doi.org/10.1016/j.gene.2006.01.031

Chakraborty A, Dorsett KA, Trummell HQ, Yang ES, Oliver PG, Bonner JA, Buchsbaum DJ, Bellis SL (2018) ST6Gal-I sialyltransferase promotes chemoresistance in pancreatic ductal adenocarcinoma by abrogating gemcitabine-mediated DNA damage. J Biol Chem 293:984–994. https://doi.org/10.1074/jbc.M117.808584

Cheng B, Tang Q, Zhang C, Chen X (2021) Glycan labeling and analysis in cells and in vivo. Annu Rev Anal Chem (Palo Alto, Calif) 14:363–387. https://doi.org/10.1146/annurev-anchem-091620-091314

Cheng Q, Hu X, Zhang X, Yang D, Zhao G, Sun L, Jiang M, Yang L, Cai J, Wang B, Zhang M, Han F, Li Y, Nie H (2023) N-glycosylation at N57/100/110 affects CD44s localization, function and stability in hepatocellular carcinoma. Eur J Cell Biol 102:151360. https://doi.org/10.1016/j.ejcb.2023.151360

Chiricolo M, Malagolini N, Bonfiglioli S, Dall'Olio F (2006) Phenotypic changes induced by expression of beta-galactoside alpha2,6 sialyltransferase I in the human colon cancer cell line SW948. Glycobiology 16:146–154. https://doi.org/10.1093/glycob/cwj045

Chu Q, Liu L, Wang W (2013) Overexpression of hCLP46 enhances Notch activation and regulates cell proliferation in a cell type-dependent manner. Cell Prolif 46:254–262. https://doi.org/10.1111/cpr.12037

Chugh S, Meza J, Sheinin YM, Ponnusamy MP, Batra SK (2016) Loss of N-acetylgalactosaminyltransferase 3 in poorly differentiated pancreatic cancer: augmented aggressiveness and aberrant ErbB family glycosylation. Br J Cancer 114:1376–1386. https://doi.org/10.1038/bjc.2016.116

Clausen H, Bennett EP (1996) A family of UDP-GalNAc: polypeptide N-acetylgalactosaminyl-transferases control the initiation of mucin-type O-linked glycosylation. Glycobiology 6:635–646. https://doi.org/10.1093/glycob/6.6.635

Coccimiglio M, Chiodo F, van Kooyk Y (2024) The sialic acid-Siglec immune checkpoint: an opportunity to enhance immune responses and therapy effectiveness in melanoma. Br J Dermatol 190:627–635. https://doi.org/10.1093/bjd/ljad517

Conze T, Carvalho AS, Landegren U, Almeida R, Reis CA, David L, Soderberg O (2010) MUC2 mucin is a major carrier of the cancer-associated sialyl-Tn antigen in intestinal metaplasia and gastric carcinomas. Glycobiology 20:199–206. https://doi.org/10.1093/glycob/cwp161

Cotton S, Azevedo R, Gaiteiro C, Ferreira D, Lima L, Peixoto A, Fernandes E, Neves M, Neves D, Amaro T, Cruz R, Tavares A, Rangel M, Silva AMN, Santos LL, Ferreira JA (2017) Targeted O-glycoproteomics explored increased sialylation and identified MUC16 as a poor prognosis biomarker in advanced-stage bladder tumours. Mol Oncol 11:895–912. https://doi.org/10.1002/1878-0261.12035

Couchman JR (2010) Transmembrane signaling proteoglycans. Annu Rev Cell Dev Biol 26:89–114. https://doi.org/10.1146/annurev-cellbio-100109-104126

Croce MV, Isla-Larrain M, Rabassa ME, Demichelis S, Colussi AG, Crespo M, Lacunza E, Segal-Eiras A (2007) Lewis x is highly expressed in normal tissues: a comparative immunohistochemical study and literature revision. Pathol Oncol Res 13:130–138. https://doi.org/10.1007/BF02893488

Cummings RD, Chiffoleau E, van Kooyk Y, McEver RP (2022a) C-type lectins. In: Varki A, Cummings RD, Esko JD, Stanley P, Hart GW, Aebi M, Mohnen D, Kinoshita T, Packer NH, Prestegard JH, Schnaar RL, Seeberger PH (eds) Essentials of glycobiology, 4th edn. Cold Spring Harbor Laboratory Press, Cold Spring Harbor, pp 455–474

Cummings RD, Liu FT, Rabinovich GA, Stowell SR, Vasta GR (2022b) Galectins. In: Varki A, Cummings RD, Esko JD, Stanley P, Hart GW, Aebi M, Mohnen D, Kinoshita T, Packer NH, Prestegard JH, Schnaar RL, Seeberger PH (eds) Essentials of glycobiology, 4th edn. Cold Spring Harbor, Cold Spring Harbor Laboratory Press, pp 491–504

da Costa V, van Vliet SJ, Carasi P, Frigerio S, Garcia PA, Croci DO, Festari MF, Costa M, Landeira M, Rodriguez-Zraquia SA, Cagnoni AJ, Cutine AM, Rabinovich GA, Osinaga E, Marino KV, Freire T (2021) The Tn antigen promotes lung tumor growth by fostering immunosuppression and angiogenesis via interaction with Macrophage Galactose-type lectin 2 (MGL2). Cancer Lett 518:72–81. https://doi.org/10.1016/j.canlet.2021.06.012

Dae HM, Kwon HY, Kang NY, Song NR, Kim KS, Kim CH, Lee JH, Lee YC (2009) Isolation and functional analysis of the human glioblastoma-specific promoter region of the human GD3 synthase (hST8Sia I) gene. Acta Biochim Biophys Sin Shanghai 41:237–245. https://doi.org/10.1093/abbs/gmp007

Dall'Olio F, Chiricolo M, D'Errico A, Gruppioni E, Altimari A, Fiorentino M, Grigioni WF (2004) Expression of beta-galactoside alpha2,6 sialyltransferase and of alpha2,6-sialylated glycoconjugates in normal human liver, hepatocarcinoma, and cirrhosis. Glycobiology 14:39–49. https://doi.org/10.1093/glycob/cwh002

Dan W, Li C, Li J, Li P, Xin M, Chen Z, Dang L, Yu Z, Li J, Shen J, Hu L, Sun S (2022) Glycoproteomic analysis reveals the effects of bisecting GlcNAc in intrahepatic cholangiocarcinoma. Glycoconj J 39:737–745. https://doi.org/10.1007/s10719-022-10085-5

de Haan N, Falck D, Wuhrer M (2020) Monitoring of immunoglobulin N- and O-glycosylation in health and disease. Glycobiology 30:226–240. https://doi.org/10.1093/glycob/cwz048

Dědová T, Braicu EI, Sehouli J, Blanchard V (2019) Sialic acid linkage analysis refines the diagnosis of ovarian cancer. Front Oncol 9:261–261. https://doi.org/10.3389/fonc.2019.00261

Denda-Nagai K, Aida S, Saba K, Suzuki K, Moriyama S, Oo-Puthinan S, Tsuiji M, Morikawa A, Kumamoto Y, Sugiura D, Kudo A, Akimoto Y, Kawakami H, Bovin NV, Irimura T (2010) Distribution and function of macrophage galactose-type C-type lectin 2 (MGL2/CD301b): efficient uptake and presentation of glycosylated antigens by dendritic cells. J Biol Chem 285:19193–19204. https://doi.org/10.1074/jbc.M110.113613

Dennis JW, Laferte S, Waghorne C, Breitman ML, Kerbel RS (1987) Beta 1-6 branching of Asn-linked oligosaccharides is directly associated with metastasis. Science 236:582–585. https://doi.org/10.1126/science.2953071

Dennis JW, Granovsky M, Warren CE (1999) Glycoprotein glycosylation and cancer progression. Biochim Biophys Acta 1473:21–34. https://doi.org/10.1016/s0304-4165(99)00167-1

de-Souza-Ferreira M, Ferreira EE, de-Freitas-Junior JCM (2023) Aberrant N-glycosylation in cancer: MGAT5 and beta1,6-GlcNAc branched N-glycans as critical regulators of tumor development and progression. Cell Oncol (Dordr) 46:481–501. https://doi.org/10.1007/s13402-023-00770-4

Dimitroff CJ, Lechpammer M, Long-Woodward D, Kutok JL (2004) Rolling of human bone-metastatic prostate tumor cells on human bone marrow endothelium under shear flow is mediated by E-selectin. Cancer Res 64:5261–5269. https://doi.org/10.1158/0008-5472.CAN-04-0691

Dimitroff CJ, Descheny L, Trujillo N, Kim R, Nguyen V, Huang W, Pienta KJ, Kutok JL, Rubin MA (2005) Identification of leukocyte E-selectin ligands, P-selectin glycoprotein ligand-1 and E-selectin ligand-1, on human metastatic prostate tumor cells. Cancer Res 65:5750–5760. https://doi.org/10.1158/0008-5472.CAN-04-4653

Ding Y, Gelfenbeyn K, Freire-de-Lima L, Handa K, Hakomori SI (2012) Induction of epithelial-mesenchymal transition with O-glycosylated oncofetal fibronectin. FEBS Lett 586:1813–1820. https://doi.org/10.1016/j.febslet.2012.05.020

Drake RR, Powers TW, Jones EE, Bruner E, Mehta AS, Angel PM (2017) MALDI mass spectrometry imaging of N-linked Glycans in cancer tissues. Adv Cancer Res 134:85–116. https://doi.org/10.1016/bs.acr.2016.11.009

Duan S, Paulson JC (2020) Siglecs as immune cell checkpoints in disease. Annu Rev Immunol 38:365–395. https://doi.org/10.1146/annurev-immunol-102419-035900

Duarte HO, Rodrigues JG, Gomes C, Hensbergen PJ, Ederveen ALH, de Ru AH, Mereiter S, Polonia A, Fernandes E, Ferreira JA, van Veelen PA, Santos LL, Wuhrer M, Gomes J, Reis CA (2021) ST6Gal1 targets the ectodomain of ErbB2 in a site-specific manner and regulates gastric cancer cell sensitivity to trastuzumab. Oncogene 40:3719–3733. https://doi.org/10.1038/s41388-021-01801-w

Duarte HO, Reis CA, Gomes J (2022) Insights on ErbB glycosylation – contributions to precision oncology. Trends Cancer 8:448–455. https://doi.org/10.1016/j.trecan.2022.02.003

Duarte HO, Gomes J, Reis CA (2023) Targeting carbohydrates in cancer–analytical and biotechnological tools. In: L ARaL (ed) Carbohydrate-based therapeutics, pp 161–200

Dube DH, Bertozzi CR (2003) Metabolic oligosaccharide engineering as a tool for glycobiology. Curr Opin Chem Biol 7:616–625. https://doi.org/10.1016/j.cbpa.2003.08.006

Duinkerken S, Li RE, van Haften FJ, de Gruijl TD, Chiodo F, Schetters STT, van Kooyk Y (2019) Chemically engineered glycan-modified cancer vaccines to mobilize skin dendritic cells. Curr Opin Chem Biol 53:167–172. https://doi.org/10.1016/j.cbpa.2019.10.001

Eisterer W, Bechter O, Hilbe W, van Driel M, Lokhorst HM, Thaler J, Bloem AC, Gunthert U, Stauder R (2001) CD44 isoforms are differentially regulated in plasma cell dyscrasias and CD44v9 represents a new independent prognostic parameter in multiple myeloma. Leuk Res 25:1051–1057. https://doi.org/10.1016/s0145-2126(01)00075-3

Elola MT, Ferragut F, Mendez-Huergo SP, Croci DO, Bracalente C, Rabinovich GA (2018) Galectins: multitask signaling molecules linking fibroblast, endothelial and immune cell programs in the tumor microenvironment. Cell Immunol 333:34–45. https://doi.org/10.1016/j.cellimm.2018.03.008

Engle DD, Tiriac H, Rivera KD, Pommier A, Whalen S, Oni TE, Alagesan B, Lee EJ, Yao MA, Lucito MS, Spielman B, Da Silva B, Schoepfer C, Wright K, Creighton B, Afinowicz L, Yu KH, Grutzmann R, Aust D, Gimotty PA, Pollard KS, Hruban RH, Goggins MG, Pilarsky C, Park Y, Pappin DJ, Hollingsworth MA, Tuveson DA (2019) The glycan CA19-9 promotes pancreatitis and pancreatic cancer in mice. Science 364:1156–1162. https://doi.org/10.1126/science.aaw3145

English NM, Lesley JF, Hyman R (1998) Site-specific de-N-glycosylation of CD44 can activate hyaluronan binding, and CD44 activation states show distinct threshold densities for hyaluronan binding. Cancer Res 58:3736–3742

Esposito M, Mondal N, Greco TM, Wei Y, Spadazzi C, Lin SC, Zheng H, Cheung C, Magnani JL, Lin SH, Cristea IM, Sackstein R, Kang Y (2019) Bone vascular niche E-selectin induces mesenchymal-epithelial transition and Wnt activation in cancer cells to promote bone metastasis. Nat Cell Biol 21:627–639. https://doi.org/10.1038/s41556-019-0309-2

Furukawa K, Ohmi Y, Hamamura K, Kondo Y, Ohkawa Y, Kaneko K, Hashimoto N, Yesmin F, Bhuiyan RH, Tajima O, Furukawa K (2022) Signaling domains of cancer-associated glycolipids. Glycoconj J 39:145–155. https://doi.org/10.1007/s10719-022-10051-1

Gagneux P, Hennet T, Varki A (2022) Biological functions of glycans. In: Varki A, Cummings RD, Esko JD, Stanley P, Hart GW, Aebi M, Mohnen D, Kinoshita T, Packer NH, Prestegard JH, Schnaar RL, Seeberger PH (eds) Essentials of glycobiology, 4th edn. Cold Spring Harbor Laboratory Press, Cold Spring Harbor, pp 79–92

Galang CK, Garcia-Ramirez J, Solski PA, Westwick JK, Der CJ, Neznanov NN, Oshima RG, Hauser CA (1996) Oncogenic Neu/ErbB-2 increases ets, AP-1, and NF-kappaB-dependent gene expression, and inhibiting ets activation blocks Neu-mediated cellular transformation. J Biol Chem 271:7992–7998. https://doi.org/10.1074/jbc.271.14.7992

Gao Y, Luan X, Melamed J, Brockhausen I (2021) Role of Glycans on key cell surface receptors that regulate cell proliferation and cell death. Cells 10. https://doi.org/10.3390/cells10051252

Garcia-Castro MI, Vielmetter E, Bronner-Fraser M (2000) N-cadherin, a cell adhesion molecule involved in establishment of embryonic left-right asymmetry. Science 288:1047–1051. https://doi.org/10.1126/science.288.5468.1047

Garcia-Vallejo JJ, van Kooyk Y (2013) The physiological role of DC-SIGN: a tale of mice and men. Trends Immunol 34:482–486. https://doi.org/10.1016/j.it.2013.03.001

Garcia-Vallejo JJ, Ilarregui JM, Kalay H, Chamorro S, Koning N, Unger WW, Ambrosini M, Montserrat V, Fernandes RJ, Bruijns SC, van Weering JR, Paauw NJ, O'Toole T, van Horssen J, van der Valk P, Nazmi K, Bolscher JG, Bajramovic J, Dijkstra CD, t Hart BA, van Kooyk Y (2014) CNS myelin induces regulatory functions of DC-SIGN-expressing, antigen-presenting cells via cognate interaction with MOG. J Exp Med 211:1465–1483. https://doi.org/10.1084/jem.20122192

Gardelli C, Russo L, Cipolla L, Moro M, Andriani F, Rondinone O, Nicotra F, Sozzi G, Bertolini G, Roz L (2021) Differential glycosylation of collagen modulates lung cancer stem cell subsets through beta1 integrin-mediated interactions. Cancer Sci 112:217–230. https://doi.org/10.1111/cas.14700

Geijtenbeek TB, Torensma R, van Vliet SJ, van Duijnhoven GC, Adema GJ, van Kooyk Y, Figdor CG (2000) Identification of DC-SIGN, a novel dendritic cell-specific ICAM-3 receptor that supports primary immune responses. Cell 100:575–585. https://doi.org/10.1016/s0092-8674(00)80693-5

Ghasempour S, Freeman SA (2023) The glycocalyx and immune evasion in cancer. FEBS J 290: 55–65. https://doi.org/10.1111/febs.16236

Gill DJ, Chia J, Senewiratne J, Bard F (2010) Regulation of O-glycosylation through Golgi-to-ER relocation of initiation enzymes. J Cell Biol 189:843–858. https://doi.org/10.1083/jcb.201003055

Goldman R, Ressom HW, Varghese RS, Goldman L, Bascug G, Loffredo CA, Abdel-Hamid M, Gouda I, Ezzat S, Kyselova Z, Mechref Y, Novotny MV (2009) Detection of hepatocellular carcinoma using glycomic analysis. Clin Cancer Res 15:1808–1813. https://doi.org/10.1158/1078-0432.CCR-07-5261

Granovsky M, Fata J, Pawling J, Muller WJ, Khokha R, Dennis JW (2000) Suppression of tumor growth and metastasis in Mgat5-deficient mice. Nat Med 6:306–312. https://doi.org/10.1038/73163

Greco B, Malacarne V, De Girardi F, Scotti GM, Manfredi F, Angelino E, Sirini C, Camisa B, Falcone L, Moresco MA, Paolella K, Di Bono M, Norata R, Sanvito F, Arcangeli S, Doglioni C, Ciceri F, Bonini C, Graziani A, Bondanza A, Casucci M (2022) Disrupting N-glycan expression on tumor cells boosts chimeric antigen receptor T cell efficacy against solid malignancies. Sci Transl Med 14:eabg3072. https://doi.org/10.1126/scitranslmed.abg3072

Gretschel S, Haensch W, Schlag PM, Kemmner W (2003) Clinical relevance of sialyltransferases ST6GAL-I and ST3GAL-III in gastric cancer. Oncology 65:139–145. https://doi.org/10.1159/000072339

Gringhuis SI, Kaptein TM, Wevers BA, van der Vlist M, Klaver EJ, van Die I, Vriend LE, de Jong MA, Geijtenbeek TB (2014) Fucose-based PAMPs prime dendritic cells for follicular T helper cell polarization via DC-SIGN-dependent IL-27 production. Nat Commun 5:5074. https://doi.org/10.1038/ncomms6074

Groux-Degroote S, Delannoy P (2021) Cancer-associated Glycosphingolipids as tumor markers and targets for cancer immunotherapy. Int J Mol Sci 22. https://doi.org/10.3390/ijms22116145

Grzeski M, Taube ET, Braicu EI, Sehouli J, Blanchard V, Klein O (2022) In situ N-glycosylation signatures of epithelial ovarian cancer tissue as defined by MALDI mass spectrometry imaging. Cancers (Basel) 14. https://doi.org/10.3390/cancers14041021

Guan F, Wang X, He F (2015) Promotion of cell migration by neural cell adhesion molecule (NCAM) is enhanced by PSA in a polysialyltransferase-specific manner. PLoS One 10: e0124237. https://doi.org/10.1371/journal.pone.0124237

Guereno M, Delgado Pastore M, Lugones AC, Cercato M, Todaro L, Urtreger A, Peters MG (2020) Glypican-3 (GPC3) inhibits metastasis development promoting dormancy in breast cancer cells by p38 MAPK pathway activation. Eur J Cell Biol 99:151096. https://doi.org/10.1016/j.ejcb.2020.151096

Guo HB, Lee I, Kamar M, Akiyama SK, Pierce M (2002) Aberrant N-glycosylation of beta1 integrin causes reduced alpha5beta1 integrin clustering and stimulates cell migration. Cancer Res 62:6837–6845

Guo HB, Lee I, Kamar M, Pierce M (2003) N-acetylglucosaminyltransferase V expression levels regulate cadherin-associated homotypic cell-cell adhesion and intracellular signaling pathways. J Biol Chem 278:52412–52424. https://doi.org/10.1074/jbc.M308837200

Guo HB, Johnson H, Randolph M, Pierce M (2009) Regulation of homotypic cell-cell adhesion by branched N-glycosylation of N-cadherin extracellular EC2 and EC3 domains. J Biol Chem 284: 34986–34997. https://doi.org/10.1074/jbc.M109.060806

Guo HB, Johnson H, Randolph M, Nagy T, Blalock R, Pierce M (2010) Specific posttranslational modification regulates early events in mammary carcinoma formation. Proc Natl Acad Sci USA 107:21116–21121. https://doi.org/10.1073/pnas.1013405107

Gupta I, Hussein O, Sastry KS, Bougarn S, Gopinath N, Chin-Smith E, Sinha Y, Korashy HM, Maccalli C (2023) Deciphering the complexities of cancer cell immune evasion: mechanisms and therapeutic implications. Adv Cancer Bio – Metastasis 8:1–27

Haltiwanger RS, Wells L, Freeze HH, Jafar-Nejad H, Okajima T, Stanley P (2022) Other classes of eukaryotic glycans. In: Varki A, Cummings RD, Esko JD, Stanley P, Hart GW, Aebi M, Mohnen D, Kinoshita T, Packer NH, Prestegard JH, Schnaar RL, Seeberger PH (eds) Essentials of glycobiology, 4th edn. Cold Spring Harbor Laboratory Press, Cold Spring Harbor, pp 155–164

Han J, Bhatta R, Liu Y, Bo Y, Elosegui-Artola A, Wang H (2023) Metabolic glycan labeling immobilizes dendritic cell membrane and enhances antitumor efficacy of dendritic cell vaccine. Nat Commun 14:5049. https://doi.org/10.1038/s41467-023-40886-7

Hanafusa K, Hotta T, Iwabuchi K (2020) Glycolipids: linchpins in the organization and function of membrane microdomains. Front Cell Dev Biol 8:589799. https://doi.org/10.3389/fcell.2020.589799

Hanahan D, Weinberg RA (2011) Hallmarks of cancer: the next generation. Cell 144:646–674. https://doi.org/10.1016/j.cell.2011.02.013

Hart GW, Housley MP, Slawson C (2007) Cycling of O-linked beta-N-acetylglucosamine on nucleocytoplasmic proteins. Nature 446:1017–1022. https://doi.org/10.1038/nature05815

Hart GW, Slawson C, Ramirez-Correa G, Lagerlof O (2011) Cross talk between O-GlcNAcylation and phosphorylation: roles in signaling, transcription, and chronic disease. Annu Rev Biochem 80:825–858. https://doi.org/10.1146/annurev-biochem-060608-102511

Hashiguchi Y, Kasai M, Fukuda T, Ichimura T, Yasui T, Sumi T (2016) Serum Sialyl-Tn (STN) as a tumor marker in patients with endometrial cancer. Pathol Oncol Res 22:501–504. https://doi.org/10.1007/s12253-015-0030-9

Hassinen A, Pujol FM, Kokkonen N, Pieters C, Kihlstrom M, Korhonen K, Kellokumpu S (2011) Functional organization of Golgi N- and O-glycosylation pathways involves pH-dependent complex formation that is impaired in cancer cells. J Biol Chem 286:38329–38340. https://doi.org/10.1074/jbc.M111.277681

Hauselmann I, Roblek M, Protsyuk D, Huck V, Knopfova L, Grassle S, Bauer AT, Schneider SW, Borsig L (2016) Monocyte induction of E-Selectin-mediated endothelial activation releases VE-cadherin junctions to promote tumor cell extravasation in the metastasis Cascade. Cancer Res 76:5302–5312. https://doi.org/10.1158/0008-5472.CAN-16-0784

He XF, Hu X, Wen GJ, Wang Z, Lin WJ (2023) O-GlcNAcylation in cancer development and immunotherapy. Cancer Lett 566:216258. https://doi.org/10.1016/j.canlet.2023.216258

Hsieh CC, Shyr YM, Liao WY, Chen TH, Wang SE, Lu PC, Lin PY, Chen YB, Mao WY, Han HY, Hsiao M, Yang WB, Li WS, Sher YP, Shen CN (2017) Elevation of beta-galactoside alpha2,6-sialyltransferase 1 in a fructoseresponsive manner promotes pancreatic cancer metastasis. Oncotarget 8:7691–7709. https://doi.org/10.18632/oncotarget.13845

Hua S, Williams CC, Dimapasoc LM, Ro GS, Ozcan S, Miyamoto S, Lebrilla CB, An HJ, Leiserowitz GS (2013) Isomer-specific chromatographic profiling yields highly sensitive and specific potential N-glycan biomarkers for epithelial ovarian cancer. J Chromatogr A 1279:58–67. https://doi.org/10.1016/j.chroma.2012.12.079

Huang HC, Chao CC, Wu PH, Chung HY, Lee HY, Suen CS, Hwang MJ, Cai BH, Kannagi R (2019) Epigenetic silencing of the synthesis of immunosuppressive Siglec ligand glycans by NF-kappaB/EZH2/YY1 axis in early-stage colon cancers. Biochim Biophys Acta Gene Regul Mech 1862:173–183. https://doi.org/10.1016/j.bbagrm.2019.01.002

Huang YL, Liang CY, Labitzky V, Ritz D, Oliveira T, Cumin C, Estermann M, Lange T, Everest-Dass AV, Jacob F (2021) Site-specific N-glycosylation of integrin alpha2 mediates collagen-dependent cell survival. iScience 24:103168. https://doi.org/10.1016/j.isci.2021.103168

Ibarlucea-Benitez I, Weitzenfeld P, Smith P, Ravetch JV (2021) Siglecs-7/9 function as inhibitory immune checkpoints in vivo and can be targeted to enhance therapeutic antitumor immunity. Proc Natl Acad Sci USA 118. https://doi.org/10.1073/pnas.2107424118

Ilarregui JM, Croci DO, Bianco GA, Toscano MA, Salatino M, Vermeulen ME, Geffner JR, Rabinovich GA (2009) Tolerogenic signals delivered by dendritic cells to T cells through a galectin-1-driven immunoregulatory circuit involving interleukin 27 and interleukin 10. Nat Immunol 10:981–991. https://doi.org/10.1038/ni.1772

Indellicato R, Trinchera M (2021) Epigenetic regulation of glycosylation in cancer and other diseases. Int J Mol Sci 22. https://doi.org/10.3390/ijms22062980

Isaji T, Gu J, Nishiuchi R, Zhao Y, Takahashi M, Miyoshi E, Honke K, Sekiguchi K, Taniguchi N (2004) Introduction of bisecting GlcNAc into integrin alpha5beta1 reduces ligand binding and down-regulates cell adhesion and cell migration. J Biol Chem 279:19747–19754. https://doi.org/10.1074/jbc.M311627200

Isaji T, Im S, Kameyama A, Wang Y, Fukuda T, Gu J (2019) A complex between phosphatidylinositol 4-kinase IIalpha and integrin alpha3beta1 is required for N-glycan sialylation in cancer cells. J Biol Chem 294:4425–4436. https://doi.org/10.1074/jbc.RA118.005208

Jian Y, Zhang L, Gong L, Ding M, Sun X, Yu X, Lv S, Li J, Yang D, Wang S (2023) CD56 polysialylation promotes the tumorigenesis and progression via the Hedgehog and Wnt/beta-catenin signaling pathways in clear cell renal cell carcinoma. Cancer Cell Int 23:319. https://doi.org/10.1186/s12935-023-03165-5

Josic D, Martinovic T, Pavelic K (2019) Glycosylation and metastases. Electrophoresis 40:140–150. https://doi.org/10.1002/elps.201800238

Ju T, Lanneau GS, Gautam T, Wang Y, Xia B, Stowell SR, Willard MT, Wang W, Xia JY, Zuna RE, Laszik Z, Benbrook DM, Hanigan MH, Cummings RD (2008) Human tumor antigens Tn and sialyl Tn arise from mutations in Cosmc. Cancer Res 68:1636–1646. https://doi.org/10.1158/0008-5472.CAN-07-2345

Kariya Y, Kawamura C, Tabei T, Gu J (2010) Bisecting GlcNAc residues on laminin-332 down-regulate galectin-3-dependent keratinocyte motility. J Biol Chem 285:3330–3340. https://doi.org/10.1074/jbc.M109.038836

Katayama Y, Hidalgo A, Furie BC, Vestweber D, Furie B, Frenette PS (2003) PSGL-1 participates in E-selectin-mediated progenitor homing to bone marrow: evidence for cooperation between E-selectin ligands and alpha4 integrin. Blood 102:2060–2067. https://doi.org/10.1182/blood-2003-04-1212

Kaur S, Kumar S, Momi N, Sasson AR, Batra SK (2013) Mucins in pancreatic cancer and its microenvironment. Nat Rev Gastroenterol Hepatol 10:607–620. https://doi.org/10.1038/nrgastro.2013.120

Khare V, Lang M, Dammann K, Campregher C, Lyakhovich A, Gasche C (2014) Modulation of N-glycosylation by mesalamine facilitates membranous E-cadherin expression in colon epithelial cells. Biochem Pharmacol 87:312–320. https://doi.org/10.1016/j.bcp.2013.10.021

Kim SK, Cho SW (2022) The evasion mechanisms of cancer immunity and drug intervention in the tumor microenvironment. Front Pharmacol 13:868695. https://doi.org/10.3389/fphar.2022.868695

Kim JH, Park CW, Um D, Baek KH, Jo Y, An H, Kim Y, Kim TJ (2014) Mass spectrometric screening of ovarian cancer with serum glycans. Dis Markers 2014:634289. https://doi.org/10.1155/2014/634289

Kizuka Y, Taniguchi N (2016) Enzymes for N-glycan branching and their genetic and nongenetic regulation in cancer. Biomol Ther 6. https://doi.org/10.3390/biom6020025

Kohler S, Ullrich S, Richter U, Schumacher U (2010) E-/P-selectins and colon carcinoma metastasis: first in vivo evidence for their crucial role in a clinically relevant model of spontaneous metastasis formation in the lung. Br J Cancer 102:602–609. https://doi.org/10.1038/sj.bjc.6605492

Komath SS, Fujita M, Hart GW, Ferguson MAJ, Kinoshita T (2022) Glycosylphosphatidylinositol anchors. In: Varki A, Cummings RD, Esko JD, Stanley P, Hart GW, Aebi M, Mohnen D, Kinoshita T, Packer NH, Prestegard JH, Schnaar RL, Seeberger PH (eds) Essentials of glycobiology, 4th edn. Cold Spring Harbor Laboratory Press, Cold Spring Harbor, pp 141–154

Krug J, Rodrian G, Petter K, Yang H, Khoziainova S, Guo W, Benard A, Merkel S, Gellert S, Maschauer S, Spermann M, Waldner M, Bailey P, Pilarsky C, Liebl A, Tripal P, Christoph J, Naschberger E, Croner R, Schellerer VS, Becker C, Hartmann A, Tuting T, Prante O, Grutzmann R, Grivennikov SI, Sturzl M, Britzen-Laurent N (2023) N-glycosylation regulates intrinsic IFN-gamma resistance in colorectal cancer: implications for immunotherapy. Gastroenterology 164(392-406):e5. https://doi.org/10.1053/j.gastro.2022.11.018

Kuo TC, Wu MH, Yang SH, Chen ST, Hsu TW, Jhuang JY, Liao YY, Tien YW, Huang MC (2021) C1GALT1 high expression is associated with poor survival of patients with pancreatic ductal adenocarcinoma and promotes cell invasiveness through integrin alpha(v). Oncogene 40:1242–1254. https://doi.org/10.1038/s41388-020-01594-4

Larsen ISB, Narimatsu Y, Joshi HJ, Siukstaite L, Harrison OJ, Brasch J, Goodman KM, Hansen L, Shapiro L, Honig B, Vakhrushev SY, Clausen H, Halim A (2017a) Discovery of an O-mannosylation pathway selectively serving cadherins and protocadherins. Proc Natl Acad Sci USA 114:11163–11168. https://doi.org/10.1073/pnas.1708319114

Larsen ISB, Narimatsu Y, Joshi HJ, Yang Z, Harrison OJ, Brasch J, Shapiro L, Honig B, Vakhrushev SY, Clausen H, Halim A (2017b) Mammalian O-mannosylation of cadherins and plexins is independent of protein O-mannosyltransferases 1 and 2. J Biol Chem 292:11586–11598. https://doi.org/10.1074/jbc.M117.794487

Lebrilla CB, Liu J, Widmalm G, Prestegard JH (2022) Oligosaccharides and polysaccharides. In: Varki A, Cummings RD, Esko JD, Stanley P, Hart GW, Aebi M, Mohnen D, Kinoshita T, Packer NH, Prestegard JH, Schnaar RL, Seeberger PH (eds) Essentials of glycobiology, 4th edn. Cold Spring Harbor Laboratory Press, Cold Spring Harbor, pp 33–42

Lee HH, Wang YN, Xia W, Chen CH, Rau KM, Ye L, Wei Y, Chou CK, Wang SC, Yan M, Tu CY, Hsia TC, Chiang SF, Chao KSC, Wistuba II, Hsu JL, Hortobagyi GN, Hung MC (2019) Removal of N-linked glycosylation enhances PD-L1 detection and predicts anti-PD-1/PD-L1 therapeutic efficacy. Cancer Cell 36:168–178 e4. https://doi.org/10.1016/j.ccell.2019.06.008
Lee TA, Tsai EY, Liu SH, Hsu Hung SD, Chang SJ, Chao CH, Lai YJ, Yamaguchi H, Li CW (2024) Post-translational modification of PD-1: potential targets for cancer immunotherapy. Cancer Res 84:800–807. https://doi.org/10.1158/0008-5472.CAN-23-2664
Lenos K, Goos JA, Vuist IM, den Uil SH, Delis-van Diemen PM, Belt EJ, Stockmann HB, Bril H, de Wit M, Carvalho B, Giblett S, Pritchard CA, Meijer GA, van Kooyk Y, Fijneman RJ, van Vliet SJ (2015) MGL ligand expression is correlated to BRAF mutation and associated with poor survival of stage III colon cancer patients. Oncotarget 6:26278–26290. https://doi.org/10.18632/oncotarget.4495
Leon F, Seshacharyulu P, Nimmakayala RK, Chugh S, Karmakar S, Nallasamy P, Vengoji R, Rachagani S, Cox JL, Mallya K, Batra SK, Ponnusamy MP (2022) Reduction in O-glycome induces differentially glycosylated CD44 to promote stemness and metastasis in pancreatic cancer. Oncogene 41:57–71. https://doi.org/10.1038/s41388-021-02047-2
Li CW, Lim SO, Chung EM, Kim YS, Park AH, Yao J, Cha JH, Xia W, Chan LC, Kim T, Chang SS, Lee HH, Chou CK, Liu YL, Yeh HC, Perillo EP, Dunn AK, Kuo CW, Khoo KH, Hsu JL, Wu Y, Hsu JM, Yamaguchi H, Huang TH, Sahin AA, Hortobagyi GN, Yoo SS, Hung MC (2018) Eradication of triple-negative breast cancer cells by targeting glycosylated PD-L1. Cancer Cell 33(187-201):e10. https://doi.org/10.1016/j.ccell.2018.01.009
Lidstrom T, Cumming J, Gaur R, Frangsmyr L, Pateras IS, Mickert MJ, Franklin O, Forsell MNE, Arnberg N, Dongre M, Patthey C, Ohlund D (2023) Extracellular galectin 4 drives immune evasion and promotes T-cell apoptosis in pancreatic cancer. Cancer Immunol Res 11:72–92. https://doi.org/10.1158/2326-6066.CIR-21-1088
Lin Y, Lubman DM (2024) The role of N-glycosylation in cancer. Acta Pharm Sin B 14:1098–1110. https://doi.org/10.1016/j.apsb.2023.10.014
Lin WL, Lin YS, Shi GY, Chang CF, Wu HL (2015) Lewisy promotes migration of oral cancer cells by glycosylation of epidermal growth factor receptor. PLoS One 10:e0120162. https://doi.org/10.1371/journal.pone.0120162
Lin MC, Chien PH, Wu HY, Chen ST, Juan HF, Lou PJ, Huang MC (2018) C1GALT1 predicts poor prognosis and is a potential therapeutic target in head and neck cancer. Oncogene 37:5780–5793. https://doi.org/10.1038/s41388-018-0375-0
Liu D, Gao Z, Yue L (2019) Fucosyltransferase 8 deficiency suppresses breast cancer cell migration by interference of the FAK/integrin pathway. Cancer Biomark 25:303–311. https://doi.org/10.3233/CBM-190209
Liu Z, Wang F, Liu X, Sang Y, Zhang L, Ren J, Qu X (2021) Cell membrane-camouflaged liposomes for tumor cell-selective glycans engineering and imaging in vivo. Proc Natl Acad Sci USA 118. https://doi.org/10.1073/pnas.2022769118
Livingston BD, Jacobs JL, Glick MC, Troy FA (1988) Extended polysialic acid chains (n greater than 55) in glycoproteins from human neuroblastoma cells. J Biol Chem 263:9443–9448
Lo Piccolo MS, Cheung NK, Cheung IY (2001) GD2 synthase: a new molecular marker for detecting neuroblastoma. Cancer 92:924–931. https://doi.org/10.1002/1097-0142(20010815)92:4<924::aid-cncr1402>3.0.co;2-o
Luca VC, Jude KM, Pierce NW, Nachury MV, Fischer S, Garcia KC (2015) Structural biology. Structural basis for Notch1 engagement of Delta-like 4. Science 347:847–853. https://doi.org/10.1126/science.1261093
Luo BH, Springer TA, Takagi J (2003) Stabilizing the open conformation of the integrin headpiece with a glycan wedge increases affinity for ligand. Proc Natl Acad Sci USA 100:2403–2408. https://doi.org/10.1073/pnas.0438060100
Lv X, Song J, Xue K, Li Z, Li M, Zahid D, Cao H, Wang L, Song W, Ma T, Gu J, Li W (2019) Core fucosylation of copper transporter 1 plays a crucial role in cisplatin-resistance of epithelial

ovarian cancer by regulating drug uptake. Mol Carcinog 58:794–807. https://doi.org/10.1002/mc.22971

Ma W, Du J, Chu Q, Wang Y, Liu L, Song M, Wang W (2011) hCLP46 regulates U937 cell proliferation via Notch signaling pathway. Biochem Biophys Res Commun 408:84–88. https://doi.org/10.1016/j.bbrc.2011.03.124

Ma L, Dong P, Liu L, Gao Q, Duan M, Zhang S, Chen S, Xue R, Wang X (2016) Overexpression of protein O-fucosyltransferase 1 accelerates hepatocellular carcinoma progression via the Notch signaling pathway. Biochem Biophys Res Commun 473:503–510. https://doi.org/10.1016/j.bbrc.2016.03.062

Macauley MS, Crocker PR, Paulson JC (2014) Siglec-mediated regulation of immune cell function in disease. Nat Rev Immunol 14:653–666. https://doi.org/10.1038/nri3737

Magalhaes A, Gomes J, Ismail MN, Haslam SM, Mendes N, Osorio H, David L, Le Pendu J, Haas R, Dell A, Boren T, Reis CA (2009) Fut2-null mice display an altered glycosylation profile and impaired BabA-mediated Helicobacter pylori adhesion to gastric mucosa. Glycobiology 19:1525–1536. https://doi.org/10.1093/glycob/cwp131

Majzner RG, Ramakrishna S, Yeom KW, Patel S, Chinnasamy H, Schultz LM, Richards RM, Jiang L, Barsan V, Mancusi R, Geraghty AC, Good Z, Mochizuki AY, Gillespie SM, Toland AMS, Mahdi J, Reschke A, Nie EH, Chau IJ, Rotiroti MC, Mount CW, Baggott C, Mavroukakis S, Egeler E, Moon J, Erickson C, Green S, Kunicki M, Fujimoto M, Ehlinger Z, Reynolds W, Kurra S, Warren KE, Prabhu S, Vogel H, Rasmussen L, Cornell TT, Partap S, Fisher PG, Campen CJ, Filbin MG, Grant G, Sahaf B, Davis KL, Feldman SA, Mackall CL, Monje M (2022) GD2-CAR T cell therapy for H3K27M-mutated diffuse midline gliomas. Nature 603:934–941. https://doi.org/10.1038/s41586-022-04489-4

Manni M, Laubli H (2021) Targeting glyco-immune checkpoints for cancer therapy. Expert Opin Biol Ther 21:1063–1071. https://doi.org/10.1080/14712598.2021.1882989

Marcos NT, Bennett EP, Gomes J, Magalhaes A, Gomes C, David L, Dar I, Jeanneau C, DeFrees S, Krustrup D, Vogel LK, Kure EH, Burchell J, Taylor-Papadimitriou J, Clausen H, Mandel U, Reis CA (2011) ST6GalNAc-I controls expression of sialyl-Tn antigen in gastrointestinal tissues. Front Biosci (Elite Ed) 3:1443–1455. https://doi.org/10.2741/e345

Marinkovich MP (2007) Tumour microenvironment: laminin 332 in squamous-cell carcinoma. Nat Rev Cancer 7:370–380. https://doi.org/10.1038/nrc2089

Marino KV, Cagnoni AJ, Croci DO, Rabinovich GA (2023) Targeting galectin-driven regulatory circuits in cancer and fibrosis. Nat Rev Drug Discov 22:295–316. https://doi.org/10.1038/s41573-023-00636-2

Marques C, Reis CA, Vives RR, Magalhaes A (2021) Heparan sulfate biosynthesis and Sulfation profiles as modulators of cancer signalling and progression. Front Oncol 11:778752. https://doi.org/10.3389/fonc.2021.778752

Marsico G, Russo L, Quondamatteo F, Pandit A (2018) Glycosylation and integrin regulation in cancer. Trends Cancer 4:537–552. https://doi.org/10.1016/j.trecan.2018.05.009

Mei Y, Wang X, Zhang J, Liu D, He J, Huang C, Liao J, Wang Y, Feng Y, Li H, Liu X, Chen L, Yi W, Chen X, Bai HM, Wang X, Li Y, Wang L, Liang Z, Ren X, Qiu L, Hui Y, Zhang Q, Leng Q, Chen J, Jia G (2023) Siglec-9 acts as an immune-checkpoint molecule on macrophages in glioblastoma, restricting T-cell priming and immunotherapy response. Nat Cancer 4:1273–1291. https://doi.org/10.1038/s43018-023-00598-9

Mendonsa AM, Na TY, Gumbiner BM (2018) E-cadherin in contact inhibition and cancer. Oncogene 37:4769–4780. https://doi.org/10.1038/s41388-018-0304-2

Mereiter S, Balmana M, Campos D, Gomes J, Reis CA (2019a) Glycosylation in the era of cancer-targeted therapy: where are we heading? Cancer Cell 36:6–16. https://doi.org/10.1016/j.ccell.2019.06.006

Mereiter S, Martins AM, Gomes C, Balmana M, Macedo JA, Polom K, Roviello F, Magalhaes A, Reis CA (2019b) O-glycan truncation enhances cancer-related functions of CD44 in gastric cancer. FEBS Lett 593:1675–1689. https://doi.org/10.1002/1873-3468.13432

Merry CLR, Lindahl U, Couchman J, Esko JD (2022) Proteoglycans and sulfated glycosaminoglycans. In: Varki A, Cummings RD, Esko JD, Stanley P, Hart GW, Aebi M, Mohnen D, Kinoshita T, Packer NH, Prestegard JH, Schnaar RL, Seeberger PH (eds) Essentials of glycobiology, 4th edn. Cold Spring Harbor Laboratory Press, Cold Spring Harbor, pp 217–232

Meyran D, Zhu J, Butler J, Macdonald S, Tantalo D, Thio N, Sek K, Ekert P, Kershaw M, Trapani J, Darcy P, Neeson P (2020) Early-phenotype Lewis Y CAR-T cells persist better in vivo and induce solid tumor regression in combination with anti-PD1. J ImmunoTherapy Cancer 8. https://doi.org/10.1136/jitc-2020-SITC2020.0126

Misra S, Heldin P, Hascall VC, Karamanos NK, Skandalis SS, Markwald RR, Ghatak S (2011) Hyaluronan-CD44 interactions as potential targets for cancer therapy. FEBS J 278:1429–1443. https://doi.org/10.1111/j.1742-4658.2011.08071.x

Mitsui Y, Yamada K, Hara S, Kinoshita M, Hayakawa T, Kakehi K (2012) Comparative studies on glycoproteins expressing polylactosamine-type N-glycans in cancer cells. J Pharm Biomed Anal 70:718–726. https://doi.org/10.1016/j.jpba.2012.06.035

Miwa HE, Song Y, Alvarez R, Cummings RD, Stanley P (2012) The bisecting GlcNAc in cell growth control and tumor progression. Glycoconj J 29:609–618. https://doi.org/10.1007/s10719-012-9373-6

Miwa HE, Koba WR, Fine EJ, Giricz O, Kenny PA, Stanley P (2013) Bisected, complex N-glycans and galectins in mouse mammary tumor progression and human breast cancer. Glycobiology 23:1477–1490. https://doi.org/10.1093/glycob/cwt075

Miyoshi E, Nishikawa A, Ihara Y, Saito H, Uozumi N, Hayashi N, Fusamoto H, Kamada T, Taniguchi N (1995) Transforming growth factor beta up-regulates expression of the N-acetylglucosaminyltransferase V gene in mouse melanoma cells. J Biol Chem 270:6216–6220. https://doi.org/10.1074/jbc.270.11.6216

Molfetta R, Zitti B, Lecce M, Milito ND, Stabile H, Fionda C, Cippitelli M, Gismondi A, Santoni A, Paolini R (2020) CD155: a multi-functional molecule in tumor progression. Int J Mol Sci 21. https://doi.org/10.3390/ijms21030922

Muhlenhoff M, Rollenhagen M, Werneburg S, Gerardy-Schahn R, Hildebrandt H (2013) Polysialic acid: versatile modification of NCAM, SynCAM 1 and neuropilin-2. Neurochem Res 38:1134–1143. https://doi.org/10.1007/s11064-013-0979-2

Mukherjee K, Chava AK, Mandal C, Dey SN, Kniep B, Chandra S, Mandal C (2008) O-acetylation of GD3 prevents its apoptotic effect and promotes survival of lymphoblasts in childhood acute lymphoblastic leukaemia. J Cell Biochem 105:724–734. https://doi.org/10.1002/jcb.21867

Munkley J, Oltean S, Vodak D, Wilson BT, Livermore KE, Zhou Y, Star E, Floros VI, Johannessen B, Knight B, McCullagh P, McGrath J, Crundwell M, Skotheim RI, Robson CN, Leung HY, Harries LW, Rajan P, Mills IG, Elliott DJ (2015) The androgen receptor controls expression of the cancer-associated sTn antigen and cell adhesion through induction of ST6GalNAc1 in prostate cancer. Oncotarget 6:34358–34374. https://doi.org/10.18632/oncotarget.6024

Munkley J, Vodak D, Livermore KE, James K, Wilson BT, Knight B, McCullagh P, McGrath J, Crundwell M, Harries LW, Leung HY, Robson CN, Mills IG, Rajan P, Elliott DJ (2016) Glycosylation is an androgen-regulated process essential for prostate cancer cell viability. EBioMedicine 8:103–116. https://doi.org/10.1016/j.ebiom.2016.04.018

Nagae M, Kizuka Y, Mihara E, Kitago Y, Hanashima S, Ito Y, Takagi J, Taniguchi N, Yamaguchi Y (2018) Structure and mechanism of cancer-associated N-acetylglucosaminyltransferase-V. Nat Commun 9:3380. https://doi.org/10.1038/s41467-018-05931-w

Nakatsura T, Nishimura Y (2005) Usefulness of the novel oncofetal antigen glypican-3 for diagnosis of hepatocellular carcinoma and melanoma. BioDrugs 19:71–77. https://doi.org/10.2165/00063030-200519020-00001

Nakatsura T, Kageshita T, Ito S, Wakamatsu K, Monji M, Ikuta Y, Senju S, Ono T, Nishimura Y (2004) Identification of glypican-3 as a novel tumor marker for melanoma. Clin Cancer Res 10: 6612–6621. https://doi.org/10.1158/1078-0432.CCR-04-0348

Natoni A, Cerreto M, De Propris MS, Petrucci MT, Fazio F, Intoppa S, Milani ML, Kirkham-McCarthy L, Henderson R, Swan D, Guarini A, O'Dwyer M, Foa R (2023) Sialofucosylation enables platelet binding to myeloma cells via P-Selectin and suppresses NK cell-mediated cytotoxicity. Cancers (Basel) 15. https://doi.org/10.3390/cancers15072154

Nishikawa A, Gu J, Fujii S, Taniguchi N (1990) Determination of N-acetylglucosaminyltransferases III, IV and V in normal and hepatoma tissues of rats. Biochim Biophys Acta 1035:313–318. https://doi.org/10.1016/0304-4165(90)90094-d

Okagawa Y, Takada K, Arihara Y, Kikuchi S, Osuga T, Nakamura H, Kamihara Y, Hayasaka N, Usami M, Murase K, Miyanishi K, Kobune M, Kato J (2016) Activated p53 with histone Deacetylase inhibitor enhances L-Fucose-mediated drug delivery through induction of Fucosyltransferase 8 expression in hepatocellular carcinoma cells. PLoS One 11:e0168355. https://doi.org/10.1371/journal.pone.0168355

Oriol R, Le Pendu J, Mollicone R (1986) Genetics of ABO, H, Lewis, X and related antigens. Vox Sang 51:161–171. https://doi.org/10.1111/j.1423-0410.1986.tb01946.x

Orr SL, Le D, Long JM, Sobieszczuk P, Ma B, Tian H, Fang X, Paulson JC, Marth JD, Varki N (2013) A phenotype survey of 36 mutant mouse strains with gene-targeted defects in glycosyltransferases or glycan-binding proteins. Glycobiology 23:363–380. https://doi.org/10.1093/glycob/cws150

Osako M, Yonezawa S, Siddiki B, Huang J, Ho JJ, Kim YS, Sato E (1993) Immunohistochemical study of mucin carbohydrates and core proteins in human pancreatic tumors. Cancer 71:2191–2199. https://doi.org/10.1002/1097-0142(19930401)71:7<2191::aid-cncr2820710705>3.0.co;2-x

Osuka RF, Yamasaki T, Kizuka Y (2024) Structure and function of N-acetylglucosaminyltransferase V (GnT-V). Biochim Biophys Acta Gen Subj 1868:130709. https://doi.org/10.1016/j.bbagen.2024.130709

Ozcan S, Barkauskas DA, Renee Ruhaak L, Torres J, Cooke CL, An HJ, Hua S, Williams CC, Dimapasoc LM, Han Kim J, Camorlinga-Ponce M, Rocke D, Lebrilla CB, Solnick JV (2014) Serum glycan signatures of gastric cancer. Cancer Prev Res (Phila) 7:226–235. https://doi.org/10.1158/1940-6207.CAPR-13-0235

Pandey A, Niknejad N, Jafar-Nejad H (2021) Multifaceted regulation of Notch signaling by glycosylation. Glycobiology 31:8–28. https://doi.org/10.1093/glycob/cwaa049

Paneque A, Fortus H, Zheng J, Werlen G, Jacinto E (2023) The Hexosamine biosynthesis pathway: regulation and function. Genes (Basel) 14. https://doi.org/10.3390/genes14040933

Park S, Colville MJ, Paek JH, Shurer CR, Singh A, Secor EJ, Sailer CJ, Huang LT, Kuo JC, Goudge MC, Su J, Kim M, DeLisa MP, Neelamegham S, Lammerding J, Zipfel WR, Fischbach C, Reesink HL, Paszek MJ (2024) Immunoengineering can overcome the glycocalyx armour of cancer cells. Nat Mater 23(3):429–438. https://doi.org/10.1038/s41563-024-01808-0

Park JJ, Yi JY, Jin YB, Lee YJ, Lee JS, Lee YS, Ko YG, Lee M (2012) Sialylation of epidermal growth factor receptor regulates receptor activity and chemosensitivity to gefitinib in colon cancer cells. Biochem Pharmacol 83:849–857. https://doi.org/10.1016/j.bcp.2012.01.007

Park SJ, Jang JY, Jeong SW, Cho YK, Lee SH, Kim SG, Cha SW, Kim YS, Cho YD, Kim HS, Kim BS, Park S, Bang HI (2017) Usefulness of AFP, AFP-L3, and PIVKA-II, and their combinations in diagnosing hepatocellular carcinoma. Medicine (Baltimore) 96:e5811. https://doi.org/10.1097/MD.0000000000005811

Park S, Lim JM, Chun JN, Lee S, Kim TM, Kim DW, Kim SY, Bae DJ, Bae SM, So I, Kim HG, Choi JY, Jeon JH (2020) Altered expression of fucosylation pathway genes is associated with poor prognosis and tumor metastasis in non-small cell lung cancer. Int J Oncol 56:559–567. https://doi.org/10.3892/ijo.2019.4953

Paszek MJ, DuFort CC, Rossier O, Bainer R, Mouw JK, Godula K, Hudak JE, Lakins JN, Wijekoon AC, Cassereau L, Rubashkin MG, Magbanua MJ, Thorn KS, Davidson MW, Rugo HS, Park

JW, Hammer DA, Giannone G, Bertozzi CR, Weaver VM (2014) The cancer glycocalyx mechanically primes integrin-mediated growth and survival. Nature 511:319–325. https://doi.org/10.1038/nature13535

Pearce OM, Laubli H (2016) Sialic acids in cancer biology and immunity. Glycobiology 26:111–128. https://doi.org/10.1093/glycob/cwv097

Peixoto A, Relvas-Santos M, Azevedo R, Santos LL, Ferreira JA (2019) Protein glycosylation and tumor microenvironment alterations driving cancer hallmarks. Front Oncol 9:380. https://doi.org/10.3389/fonc.2019.00380

Peng N, Liu J, Hai S, Liu Y, Zhao H, Liu W (2024) Role of post-translational modifications in colorectal cancer metastasis. Cancers (Basel) 16. https://doi.org/10.3390/cancers16030652

Perdicchio M, Ilarregui JM, Verstege MI, Cornelissen LA, Schetters ST, Engels S, Ambrosini M, Kalay H, Veninga H, den Haan JM, van Berkel LA, Samsom JN, Crocker PR, Sparwasser T, Berod L, Garcia-Vallejo JJ, van Kooyk Y, Unger WW (2016) Sialic acid-modified antigens impose tolerance via inhibition of T-cell proliferation and de novo induction of regulatory T cells. Proc Natl Acad Sci USA 113:3329–3334. https://doi.org/10.1073/pnas.1507706113

Pinho SS, Reis CA (2015) Glycosylation in cancer: mechanisms and clinical implications. Nat Rev Cancer 15:540–555. https://doi.org/10.1038/nrc3982

Pinho S, Marcos NT, Ferreira B, Carvalho AS, Oliveira MJ, Santos-Silva F, Harduin-Lepers A, Reis CA (2007) Biological significance of cancer-associated sialyl-Tn antigen: modulation of malignant phenotype in gastric carcinoma cells. Cancer Lett 249:157–170. https://doi.org/10.1016/j.canlet.2006.08.010

Pinho SS, Figueiredo J, Cabral J, Carvalho S, Dourado J, Magalhaes A, Gartner F, Mendonfa AM, Isaji T, Gu J, Carneiro F, Seruca R, Taniguchi N, Reis CA (2013) E-cadherin and adherens-junctions stability in gastric carcinoma: functional implications of glycosyltransferases involving N-glycan branching biosynthesis, N-acetylglucosaminyltransferases III and V. Biochim Biophys Acta 1830:2690–2700. https://doi.org/10.1016/j.bbagen.2012.10.021

Pocas J, Marques C, Gomes C, Otake AH, Pinto F, Ferreira M, Silva T, Faria-Ramos I, Matos R, Ribeiro AR, Senra E, Cavadas B, Batista S, Maia J, Macedo JA, Lima L, Afonso LP, Ferreira JA, Santos LL, Polonia A, Osorio H, Belting M, Reis CA, Costa-Silva B, Magalhaes A (2023) Syndecan-4 is a maestro of gastric cancer cell invasion and communication that underscores poor survival. Proc Natl Acad Sci USA 120:e2214853120. https://doi.org/10.1073/pnas.2214853120

Pochec E, Litynska A, Bubka M, Amoresano A, Casbarra A (2006) Characterization of the oligosaccharide component of alpha3beta1 integrin from human bladder carcinoma cell line T24 and its role in adhesion and migration. Eur J Cell Biol 85:47–57. https://doi.org/10.1016/j.ejcb.2005.08.010

Posey AD Jr, Schwab RD, Boesteanu AC, Steentoft C, Mandel U, Engels B, Stone JD, Madsen TD, Schreiber K, Haines KM, Cogdill AP, Chen TJ, Song D, Scholler J, Kranz DM, Feldman MD, Young R, Keith B, Schreiber H, Clausen H, Johnson LA, June CH (2016) Engineered CAR T cells targeting the cancer-associated Tn-Glycoform of the membrane Mucin MUC1 control adenocarcinoma. Immunity 44:1444–1454. https://doi.org/10.1016/j.immuni.2016.05.014

Potapenko IO, Haakensen VD, Luders T, Helland A, Bukholm I, Sorlie T, Kristensen VN, Lingjaerde OC, Borresen-Dale AL (2010) Glycan gene expression signatures in normal and malignant breast tissue; possible role in diagnosis and progression. Mol Oncol 4:98–118. https://doi.org/10.1016/j.molonc.2009.12.001

Powers TW, Neely BA, Shao Y, Tang H, Troyer DA, Mehta AS, Haab BB, Drake RR (2014) MALDI imaging mass spectrometry profiling of N-glycans in formalin-fixed paraffin embedded clinical tissue blocks and tissue microarrays. PLoS One 9:e106255. https://doi.org/10.1371/journal.pone.0106255

Prapa M, Chiavelli C, Golinelli G, Grisendi G, Bestagno M, Di Tinco R, Dall'Ora M, Neri G, Candini O, Spano C, Petrachi T, Bertoni L, Carnevale G, Pugliese G, Depenni R, Feletti A, Iaccarino C, Pavesi G, Dominici M (2021) GD2 CAR T cells against human glioblastoma. NPJ Precis Oncol 5:93. https://doi.org/10.1038/s41698-021-00233-9

Radhakrishnan P, Dabelsteen S, Madsen FB, Francavilla C, Kopp KL, Steentoft C, Vakhrushev SY, Olsen JV, Hansen L, Bennett EP, Woetmann A, Yin G, Chen L, Song H, Bak M, Hlady RA, Peters SL, Opavsky R, Thode C, Qvortrup K, Schjoldager KT, Clausen H, Hollingsworth MA, Wandall HH (2014) Immature truncated O-glycophenotype of cancer directly induces oncogenic features. Proc Natl Acad Sci USA 111:E4066–E4075. https://doi.org/10.1073/pnas.1406619111

Raglow Z, McKenna MK, Bonifant CL, Wang W, Pasca di Magliano M, Stadlmann J, Penninger JM, Cummings RD, Brenner MK, Markovitz DM (2022) Targeting glycans for CAR therapy: the advent of sweet CARs. Mol Ther 30:2881–2890. https://doi.org/10.1016/j.ymthe.2022.07.006

Raiter A, Lipovetsky J, Stenbac A, Lubin I, Yerushalmi R (2023) TNBC-derived Gal3BP/Gal3 complex induces immunosuppression through CD45 receptor. Onco Targets Ther 12:2246322. https://doi.org/10.1080/2162402X.2023.2246322

Recchi MA, Hebbar M, Hornez L, Harduin-Lepers A, Peyrat JP, Delannoy P (1998) Multiplex reverse transcription polymerase chain reaction assessment of sialyltransferase expression in human breast cancer. Cancer Res 58:4066–4070

Reily C, Stewart TJ, Renfrow MB, Novak J (2019) Glycosylation in health and disease. Nat Rev Nephrol 15:346–366. https://doi.org/10.1038/s41581-019-0129-4

Reis CA, David L, Correa P, Carneiro F, de Bolos C, Garcia E, Mandel U, Clausen H, Sobrinho-Simoes M (1999) Intestinal metaplasia of human stomach displays distinct patterns of mucin (MUC1, MUC2, MUC5AC, and MUC6) expression. Cancer Res 59:1003–1007

Reis CA, Osorio H, Silva L, Gomes C, David L (2010) Alterations in glycosylation as biomarkers for cancer detection. J Clin Pathol 63:322–329. https://doi.org/10.1136/jcp.2009.071035

Ren LR, Zhang LP, Huang SY, Zhu YF, Li WJ, Fang SY, Shen L, Gao YL (2016) Effects of sialidase NEU1 siRNA on proliferation, apoptosis, and invasion in human ovarian cancer. Mol Cell Biochem 411:213–219. https://doi.org/10.1007/s11010-015-2583-z

Ribatti D (2017) The concept of immune surveillance against tumors. The first theories. Oncotarget 8:7175–7180. https://doi.org/10.18632/oncotarget.12739

Riley RS, June CH, Langer R, Mitchell MJ (2019) Delivery technologies for cancer immunotherapy. Nat Rev Drug Discov 18:175–196. https://doi.org/10.1038/s41573-018-0006-z

Rini JM, Moremen KW, Davis BG, Esko JD (2022) Glycosyltransferases and glycan-processing enzymes. In: Varki A, Cummings RD, Esko JD, Stanley P, Hart GW, Aebi M, Mohnen D, Kinoshita T, Packer NH, Prestegard JH, Schnaar RL, Seeberger PH (eds) Essentials of glycobiology, 4th edn. Cold Spring Harbor Laboratory Press, Cold Spring Harbor, pp 67–78

Ritchie DS, Neeson PJ, Khot A, Peinert S, Tai T, Tainton K, Chen K, Shin M, Wall DM, Honemann D, Gambell P, Westerman DA, Haurat J, Westwood JA, Scott AM, Kravets L, Dickinson M, Trapani JA, Smyth MJ, Darcy PK, Kershaw MH, Prince HM (2013) Persistence and efficacy of second generation CAR T cell against the LeY antigen in acute myeloid leukemia. Mol Ther 21:2122–2129. https://doi.org/10.1038/mt.2013.154

Rodrigues JG, Duarte HO, Gomes C, Balmana M, Martins AM, Hensbergen PJ, de Ru AH, Lima J, Albergaria A, van Veelen PA, Wuhrer M, Gomes J, Reis CA (2021) Terminal alpha2,6-sialylation of epidermal growth factor receptor modulates antibody therapy response of colorectal cancer cells. Cell Oncol (Dordr) 44:835–850. https://doi.org/10.1007/s13402-021-00606-z

RodrIguez E, Schetters STT, van Kooyk Y (2018) The tumour glyco-code as a novel immune checkpoint for immunotherapy. Nat Rev Immunol 18:204–211. https://doi.org/10.1038/nri.2018.3

Rodriguez E, Boelaars K, Brown K, Madunic K, van Ee T, Dijk F, Verheij J, Li RJE, Schetters STT, Meijer LL, Le Large TYS, Driehuis E, Clevers H, Bruijns SCM, O'Toole T, van Vliet SJ, Bijlsma MF, Wuhrer M, Kazemier G, Giovannetti E, Garcia-Vallejo JJ, van Kooyk Y (2022) Analysis of the glyco-code in pancreatic ductal adenocarcinoma identifies glycan-mediated immune regulatory circuits. Commun Biol 5:41. https://doi.org/10.1038/s42003-021-02934-0

Romer TB, Aasted MKM, Dabelsteen S, Groen A, Schnabel J, Tan E, Pedersen JW, Haue AD, Wandall HH (2021) Mapping of truncated O-glycans in cancers of epithelial and non-epithelial origin. Br J Cancer 125:1239–1250. https://doi.org/10.1038/s41416-021-01530-7

Roth J, Zuber C, Wagner P, Taatjes DJ, Weisgerber C, Heitz PU, Goridis C, Bitter-Suermann D (1988) Reexpression of poly(sialic acid) units of the neural cell adhesion molecule in Wilms tumor. Proc Natl Acad Sci USA 85:2999–3003. https://doi.org/10.1073/pnas.85.9.2999

Ruckhaberle E, Karn T, Rody A, Hanker L, Gatje R, Metzler D, Holtrich U, Kaufmann M (2009) Gene expression of ceramide kinase, galactosyl ceramide synthase and ganglioside GD3 synthase is associated with prognosis in breast cancer. J Cancer Res Clin Oncol 135:1005–1013. https://doi.org/10.1007/s00432-008-0536-6

Ruhaak LR, Taylor SL, Stroble C, Nguyen UT, Parker EA, Song T, Lebrilla CB, Rom WN, Pass H, Kim K, Kelly K, Miyamoto S (2015) Differential N-glycosylation patterns in lung adenocarcinoma tissue. J Proteome Res 14:4538–4549. https://doi.org/10.1021/acs.jproteome.5b00255

Saldova R, Royle L, Radcliffe CM, Hamid UMA, Evans R, Arnold JN, Banks RE, Hutson R, Harvey DJ, Antrobus R, Petrescu SM, Dwek RA, Rudd PM (2007) Ovarian cancer is associated with changes in glycosylation in both acute-phase proteins and IgG. Glycobiology 17:1344

Sanji AS, Manasa J, Gurav MJ, Batra SK, Chachadi VB (2024) Cancer snap-shots: biochemistry and glycopathology of O-glycans: a review. Int J Biol Macromol 260:129318. https://doi.org/10.1016/j.ijbiomac.2024.129318

Schjoldager KT, Narimatsu Y, Joshi HJ, Clausen H (2020) Global view of human protein glycosylation pathways and functions. Nat Rev Mol Cell Biol 21:729–749. https://doi.org/10.1038/s41580-020-00294-x

Schnaar RL (2023) Gangliosides as Siglec ligands. Glycoconj J 40:159–167. https://doi.org/10.1007/s10719-023-10101-2

Schnaar RL, Sandhoff R, Tiemeyer M, Kinoshita T (2022) Glycosphingolipids. In: Varki A, Cummings RD, Esko JD, Stanley P, Hart GW, Aebi M, Mohnen D, Kinoshita T, Packer NH, Prestegard JH, Schnaar RL, Seeberger PH (eds) Essentials of glycobiology, 4th edn. Cold Spring Harbor Laboratory Press, Cold Spring Harbor, pp 129–140

Seidenfaden R, Krauter A, Schertzinger F, Gerardy-Schahn R, Hildebrandt H (2003) Polysialic acid directs tumor cell growth by controlling heterophilic neural cell adhesion molecule interactions. Mol Cell Biol 23:5908–5918. https://doi.org/10.1128/MCB.23.16.5908-5918.2003

Senbanjo LT, Chellaiah MA (2017) CD44: a multifunctional cell surface adhesion receptor is a regulator of progression and metastasis of cancer cells. Front Cell Dev Biol 5:18. https://doi.org/10.3389/fcell.2017.00018

Shi Y, Wang Y, Yang R, Zhang W, Zhang Y, Feng K, Lv Q, Niu K, Chen J, Li L, Zhang Y (2022) Glycosylation-related molecular subtypes and risk score of hepatocellular carcinoma: novel insights to clinical decision-making. Front Endocrinol (Lausanne) 13:1090324. https://doi.org/10.3389/fendo.2022.1090324

Shu L, Lin S, Zhou S, Yuan T (2024) Glycan-Lectin interactions between platelets and tumor cells drive hematogenous metastasis. Platelets 35:2315037. https://doi.org/10.1080/09537104.2024.2315037

Singh C, Shyanti RK, Singh V, Kale RK, Mishra JPN, Singh RP (2018) Integrin expression and glycosylation patterns regulate cell-matrix adhesion and alter with breast cancer progression. Biochem Biophys Res Commun 499:374–380. https://doi.org/10.1016/j.bbrc.2018.03.169

Smith BAH, Bertozzi CR (2021) The clinical impact of glycobiology: targeting selectins, Siglecs and mammalian glycans. Nat Rev Drug Discov 20:217–243. https://doi.org/10.1038/s41573-020-00093-1

Soares da Costa D, Reis RL, Pashkuleva I (2017) Sulfation of glycosaminoglycans and its implications in human health and disorders. Annu Rev Biomed Eng 19:1–26. https://doi.org/10.1146/annurev-bioeng-071516-044610

Song Y, Aglipay JA, Bernstein JD, Goswami S, Stanley P (2010) The bisecting GlcNAc on N-glycans inhibits growth factor signaling and retards mammary tumor progression. Cancer Res 70:3361–3371. https://doi.org/10.1158/0008-5472.CAN-09-2719

Stanczak MA, Laubli H (2023) Siglec receptors as new immune checkpoints in cancer. Mol Asp Med 90:101112. https://doi.org/10.1016/j.mam.2022.101112

Stanley P (2016) What have we learned from Glycosyltransferase knockouts in mice? J Mol Biol 428:3166–3182. https://doi.org/10.1016/j.jmb.2016.03.025

Stanley P, Moremen KL, Lewis NE, Taniguchi N, Aebi M (2022) N-Glycans. In: Varki A, Cummings RD, Esko JD, Stanley P, Hart GW, Aebi M, Mohnen D, Kinoshita T, Packer NH, Prestegard JH, Schnaar RL, Seeberger PH (eds) Essentials of glycobiology, 4th edn. Cold Spring Harbor Laboratory Press, Cold Spring Harbor

Steentoft C, Vakhrushev SY, Joshi HJ, Kong Y, Vester-Christensen MB, Schjoldager KT, Lavrsen K, Dabelsteen S, Pedersen NB, Marcos-Silva L, Gupta R, Bennett EP, Mandel U, Brunak S, Wandall HH, Levery SB, Clausen H (2013) Precision mapping of the human O-GalNAc glycoproteome through SimpleCell technology. EMBO J 32:1478–1488. https://doi.org/10.1038/emboj.2013.79

Stowell SR, Ju T, Cummings RD (2015) Protein glycosylation in cancer. Annu Rev Pathol 10:473–510. https://doi.org/10.1146/annurev-pathol-012414-040438

Sun L, Li CW, Chung EM, Yang R, Kim YS, Park AH, Lai YJ, Yang Y, Wang YH, Liu J, Qiu Y, Khoo KH, Yao J, Hsu JL, Cha JH, Chan LC, Hsu JM, Lee HH, Yoo SS, Hung MC (2020) Targeting glycosylated PD-1 induces potent antitumor immunity. Cancer Res 80:2298–2310. https://doi.org/10.1158/0008-5472.CAN-19-3133

Sun K, Feng Z, Fan C, Min X, Zhang P, Xia L (2023) A glycosylation signature for predicting the progression and immunotherapeutic response of prostate cancer. J Gene Med 25:e3489. https://doi.org/10.1002/jgm.3489

Sweeney JG, Liang J, Antonopoulos A, Giovannone N, Kang S, Mondala TS, Head SR, King SL, Tani Y, Brackett D, Dell A, Murphy GF, Haslam SM, Widlund HR, Dimitroff CJ (2018) Loss of GCNT2/I-branched glycans enhances melanoma growth and survival. Nat Commun 9:3368. https://doi.org/10.1038/s41467-018-05795-0

Szczykutowicz J (2023) Ligand recognition by the macrophage Galactose-type C-type Lectin: self or non-self?-a way to trick the host's immune system. Int J Mol Sci 24. https://doi.org/10.3390/ijms242317078

Tahara S, Okumura G, Matsuo T, Shibuya A, Shibuya K (2024) Essential role of CD155 glycosylation in functional binding to DNAM-1 on natural killer cells. Int Immunol 36:317–325. https://doi.org/10.1093/intimm/dxae005

Tanaka F, Otake Y, Nakagawa T, Kawano Y, Miyahara R, Li M, Yanagihara K, Nakayama J, Fujimoto I, Ikenaka K, Wada H (2000) Expression of polysialic acid and STX, a human polysialyltransferase, is correlated with tumor progression in non-small cell lung cancer. Cancer Res 60:3072–3080

Taniguchi N, Honke K, Fukuda M (2011) Handbook of glycosyltransferases and related genes. Springer Science & Business Media

Taniguchi N, Ohkawa Y, Maeda K, Harada Y, Nagae M, Kizuka Y, Ihara H, Ikeda Y (2021) True significance of N-acetylglucosaminyltransferases GnT-III, V and alpha1,6 fucosyltransferase in epithelial-mesenchymal transition and cancer. Mol Asp Med 79:100905. https://doi.org/10.1016/j.mam.2020.100905

Tao Y, Li H, Huang R, Mo D, Zeng T, Fang M, Li M (2018) Clinicopathological and prognostic significance of cancer stem cell markers in ovarian cancer patients: evidence from 52 studies. Cell Physiol Biochem 46:1716–1726. https://doi.org/10.1159/000489586

Theocharis AD, Karamanos NK (2019) Proteoglycans remodeling in cancer: underlying molecular mechanisms. Matrix Biol 75-76:220–259. https://doi.org/10.1016/j.matbio.2017.10.008

Theocharis AD, Tsolakis I, Tzanakakis GN, Karamanos NK (2006) Chondroitin sulfate as a key molecule in the development of atherosclerosis and cancer progression. Adv Pharmacol 53:281–295. https://doi.org/10.1016/S1054-3589(05)53013-8

Thomas D, Rathinavel AK, Radhakrishnan P (2021) Altered glycosylation in cancer: a promising target for biomarkers and therapeutics. Biochim Biophys Acta Rev Cancer 1875:188464. https://doi.org/10.1016/j.bbcan.2020.188464

Trinchera M, Aronica A, Dall'Olio F (2017) Selectin Ligands Sialyl-Lewis a and Sialyl-Lewis x in gastrointestinal cancers. Biology (Basel) 6. https://doi.org/10.3390/biology6010016

Tu CF, Wu MY, Lin YC, Kannagi R, Yang RB (2017) FUT8 promotes breast cancer cell invasiveness by remodeling TGF-beta receptor core fucosylation. Breast Cancer Res 19:111. https://doi.org/10.1186/s13058-017-0904-8

Valle-Argos B, Chiodin G, Bryant DJ, Taylor J, Lemm E, Duriez PJ, Rock PJ, Strefford JC, Forconi F, Burack RW, Packham G, Stevenson FK (2021) DC-SIGN binding to mannosylated B-cell receptors in follicular lymphoma down-modulates receptor signaling capacity. Sci Rep 11:11676. https://doi.org/10.1038/s41598-021-91112-7

van Gisbergen KP, Aarnoudse CA, Meijer GA, Geijtenbeek TB, van Kooyk Y (2005) Dendritic cells recognize tumor-specific glycosylation of carcinoembryonic antigen on colorectal cancer cells through dendritic cell-specific intercellular adhesion molecule-3-grabbing nonintegrin. Cancer Res 65:5935–5944. https://doi.org/10.1158/0008-5472.CAN-04-4140

van Kooyk Y, Ilarregui JM, van Vliet SJ (2015) Novel insights into the immunomodulatory role of the dendritic cell and macrophage-expressed C-type lectin MGL. Immunobiology 220:185–192. https://doi.org/10.1016/j.imbio.2014.10.002

van Vliet SJ, Gringhuis SI, Geijtenbeek TB, van Kooyk Y (2006) Regulation of effector T cells by antigen-presenting cells via interaction of the C-type lectin MGL with CD45. Nat Immunol 7: 1200–1208. https://doi.org/10.1038/ni1390

van Vliet SJ, Saeland E, van Kooyk Y (2008) Sweet preferences of MGL: carbohydrate specificity and function. Trends Immunol 29:83–90. https://doi.org/10.1016/j.it.2007.10.010

van Vliet SJ, Bay S, Vuist IM, Kalay H, Garcia-Vallejo JJ, Leclerc C, van Kooyk Y (2013) MGL signaling augments TLR2-mediated responses for enhanced IL-10 and TNF-alpha secretion. J Leukoc Biol 94:315–323. https://doi.org/10.1189/jlb.1012520

Vazquez-Martin C, Gil-Martin E, Fernandez-Briera A (2005) Elevation of ST6Gal I activity in malignant and transitional tissue in human colorectal cancer. Oncology 69:436–444. https://doi.org/10.1159/000089999

Verkerke H, Dias-Baruffi M, Cummings RD, Arthur CM, Stowell SR (2022) Galectins: An ancient family of carbohydrate binding proteins with modern functions. Methods Mol Biol 2442:1–40. https://doi.org/10.1007/978-1-0716-2055-7_1

Walmod PS, Kolkova K, Berezin V, Bock E (2004) Zippers make signals: NCAM-mediated molecular interactions and signal transduction. Neurochem Res 29:2015–2035. https://doi.org/10.1007/s11064-004-6875-z

Wang PH, Lee WL, Lee YR, Juang CM, Chen YJ, Chao HT, Tsai YC, Yuan CC (2003) Enhanced expression of alpha 2,6-sialyltransferase ST6Gal I in cervical squamous cell carcinoma. Gynecol Oncol 89:395–401. https://doi.org/10.1016/s0090-8258(03)00127-6

Wang PH, Lee WL, Juang CM, Yang YH, Lo WH, Lai CR, Hsieh SL, Yuan CC (2005) Altered mRNA expressions of sialyltransferases in ovarian cancers. Gynecol Oncol 99:631–639. https://doi.org/10.1016/j.ygyno.2005.07.016

Wang JL, Su WY, Lin YW, Xiong H, Chen YX, Xu J, Fang JY (2017) CD44v6 overexpression related to metastasis and poor prognosis of colorectal cancer: a meta-analysis. Oncotarget 8: 12866–12876. https://doi.org/10.18632/oncotarget.14163

Wang J, Sun J, Liu LN, Flies DB, Nie X, Toki M, Zhang J, Song C, Zarr M, Zhou X, Han X, Archer KA, O'Neill T, Herbst RS, Boto AN, Sanmamed MF, Langermann S, Rimm DL, Chen L (2019) Siglec-15 as an immune suppressor and potential target for normalization cancer immunotherapy. Nat Med 25(4):656–666. https://doi.org/10.1038/s41591-019-0374-x

Wang S, Zhang X, Yang C, Xu S (2019) MicroRNA-198-5p inhibits the migration and invasion of non-small lung cancer cells by targeting fucosyltransferase 8. Clin Exp Pharmacol Physiol 46: 955–967. https://doi.org/10.1111/1440-1681.13154

Wegscheider AS, Wojahn I, Gottheil P, Spohn M, Kas JA, Rosin O, Ulm B, Nollau P, Wagener C, Niendorf A, Wolters-Eisfeld G (2024) CD301 and LSECtin glycan-binding receptors of innate immune cells serve as prognostic markers and potential predictors of immune response in breast cancer subtypes. Glycobiology 34. https://doi.org/10.1093/glycob/cwae003

Wei A, Fan B, Zhao Y, Zhang H, Wang L, Yu X, Yuan Q, Yang D, Wang S (2016) ST6Gal-I overexpression facilitates prostate cancer progression via the PI3K/Akt/GSK-3beta/beta-catenin signaling pathway. Oncotarget 7:65374–65388. https://doi.org/10.18632/oncotarget.11699
Weiz S, Wieczorek M, Schwedler C, Kaup M, Braicu EI, Sehouli J, Tauber R, Blanchard V (2016) Acute-phase glycoprotein N-glycome of ovarian cancer patients analyzed by CE-LIF. Electrophoresis 37:1461–1467. https://doi.org/10.1002/elps.201500518
Wu X, Zhao J, Ruan Y, Sun L, Xu C, Jiang H (2018) Sialyltransferase ST3GAL1 promotes cell migration, invasion, and TGF-beta1-induced EMT and confers paclitaxel resistance in ovarian cancer. Cell Death Dis 9:1102. https://doi.org/10.1038/s41419-018-1101-0
Xiao H, Woods EC, Vukojicic P, Bertozzi CR (2016) Precision glycocalyx editing as a strategy for cancer immunotherapy. Proc Natl Acad Sci USA 113:10304–10309. https://doi.org/10.1073/pnas.1608069113
Yamamoto F, Clausen H, White T, Marken J, Hakomori S (1990) Molecular genetic basis of the histo-blood group ABO system. Nature 345:229–233. https://doi.org/10.1038/345229a0
Yamamoto H, Kaneko Y, Rebbaa A, Bremer EG, Moskal JR (1997) alpha2,6-Sialyltransferase gene transfection into a human glioma cell line (U373 MG) results in decreased invasivity. J Neurochem 68:2566–2576. https://doi.org/10.1046/j.1471-4159.1997.68062566.x
Yamamoto H, Oviedo A, Sweeley C, Saito T, Moskal JR (2001) Alpha2,6-sialylation of cell-surface N-glycans inhibits glioma formation in vivo. Cancer Res 61:6822–6829
Yang X, Tang J, Rogler CE, Stanley P (2003) Reduced hepatocyte proliferation is the basis of retarded liver tumor progression and liver regeneration in mice lacking N-acetylglucosaminyltransferase III. Cancer Res 63:7753–7759
Yang HF, Yu M, Jin HD, Yao JQ, Lu ZL, Yabasin IB, Yan Q, Wen QP (2017) Fentanyl promotes breast cancer cell Stemness and epithelial-Mesenchymal transition by Upregulating alpha1, 6-Fucosylation via Wnt/beta-catenin signaling pathway. Front Physiol 8:510. https://doi.org/10.3389/fphys.2017.00510
Yang D, Han F, Cai J, Sun H, Wang F, Jiang M, Zhang M, Yuan M, Zhou W, Li H, Yang L, Bai Y, Xiao L, Dong H, Cheng Q, Mao H, Zhou L, Wang R, Li Y, Nie H (2023) N-glycosylation by N-acetylglucosaminyltransferase IVa enhances the interaction of integrin beta1 with vimentin and promotes hepatocellular carcinoma cell motility. Biochim Biophys Acta, Mol Cell Res 1870:119513. https://doi.org/10.1016/j.bbamcr.2023.119513
Yoshida H, Koodie L, Jacobsen K, Hanzawa K, Miyamoto Y, Yamamoto M (2020) B4GALNT1 induces angiogenesis, anchorage independence growth and motility, and promotes tumorigenesis in melanoma by induction of ganglioside GM2/GD2. Sci Rep 10:1199. https://doi.org/10.1038/s41598-019-57130-2
Zachara NE, Akimoto Y, Boyce M, Hart GW (2022) The O-GlcNAc modification. In: Varki A, Cummings RD, Esko JD, Stanley P, Hart GW, Aebi M, Mohnen D, Kinoshita T, Packer NH, Prestegard JH, Schnaar RL, Seeberger PH (eds) Essentials of glycobiology, 4th edn. Cold Spring Harbor Laboratory Press, Cold Spring Harbor, pp 251–264
Zahradnikova M, Ihnatova I, Lattova E, Uhrik L, Stuchlikova E, Nenutil R, Valik D, Nalezinska M, Chovanec J, Zdrahal Z, Vojtesek B, Hernychova L, Novotny MV (2021) N-Glycome changes reflecting resistance to platinum-based chemotherapy in ovarian cancer. J Proteome 230:103964. https://doi.org/10.1016/j.jprot.2020.103964
Zeng G, Gao L, Birkle S, Yu RK (2000) Suppression of ganglioside GD3 expression in a rat F-11 tumor cell line reduces tumor growth, angiogenesis, and vascular endothelial growth factor production. Cancer Res 60:6670–6676
Zhang G, Isaji T, Xu Z, Lu X, Fukuda T, Gu J (2019) N-acetylglucosaminyltransferase-I as a novel regulator of epithelial-mesenchymal transition. FASEB J 33(2):2823–2835. https://doi.org/10.1096/fj.201801478R
Zhang H, Meng F, Wu S, Kreike B, Sethi S, Chen W, Miller FR, Wu G (2011) Engagement of I-branching beta-1, 6-N-acetylglucosaminyltransferase 2 in breast cancer metastasis and TGF-beta signaling. Cancer Res 71:4846–4856. https://doi.org/10.1158/0008-5472.CAN-11-0414

Zhang L, Chen Y, Wang K (2019) Comparison of CA125, HE4, and ROMA index for ovarian cancer diagnosis. Curr Probl Cancer 43:135–144. https://doi.org/10.1016/j.currproblcancer.2018.06.001

Zhao Y, Nakagawa T, Itoh S, Inamori K, Isaji T, Kariya Y, Kondo A, Miyoshi E, Miyazaki K, Kawasaki N, Taniguchi N, Gu J (2006) N-acetylglucosaminyltransferase III antagonizes the effect of N-acetylglucosaminyltransferase V on alpha3beta1 integrin-mediated cell migration. J Biol Chem 281:32122–32130. https://doi.org/10.1074/jbc.M607274200

Zhao Y, Sato Y, Isaji T, Fukuda T, Matsumoto A, Miyoshi E, Gu J, Taniguchi N (2008) Branched N-glycans regulate the biological functions of integrins and cadherins. FEBS J 275:1939–1948. https://doi.org/10.1111/j.1742-4658.2008.06346.x

Zhou X, Chi K, Zhang C, Liu Q, Yang G (2023) Sialylation: a cloak for tumors to trick the immune system in the microenvironment. Biology (Basel) 12. https://doi.org/10.3390/biology12060832

The manufacturer's authorised representative in the EU is Springer Nature Customer Service Centre GmbH, Europaplatz 3, 69115 Heidelberg, Germany. If you have any concerns regarding our products, please contact ProductSafety@springernature.com

Printed and bound by CPI Group (UK) Ltd, Croydon, CR0 4YY
01/10/2026
02225866-0006